The Devastating Delusions Of The Contemporary Mind

The Devastating Delusions Of The Contemporary Mind

A Scientific Exposition

By
Henry E. Jones, M.D.
2010

Copyright © 2010 by Henry E. Jones, M.D.

All rights reserved. No part of this book may be reproduced, stored, or transmitted by any means—whether auditory, graphic, mechanical, or electronic—without written permission of both publisher and author, except in the case of brief excerpts used in critical articles and reviews. Unauthorized reproduction of any part of this work is illegal and is punishable by law.

ISBN: 978-0-557-58479-6

PREFACE

In order to fully appreciate the contemporary mind it is necessary to understand how it developed. Psychological evolution provides a means of accomplishing this by revealing man's original mental state and the intermediary stages in the development of the contemporary mind. So before we began the discussion of the mind of modern humans we need to understand the process of psychological evolution and how this process has selected the psychological traits which characterize contemporary psychology.

We will evaluate four different psychological types that have developed during mankind's existence. By studying and understanding the pre-modern psychological adaptations a clear picture of the contemporary mind will emerge. We can then appreciate the strengths and liabilities of the modern mind giving us insight into why we behave as we do, and perhaps where we are headed.

Table of Contents

Psychological Evolution. ix

SECTION I Hunter-Gatherers 1
   Chapter 1. Genesis 3
   Chapter 2 The Mind of Hunter-Gatherers 13
   Chapter 3 How Human Psychology May be Changed 45

SECTION II Tribalism 65
   Chapter 4 Problems in Paradise 67
   Chapter 5 THEISM 91
   Chapter 6 Tribe 125
   Chapter 7 The Reification Error 139

SECTION III Statism 177
   Chapter 8 The Human Environment 179
   Chapter 9 Disclosure 193
   Chapter 10 The Slow Advancement Of Freedom 365
   Chapter 11 Modern Religions 377
   Chapter 12 The State 381
   Chapter 13 The Contemporary Mind 399

SECTION IV Deism 445
   Chapter 14 The Present 447
   Chapter 15 Conceptual Model VS *Super*concept 469

References 499

Glossary. 507

About The Author 511

Psychological Evolution

Beginning in the 18th century James Hutton[1], Charles Lyell[2] and Alfred Wegener[3] awakened mankind to the fact that the earth is not a static lump of dirt as believed for millennia. They identified processes operating over enormous periods of time which constantly change the oceans and continents. Ice ages, warming periods with melting polar ice, continental drift, tectonic plates colliding causing earth quakes and volcanoes, and many other phenomena reveal our planet to be a constantly changing, dynamic entity that never rests.

Charles Darwin[4], following Hutton[1], then shook up the establishment in the 19th century with his proposal that biological species are not fixed and static either. Darwin's theory of biological evolution holds that genetic mutation provides a means of adaptation. The processes of natural selection allow the environment to favor some mutations over others. By this means new species develop as others become extinct.

Then Darwin carried his theory further. He stated that the same principles which determined the origin and survival of the other biological species also applied to man. Homo sapiens were the result of biological evolution just like other living creatures.

Darwin held that success for a biological species was to avoid extinction and increase its population. He further postulated that a successful adaptation in one environment could be a liability in another environment. Darwin showed that it is not strength, nor speed, or even intelligence that most promote survival, but adaptability.

In the 1970's Michael Ghiselin[5], Jerome Barkow[6], Leda Cosmides[7], and John Tooby[8] raised the possibility that the basic psychology of human beings may adapt and evolve. They hold that much of human behavior is generated by psychological adaptations that evolve to solve recurrent problems in human environments.

The supposition is that as the environment changes, at some point the accumulated changes render man's current psychology less effective at coping with the survival challenges. There is furthermore the assumption that humans have the capacity through some mechanism, to change their basic psychology. Then through the process of natural selection the fittest psychology survives. The old psychology which had become poorly adaptive is thereby replaced by a new psychology which provides a more effective coping strategy.

Are there any events or 'common sense' observations that might suggest that this type of evolution may actually occur? A review of the pre-history and history of our species suggests that human psychology can and does evolve. Our oldest ancestors for example were entirely family centered. They lived a nomadic lifestyle for many millenniums without ever establishing population centers. Over 9000 generations of our first ancestors failed to establish permanent villages or cities. More than 90% of man's existence has been lived in this manner.

Then over a relatively short span of time these nomadic peoples changed. They suddenly adopted a communal lifestyle with their creation of the tribe. The tribal structure spread around the world and for millennia was the most common type of societal organization. Then after another huge time span tribal man was able to change his psychology again creating the democratic city-state. Democracy failed to push the tribe out and both continue in existence. Then more recently mankind fashioned a psychology that invented and embraced the modern nation-state.

Many types of societal organization have always been theoretically possible. That which limits the type of societal structure selected by humans at any given time is the physical and psychological environment. Within the parameters set by the physical environment it is man's psychology which limits the possibilities for adaptation. How elastic or malleable is human psychology? Each dramatic change in man's societal structure would have first required a dramatic change in man's thinking. Such epochal changes in the way people think might first have required a change in man's basic psychology.

Although exceedingly rare, under extreme circumstances it appears that man is able to alter his psychology and adapt to unusual environmental pressures. I will present here what I believe to be a credible theory of psychological evolution. This theory explains how humans could alter their basic psychology and adapt to changes in the physical and social environment. I will outline the psychological types that I believe have evolved into dominance over the centuries. I will describe what created the need for these psychological adaptations, how these psychological epochs were accomplished and the success these psychological adaptations have achieved. I will outline how I think this has occurred on four occasions during the existence of our species. I will explain what I believe triggers such psychological changes and how the principles of evolution select which changes survive and proliferate.

Finally, no rule of evolution places a moral requirement on a survival strategy. <u>In the struggle for survival the only rule is that a mutation, or a psychological adaptation, provides a means for the species to avoid extinction.</u>

Nomadic
Tribal
Democratic City, State.
Nation - State

SECTION I

Hunter-Gatherers

Man's First Psychological Adaptation
200,000 BC----20,000 BC

Chapter 1.
Genesis

One day, over 200,000[9] years ago, probably in what today is South Africa[10], a remarkable event was in progress. Inside a structure that was so rudimentary that it can hardly be called a structure, a miracle was taking place. Inside the poorly constructed hut made of tree limbs and branches, a world shaking event was unfolding. While others of her kind milled about outside, a female primate was struggling to give birth to her infant. The baby's extra large head was making delivery very difficult. No one was in the hut to help her as she painstakingly succeeded where perhaps others before her had failed. She succeeded at giving birth to a very special child.

The baby's mother may have been a member of the species Homo erectus. Her infant was born with a number of genetic mutations. We do not know what caused this baby to have these mutations. We do know that these mutated genes caused this infant's brain to grow much larger than that of its predecessors. These altered genes may have also reshaped the infant's larynx and enlarged its vocal cords. When this infant matured and reproduced, these profound genetic alterations passed to its offspring and to all subsequent descendants. Thus with the birth of this child a new species, our species, Homo sapiens was born into existence.

Before the birth of this child, our pre-Homo sapien ancestors had already made some important inventions and discoveries. They had learned to make stone and bone tools, to utilize fire and they may have already domesticated the wolf. Our species inherited these advances and adopted a lifestyle similar to our primate ancestors. Living and traveling in small family groups, they followed the game and the seasons. With bigger, more capable brains we excelled as hunters, fishermen, foragers and food gatherers. We refer to these ancient ancestors as 'hunter-gatherers'.

Hunter-gatherers spent a great deal of their time on the move. Sometimes they walked long distances everyday for weeks. They followed the migration of animals, and searched for berries, nuts, fruits and vegetables[11]. If they found a lush crop of fruit or nuts, or a particularly good fishing site, they might camp for days or weeks.

These nomadic families accumulated few possessions because they moved often and had to carry all that they owned with them. Yet edible plants and animals were so plentiful that starvation was practically unknown. Anthropologists estimate that the adults in a hunter-gatherer family had to work finding food and hunting 2 or 3 six hour days per week[11]. One adult's labor at gathering and hunting supported 4 or 5 people. 35 percent of the population didn't work at all! [11] Even with all the walking and moving hunter-gatherers had a great deal of free time.

Leisure was spent resting in camp, doing handcrafts, visiting other camps, entertaining visitors from other camps, telling stories, playing games, playing music, singing and dancing[11.] Today people equate affluence with the accumulation of material goods. By that standard hunter-gatherers were very poor. But from the standpoint of material want, hunger and leisure time, our hunter-gatherer ancestors enjoyed a level of affluence rarely achieved in societies today.

Researchers have studied the few hunter-gatherer groups that existed into modern times. Several of these scientist have mentioned how unconcerned these people are about the source of their next meal[11]. They do not worry about food. They are sure that a food source will appear and self-assured that they will be able to procure it. Where scarcity and hardship mark the world view of modern man, abundance and ease of life seems to be the orientation of our hunter-gatherer ancestors. This experience of plenty and abundance may help to explain the tranquil and peaceful nature of these ancient peoples.

Hunter-gatherers were indeed remarkably peaceful[11]. Murder and war-making was practically unknown to them. Most anthropologists have commented on this lack of aggression as characteristic of the hunter-gatherers. I suppose their sense of abundance made fighting over things seem ridiculous.

Occasionally when the fish were actively spawning, or hunting unusually good, or the crop of wild nuts, fruits, or vegetables especially abundant, several families might camp nearby and close to one another. Maybe at the bend of a river, in a large meadow at the base of a mountain, or at the sea shore, a group of families would congregate and cooperate in the harvest. At such times people engaged in the trade of simple tools and commodities. And at such times young people found each other and new families were established. But as quickly as such events convened they disbanded. As soon as the available food was utilized it was back to trekking again. Soon each family was back going its nomadic way in search of game and edible plants.

The Hollow Tree

Kati was 15 springs and Lade, her brother 9. They suddenly stopped their running along the river path and shouted back at their family. "We found the hollow tree!"

Exhausted they leaned against a huge oak tree. Their two dogs, Pete and Shobo had been running with them. Now they lay down. Surely this was the tree, it had a large hollow, but the hollow was empty! Soon Papa and Granddad arrived, then Mom and Grandma, holding little keche. Dad said, "This is the tree. We hid the fishing poles, string and hooks right in this hollow. It was well camouflaged."

As he pointed to the hollow in the tree Granddad and Grandma shook their heads in agreement. "Well", Mom said, "We can find some more poles around here. I have some extra bone hooks."

Granddad Haddi knew there was a canebrake not too far away. "I will get us some poles" he said as he turned and walked away from the river's edge. Kati and Lade followed Granddad. Shobo went with them too. While they were gone Mom and Grandma built a fire and laid out some animal skins. Pete smelled around the campsite. Grandma bounced Keche on her knee.

After a while Haddi, the kids and Shobo returned with many reeds and long canes. Haddi cut 6 long straight cane poles. Papa Jos joined them as they were returning to the river. Haddi, Jos and Lade worked with flint knives making the poles ready. By nightfall they had 6 fine fishing poles.

Everyone slept near the fire as the night was cool. The dogs slept sound. They were good watch dogs so as long as they were quiet everyone felt safe.

The family was up at daybreak the next morning. Kati and Lade found many big earth worms for bait. They drank some heated bean juice before going down to the river's edge'. Soon everyone was fishing. But the fishing was poor and it did not improve all morning. Only a few tiny fish were caught by noon.

Mom cooked while little Keche watched. Mom stuck a pointed stick through the small fish and roasted them over the hot embers. Keche watched Pete and Pete watched the cooking fish. Kati and Lade wanted to go further down the river and try to find a better fishing spot. "Here take these and eat them on your way," Mom said.

"We will follow along soon" Haddi said "Take Shobo with you."

Lade and Kati began running along a path on the edge of the river with their fishing poles and cooked fish. Shobo ran out ahead of them. Grandma and Mom called to the men and started breaking camp.

The family walked for a good while when they suddenly heard someone approaching from up ahead. They stopped walking and waited quietly. It was the children returning from their forward exploration. They saw Shobo first. Then Lade and Kati appeared. Kati said, "We came upon some people up ahead busy fishing."

Lade broke in, "They are fishing with our missing poles!"

Haddi asked, "Are they catching any fish?"

"Yes, they are catching lots of fish, and some really big ones too!" Lade expressed.

"Let's go confront them" Grandma said. And with that they were all off down the path toward the strangers. Shobo led the way with Kati and the others right behind.

Not too long after a brisk walk and they came upon a group of people fishing. They had already been successful as a long string of fish was cooking over a big fire.

Granddad shouted out to the group of fishermen, "That's a fine mess of fish you have caught. Are the fish still biting?"

An older man shouted back, "The fish are biting great today. Come on down and throw in a hook".

Haddi responded, "Thanks we will join you."

The family took out their fishing gear, laid their skins and tools aside and joined the strangers. The fishing was great. These people also had two dogs. Shobo and Pete made themselves at home.

In the late evening as the sun went down everyone quit fishing. Then in the glow of the camp fire all the fish were dressed and impaled on a pointed stick. Then they were set near the heat of the fire to dry and slowly cook. Everyone took part in preparing the fish.

As they worked they talked of Ytilaer the Great Spirit. Yes it was the Great Spirit of the Forest that produced the fish. It was the Great Spirit Ytilaer that had brought the two families together at this place in the river where so many fine fish could be caught. Ytilaer had shown people how to fish. Yes, this abundance of food had been provided through the generous love of Ytilaer the Great Spirit. Hunter-gatherers believed that Ytilaer talked to them constantly, but in a voice not always easy to understand. These people believed it their job to listen carefully to the voice of Ytilaer and try to understand his message. The older members of the family, those who had listened to Ytilaer the longest, and who had tried the hardest to understand him, had learned more of his secrets. Grandfather Deneen had learned to know where the fish were biting from Ytilaer. Grandmother had learned which plants were edible by listening to her parents and to Ytilaer. The hunter-gatherers believed that all knowledge ultimately came from Ytilaer. It was therefore everyone's duty to listen and learn as much as possible from the Great Spirit of the Forest. There was nothing that Ytilear did not know if humans would but listen, watch and observe. Ytilaer's language was not easy to understand, but with study and devotion to him one could learn and through such learning, one could know him. Ytilaer was their God. They sought to listen to him and understand their world.

The Haddi family camped with the Deneen family for three weeks. Many dried fish were prepared. But finally the good fishing subsided and it was time to part. The Deneens were to follow a branch of the river like they did every year while Haddi and his family would continue down the main river past a salt lick and on to the ocean shore.

Kati had made a boyfriend of the Deneen boy Justi.

They promised to meet up in 12 moons at the old hollow tree. Papa, Mom and Grandma talked with the boy's family and all agreed to meet-up next spring. Kati and Justi wanted to start their own family and both families agreed to help them.

No one in the family mentioned the missing fishing poles. But when the family returned to their old fishing spot the following year they found their gear back in the tree hollow along with some fresh string, newly carved hooks, and a new flint knife. This knife was really pretty with a deer antler handle and deer skin sheath.

When Kati left the family the following year to go with Justi, staying with his family, Haddi gave the new flint knife to Kati and Justi. For many generations meetings at the hollow oak tree at the river's edge were an annual tradition. For many years Kati's children and grandchildren told and re-told the story of the hollow tree and the missing fishing poles.

Hunter-Gatherer Lifestyle

Human beings lived the hunter-gatherer lifestyle, with no appreciable change, for more than 180,000 years! If we allow 5 generations per century, then 9000 generations of your ancestors, and mine, spent their lives in this way. During this vast 180,000 year period our people suffered almost no poverty and little hunger. This was a time before there were any social organizations or institutions except for the biological family. Monogamous sexual relationships appear to have been the usual arrangement[11]. Whatever the sexual relationships were, they appear to have been voluntary and non-violent[11]. There were no social class divisions, no warfare, no human or blood sacrifices, and no enslavement! There were no kings or presidents, and no priest or shamans. There were no permanent settlements and no governments! Perhaps there were some larger extended families or clans, but there were no tribes or chiefs, no schools or churches, no villages, or cities, and no recognized authority outside the family elders. There apparently were a few 'rules', but enforcement of those rules was up to the individuals. Yet *many generations of people may not have witnessed a single episode of aggressive, violent, murderous behavior!*

This hunter-gatherer lifestyle and belief system must have been a very sustainable, accommodating, and satisfying way of life because for over 90% of man's existence he lived in this manner[11]. For a very long time the family and the hunter-gatherer ruled supreme.

You may have been taught that our pre-historic ancestors were brutish cavemen that bashed heads with clubs and pulled struggling women off to caves by their hair. This is the spin given the anthropological and archaeological data by the biases and prejudices of modern man. Commentators from a Judeo-Christian background and academics schooled in Freudian psychology, apply those views of human nature to our ancestors. There is almost no data to uphold that opinion. There is more evidence to support the view expressed here.

At the dawn of recorded history, about 5000 years ago, many people's still lived just like this, as hunter-gatherers[11]. Even today in isolated areas of the earth a few bands of hunter-gatherers remain[11]. Thus this hunter-gatherer lifestyle represents the most typical way for humans to live. Every other way of life that mankind has tried, or is trying, is relatively new, recent, relatively untested and experimental!

When modern contemporary humans take a vacation from their hectic work-a-day world, they often resort to hunter-gatherer activities to relax and refresh themselves. We play with our dogs. We like to barbecue or grill in the backyard. We like to go camping, hunting or have a fish fry. We love to take our recreational vehicles out for a trip. We enjoy, for a while, fishing, backpacking and hiking, the daily lifestyle of our ancient ancestors. We too find some aspects of the hunter-gatherer way of life very relaxing and satisfying.

Chapter 2
The Mind of Hunter-Gatherers

We need to understand the psychology of our very earliest ancestors. How did the new Homo sapien brain work[12]? How did our earliest ancestors manage such a non-violent lifestyle? What was the psychology of these hunter-gatherers? And did this hunter-gatherer psychology provide some mechanism by which man could adapt to changing environments over subsequent millenniums?

We will gather what information we can from anthropology, clinical psychiatry, neuro-anatomy, neuro-chemistry and neuro-physiology. We will also gather what information we can from the few hunter-gatherers that are still around. Utilizing all this scientific data we will now construct a model of the hunter-gather mind!

'Mind' is a manifestation of the organization and activity of the human brain[12]. Thinking, contemplating, imagining, remembering, reminiscing, fantasizing, calculating, and changing our mind, all involve active brain processes that result in real structural changes in the brain. Atomic, molecular, electric, hormonal and chemical changes follow from every mental 'act'[12]. To fully understand the hunter-gatherer psychology and entertain a theory of psychological evolution we need to know some fundamentals of brain anatomy and function.

The human brain can be divided into two major anatomically areas, the Forebrain and the Hindbrain. These areas not only differ anatomically but also functionally[12]. The Forebrain includes the left and right cerebral hemispheres, each with its four lobes. The frontal, parietal, occipital and temporal lobes each with its 'gray matter' are the largest and most obvious parts of the human brain.

The Forebrain also includes the thalamus, hypothalamus, amygdala, hippocampus, and limbic system. The scientific name for this Forebrain area of the brain is: Prosencephalon. This area is believed to be the part of the brain most involved with abstract conceptual thought[12]. This is the area that scientists believe is programmable through experience and learning. This is probably the area of the brain that gives us an imagination! This Forebrain area is also called the new-brain because it is an area that is poorly developed in the lower animals, and much better developed in the higher animals. The Forebrain reaches its zenith of development with the brain of man.

Then there is the Hindbrain, or Rhombencephalon. This area is referred to as the old-brain because it is present in primitive and ancient creatures, as well as in the more advanced animals. The Hindbrain of humans closely resembles that found in animals. Connecting the Forebrain and the Hindbrain is a small area called the Midbrain or Mesencephalon. Most authorities believe this area functions mainly as a communications link or nerve crossroads, between the two larger areas of the brain.

Prosencephalon

The Forebrain of Homo sapiens is sometimes referred to as the new brain because it evolved into being in its present form just 200,000 years ago. When discussing evolution, we generally speak in increments of millions of years. So we human beings with our new and enlarged Forebrains are very new on the scene. And with this bigger and better Forebrain, we are able to do some marvelous things.

To more easily understand what the human Forebrain can do, and how it does it, we can compare it to the personal computer. Computer hardware is the mechanical, magnetic, electronic, and electric components that make up a computer system. Brain tissues are analogous to this computer hardware[12]. Computer software is the programs, routines, and instructions that control the functioning of the hardware and direct its operation. Software in the brain, like in a computer, is the organization of electrical connections. We are born with some of our software already in place, but much of it must be learned. Homo sapiens are therefore not born with a fully programmed and functioning mental computer.

After birth the human mind must develop through three levels of programming[13]. The most fundamental 'programming' is that 'software' with which we are born, that provided by genetics. This 'programming' is wired-in or hard-wired and does not require learning or experience to acquire. Much of this genetic 'programming' is contained in the Hindbrain which we will discuss presently. This genetic or innate software forms the basic level of the mind. Then through experience and learning the brain continues with what genetics provides, adding to the innate programming. The first programming to be added to the genetic 'software' is derived through perception. Then lastly the conceptual process develops and contributes concepts which become the premier programming tool and building block of the adult human mind.

The process of acquiring mental programming can therefore be described as developing[13] chronologically through three stages. An infant's mind develops through: (1) the stage of sensations, (2) the perceptual stage, and (3) the conceptual stage. We need examine each of these three stages of programming.

At birth the infant is primarily in the stage of sensations. Here the brain acts as a measuring device. It measures light waves, then transduces them into electric impulses; it measures sound waves and transduces them into electric impulses; and it measures minute amounts of chemical in the air and in food, and transduces them into electric impulses. It measures pressure and temperature against the skin and transduces them into electric impulses. The human brain does this measuring and transducing into electric impulses inherently and automatically. Because of the way the brain is constructed a human being automatically experiences these electric impulses, depending on where in the brain they arrive, as sight, sound, smell, touch , and taste.

From these sensations, the brain automatically forms percepts[13]. A percept is a group of sensations automatically bundled and integrated by the brain. Percepts are formed using innate, wired-in, genetically derived 'software'. It is in this percept form that animals, and human beings, grasp the evidence of their senses and apprehend the external world. Functionally, the human brain begins processing information beginning at this perceptual stage[13]. A newborn human being begins forming percepts immediately upon birth. Some studies even suggest that some rudimentary percepts may be formed before birth while the baby is still in the womb[21]!

Thus the human brain automatically takes sensations and bundles and organizes them into percepts. While this process is automatic the brain must perform this operation with information provided by the environment. Environments contain universal as well as unique components. Therefore every infants perceptual experience is unique. These early childhood **percepts are the foundation of the individual's knowledge**[13]. Percepts are organized and integrated by the infant's earliest learning experiences to form his unique perceptual mind. Thus a child begins his lifelong process of building his mind starting first with percepts. Then he adds first, second, third and higher order concepts. The infant's brain is making tens of thousands of connections every second! These connections proceed at a furious pace even as the infant sleeps[22]!

This early, perceptual stage of mental programming is extremely powerful. These first percepts connect directly with the genetically inherited programs available from birth. Percepts are the first elements of the mind to result from experience and therefore lie at the very foundation of the individuals mental development. The perceptual software created during this time of an infant's mental development will become the most tenacious and most difficult to change in later life. Like the very first experiences of some lower animals this perceptual programming acts like psychological imprinting.

Human Imprinting

While the newborn and very young infant spends a large percentage of his time sleeping, sleeping is not the dormant activity some assume. The seemingly passive infant sleeping soundly in his crib has a brain which is bursting with lighting-fast activity! His little brain is making tens of thousands of connections every second! His human and physical environment is programming his mind.

'Imprint programming'[14] is simply the earliest programming that is the result of experience as opposed to genetically derived programming. This early programming is largely perceptual, non-verbal, and unconscious. One of the reasons imprint programming is so difficult to change in later life is that it 'feels' right. The emotion connected to this early programming is very powerful and feels 'normal' and 'correct'. Emotion connected to this early imprint programming is so strong that it can usually over-rule conceptual thought. This earliest perceptual programming 'feels' much like genetically endowed programming, it is intuitive. This makes modification and change of this 'imprinted' software later in life very difficult.

Another reason imprint programming is so difficult to change is the *super*concept or conceptual standard of truth which I will explain momentarily. This is the unifying concept which ties all of a person's concepts together into a non-contradictory whole. To change a few concepts within a matrix of concepts held together by non-contradictory associations is very difficult to accomplish, and in time tend to be reversed to once again fit into the theme established by this super unifying concept which defines the individual's mind.

Reason, conceptualization, and therefore counseling and psychotherapy are less effective than we might wish at altering this imprinted software. Imprint software thereby has the power to effectively hold the individual's behavior in-line with those experiences and concepts which comply with their 'imprinted' *super*concept. Thus the importance of early childhood experience on lifelong belief and behavior.

Software that forms the *super*concept shapes a person's identity. So once it is formed it is extremely difficult to change. You can appreciate the difficulty involved in a person using his conceptual processes to try and change the largely perceptual software of the imprint period. If a person believes something intuitively, it is very difficult to change their opinion through reason and logic. Conceptual programming utilizes language, the perceptual programming of early childhood is non-verbal and highly emotional. In practical terms the imprint software that an infant and young child develops during this period is essentially permanent. It can be changed but only through a very difficult and painful personal odyssey with or without professional assistance. Therefore many of the first beliefs and behaviors incorporated into an infant's mind are essentially fixed. Fixed and permanent that is except for the mechanism of change we are coming to presently.

Concept Formation

The human mind is apparently unique in its ability to manipulate and play with percepts. The infant 'plays' with these images in his mind measuring, comparing, contrasting, analyzing and examining each percept in minute detail. This use of imagination and playfulness, absent any fear or inhibition, in a mental environment of total freedom, has no parallel in the mental operations of any other living creature on earth!

After the perceptual stage of infant mental development comes this third and final stage: concept formation. From his intrapsychic play the infant notes differences and similarities among percepts. Out of this playful imagining he forms his first concept.

The first concepts are the most simple of the conceptual building blocks and are derived from underlying percepts[13]. The first concept a child forms may be the idea of 'thing'. The fact that something exists is implicit in every percept. A child may grasp his first concept 'thing' implicitly directly from the perceptual level (to perceive a thing is to perceive that it exists). This process of abstracting first order concepts from underlying percepts begins the conceptual building of the individual's knowledge and understanding of the world[13].

Next the brain, through its ability at abstraction through relative measurement, forms secondary concepts by dividing 'things' into different kinds of 'things'. These secondary concepts are of concrete objects like 'mama', 'toy', 'table', 'chair', as well as secondary concepts of attributes and actions. *The key to human cognition is this ability to conceive of entities as **units** of a larger group*[13]. This is the human beings' distinctive method of thinking. Thus a growing child develops the concept of, for example 'chair', recognizing the chair he uses at the table as one example, one unit, of a whole class of similar objects. This class of objects, this concept, has a name, 'chair'. So language becomes a method of labeling concepts, and then the tool of further conceptual thought. 'Chair', is further refined into 'high chair', 'rocking chair' 'my chair', 'daddy's chair', and so on. 'Table' goes through the same process, 'kitchen table', 'dining table', etcetera. Every object and every action in the child's environment undergoes similar cognitive definition, classification and labeling[13]. Imagination plays a big part in concept creation.

The developing mind, through experience, accumulates an ever increasing store of concepts. Thus from a base of percepts, an ever-growing matrix of concepts, and associative connections between concepts, grows over time. This creates an ever expanding four-dimensional matrix of percepts, concepts and their connections which forms the human mind.

As this matrix of concepts, and their associations expand, two interacting cognitive processes are taking place. From a base of percepts, then primary, secondary, and higher order concepts, cognitive development moves toward[13] (a) more extensive knowledge, and (b) more intensive knowledge. That is to say (a) toward wider integrations, and (b) more precise differentiations[13]. Following this process and in accordance with cognitive evidence, earlier-formed concepts are integrated into wider ones or subdivided into narrower ones. The process of forming a concept is not complete until its constituent units have been integrated into a single mental unit and defined by a specific word[13].

The process of abstracting from an abstraction is not simply memorizing a word. Mental work must be expended and it is not an arbitrary selection. Each new concept must stand up to rigorous analysis, it must 'make sense' within the context of all the individual's concepts and associations. Every new concept must be integrated without contradiction into the total of one's mind[13]. "Mind" being all of the accumulated concepts, integrated from percept-derived concepts at the base, through all levels of abstraction[13].

It is believed that as a person sleeps each night his mind cycles through its entire databank of concepts at least 6 times requiring at least 15 minutes for each cycle[23]! Sleep may actually be more about giving the brain computer time to 'catch up' than for the physical body to rest up! This is why we often awaken in the morning with the answer to some vexing problem from the day before.

Animals have the ability to form percepts. Some animals can form a few basic, first order concepts. But no creature can match the human ability to easily and rapidly form and store massive numbers of second, third and higher order concepts. This ability of the human brain at conceptual thought appears to have no limitation. The human imagination seems boundless and able to conceive anything regardless of whether or not it is real or even possible. This ability to imagine and conceive the seemingly impossible is man's greatest strength.

Contradictions

As the human mind pursues its relentless cognition and continually tries to integrate concepts and make logical associations, it is trying to eliminate contradictions[13]. This process is called 'thinking' and it proceeds automatically if not sabotaged or derailed by some adverse condition. Thinking seeks to eliminate contradictions because they prevent integration. The human mind therefore abhors a contradiction.

When the mind is confronted with a concept that cannot be integrated into its matrix of fully integrated concepts, that cannot be connected by logical links or associations all the way back and down to the level of perceptions, the mind will try to kick the concept out. The easiest thing for the human mind to do with a contradiction is to reject the contradicting concept as untrue and throw it out.

We hear people today say things like; "I'm not falling for that bull", "I'm not about to swallow that" and "that sounds untrue to me". These statements, and many others like them, are used to communicate to others that the concept does not logically integrate into the speaker's knowledge bank and is therefore rejected. An idea may not be thrown out immediately. The concept may even be 'kept on file' and rechecked from time to time. The person may ponder the idea for a period of time, but typically humans will not accept ideas that cannot be fully integrated into their mind. A person may ponder the validity of a belief or concept until he is able to integrate it into his overall knowledge, or he will reject it and refuse to accept the idea as true.

This rejection of floating abstractions, concepts that make no sense, and beliefs that cannot be integrated goes on continuously. Accepting beliefs and forming concepts is parsimonious. The human mind seeks the most simple and least concepts required to explain its observations[13]. Conceptual integration, without contradiction, from the ground up, has epistemological primacy. This raises the question of what shall serve as the reference standard of non-contradiction? We will return to this shortly.

Rhombencephalon

Now we come to the second large area of the human brain. From an evolutionary standpoint this area of the human brain is much older than the Forebrain we have been discussing. This area of the nervous system has been evolving and changing literally since life began on earth at least 3.5 billion years ago! Seriously, the first creatures to leave the seas and live on land 360 million years ago had Hindbrains not so different from us modern humans! Our Forebrains are the latest greatest, but our Hindbrains are as old and stogy as they come!

The Hindbrain includes the cerebellum, pons, and medulla oblongata. This area of the brain functions in humans principally the same as it does in all animals. It supports vital bodily functions. It regulates heart rate, respiration rate, sleep, digestion and most of the other automatic, so called 'vegetative' functions. These functions in the Hindbrain of man and animals are fixed by genetics. They are wired-in, and largely inalterable and unchangeable. While this area can be trained to some extent, it is not under direct volitional control. Mostly this area operates automatically, and unconsciously.

Lying within the human Hindbrain at birth is a structure that will require some conceptual development before it becomes fully functional. This structure, like the others in the Hindbrain, has its basic functions wired-in and fixed by genetics. The individual has little control over this structure, or how it works, yet it is this structure more than any other that will determine the course of a person's life. This structure is the **SELF**.

The **self** emanates from Hindbrain tissue and has the attributes of this area of the brain. The **self** is not an abstract creation of computer cognition, it is hardware, and it is fundamentally inalterable and unchangeable. But the **self** will require some programming by the Forebrain in order to **conceive** of its **self**. Thus

self-awareness is a Forebrain phenomenon. Thus the **self** as we are aware of it, exists and is represented in both Hindbrain and Forebrain. So we must now return to the Forebrain to continue the conceptual programming of self-awareness.

Identity Formation

The **self**, with some additional Forebrain programming, will become our experience of who we are. The **Self** will contain our deepest hopes, fears, ambitions and desires. It is our inner life. The **self** is that small voice down deep inside that represents our true being. It is through the **self** that we experience pleasure, pain, hope and disappointment. It is the **self** that is the executer of thought and action. This awareness of one's **self** will require some additional programming, the creation of additional software which will create an **identity** for the **self**. Identity is composed of those concepts which define our **self** to our **self**.

We become aware of our **self** through our **identity** which is both a perceptual and a conceptual process. The programming which creates an identity for the **self** will result in self-awareness. This takes place at the juncture of genetic programming with the earliest imprint programming by the environment. The **self** is programmed by genes and by the infant's earliest percepts and concepts. Self-awareness is therefore the result of some of the very earliest programming in the mind. This takes place during the first few months of life! It is therefore some of the most powerful and tenacious programming in the human mind.

There are many aspects to the **self** acquiring an identity. There is personal identity, moral identity, sexual identity and many other aspects of identity which are necessary in defining the **self**. The most basic aspect of identity to develop is the individual's **epistemological** orientation. This programming will determine how the person's cognition will work. We will focus on this part of **identity** because it is this part which is most important to understanding psychological evolution.

*Super*concept

The source of all the data available to every human being is the environment. The environment can be divided into two great areas which both newborns and adult humans must face. There is the non-human realm we call external reality or nature. And there is the human environment made up of people and their perceptual and conceptual products. Sensations from both these areas of the environment immediately begin flooding the infant's brain from the moment of birth. Very soon the infant begins assembling the incoming sensations into percepts. Using these percepts the infant begins constructing a fantasy world inside his mind. Soon he is able to conceptualize an identity and thereby becomes conscious of his **self**.

The purpose of identity is to conceptualize the **self** to the **self** as self-awareness develops. In the process of forming concepts *where the self is involved* it is necessary that the dimensions or boundaries *of the self* be accurately demarcated. Epistemological concepts are developed by the infant to demarcate and separate his **self** from non-self. Then these concepts are used to provide the conceptual basis to differentiate the individual's **self** from the human and non-human environment around him. This is an important part of self identity as it also provides the concepts needed to think clearly and categorize appropriately in mental operations pertaining to the both aspects of the environment.

As soon as an infant has a beginning **identity** and can conceive of his **self** he becomes **the** important entity, the main actor in his evolving mental construction, his fantasy world. Throughout his life the individual will take information from the two aspects of the environment into his fantasy world. He applies his imagination to the incoming information to constantly digest, understand, modify and weave it into his ever-evolving mental construction. During childhood a person's **identity** is continually enhanced. Usually by adolescence a person has formed an overarching, unifying philosophy or theme for his fantasy world and his **self** within it. This world view contains a standard of truth for non-contradictory cognition. This is the person's *super***concept**.

This world view, fantasy world or 'philosophy' which the individual creates in his mind is probably not his last. He will probably build and discard many belief systems or imaginary constructs over the course of his life. But this initial or **first** mental construction, this one with perceptual attachments that go back to the beginning of consciousness, this world view and concept of his **self**, which is anchored in imprint programming, ***and which contains the cognitive standard for non-contradiction*** remains established and important to him all his life. Some psychologists have referred to this entity as the **true self**. This original world view and identity with its reference standard, is the individual's *super*concept and it typically remains his lifelong default standard of truth for important decisions. This is the person's original imprinted epistemological reference standard of non-contradiction. This is his *super***concept.** The *super*concept is the earliest part of identity. Later experiences from family, society, culture, ethnic background, religion and occupation will add elements to one's **identity**. But the *super*concept will remain at the core of that **identity**. While the more superficial aspects of identity may change over a person's lifetime, this core portion of their identity is very difficult to change and tends to remain constant throughout their life!

This imprinted epistemological standard, this *super*concept is the most important part of identity and it is the most important aspect of the human mind because it is the most essential to survival. Epistemology is a branch of philosophy. It is the academic study, or the science, of how we know the truth. Philosophers have debated this question for thousands of years. A newborn however does not have the luxury of dispassionate and endless debate. Survival for an infant is not a philosophical indulgence; it is a highly emotional emergency! While this super belief will probably last a lifetime it must be adopted very soon after birth!

In order to think, in order to think the efficient way humans think, an infant must incorporate into his mind a standard of non-contradiction. Thinking requires non-contradictory integration of concepts so he must have **a standard reference of truth** to accomplish this and he must have it very quickly. A person may learn half of what he will ever know within the first five years of life. For his cognitive processes to work properly, so that he can learn the massive amount he needs to know as quickly as possible, he must have the key which unlocks the enormous potential of his conceptual mind! To do this he must have a key concept, a ***super*concept**, a world view of the way things work, a reference standard of truth with which to calibrate his thinking.

A *super*concept is necessary because there must be a standard of reference against which each concept is gauged or judged for non-contradiction. This concept must rank above, and determine the order and organization of all the individual's concepts. The human mind is a four dimensional matrix of concepts associated in a non-contradictory manner. The roots of this matrix are in perception. Arching over this entire perceptual-conceptual system and serving as the reference standard of truth is this ***super*concept**. Perhaps it should be labeled the *super*percept because it originates in perception and for many people it is never consciously conceptualized. The *super*concept is necessarily singular. By its nature there is no such thing as dual or multiple *super*concepts.

The first thing that should be done prior to using a precision measuring instrument is to check its adjustment and make sure that it is precisely calibrated. This requires that some standard weight or measure be used as a reference. So it is with cognition. A concept contradicts what? What represents 'truth'? What truth is contradicted by an idea which is accordingly rejected? Which beliefs stand in non-contradictory association to the other concepts in one's mind? By what ultimate standard of truth are they integrated? A *super*concept is required in order to accomplish this non-contradictory integration. The integration of concepts requires such an over-arching *super*concept, under which all concepts are subsumed. This *super*concept or umbrella belief is the reference standard by which contradiction is measured or guaged. This super reference ties all other concepts together in a unifying whole creating a singular mind with a single standard of non-contradiction. A *super*concept is required to organize and facilitate continuous non-contradictory cognition and integration.

A *super*concept by which all concepts are judged is a survival necessity. Effective thinking, including decision- making relies upon it. Without an effective cognitive reference the mind may cogitate endlessly, excessively and exhaustively without being able to reach any conclusion. Much of the time spent in sleep is necessary because of the mind's requirement that the total of its contents be repeatedly checked for integration[23]. Several times every night one's entire conceptual inventory is examined, evaluated and brought into non-contradictory agreement with this *super*concept! The inability to achieve non-contradictory integration causes fitful dreaming and a poor night's sleep. Often we awaken with the solution to our previously unsuccessful efforts to integrate a concept into our mind. Such integration requires a *super*concept!

The process of forming a *super*concept begins in perception; it is conceptualized and carried forward over the course of childhood through all levels of mental integration and during adolescence finalized. This reference standard of truth begins as some of the earliest programming in a person's mind. This is the one concept which must exist and must link to every other percept and concept in the individual's mind. This programming directs the basic operating system of cognition and provides the standard by which all subsequent data is appraised and judged. The beginnings of this programming falls solidly into the category of 'imprint programming', it is non-verbal and unconscious. It begins as a percept, but it is conceptualized with increasing clarity through childhood and completed in the adult with a firmly established *super*concept.

Many philosophies, religions, and ideologies contain a reference standard of truth which is internally consistent to one degree or another. These reference standards are **conceptual.** An individual may intellectually subscribe to any number of theories, ideologies, theologies, philosophies or other belief systems. Contained within each belief system may be an explicit or implicit reference standard of truth. The standards of truth underpinning these belief systems may correlate or conflict. A person may adhere to such beliefs or discard them with relative ease and then adopt others or not, without much emotional difficulty. A conceptual reference standard of truth that cannot be integrated into the imprinted epistemological perceptions of early infancy, the person's *super*concept, cannot reinforce or authenticate their *superconcept!* Such intellectual constructs as philosophies, religious beliefs, ideologies, theories, and other belief systems can become a portion of a person's identity later in childhood or adolescence, if it can be successfully integrated into their *super*concept. The more successfully a conceptual theology or ideology can be integrated with these early experiences of infancy, the more likely such a theology or ideology will be adopted as part of their **identity**. **Thus the *super*concept a person has was determined by the epistemological *percepts* incorporated during infancy!** Their *super*concept will then determine what type of ideologies and theologies they will find meaningful later in life and then perhaps adopt as a conceptual part of their **identity**! This explains the importance of infant and early childhood experience on a person's philosophical and religious preferences later in life.

To further clarify the importance of the *super*concept to cognition let me use this analogy. Imagine a number of carpenters building a wood frame house. Normally each worker carries on his person a tape measure calibrated in an acceptable standard of identical lengths. Imagine the chaos that would occur on the work site if each builder used a measuring device calibrated to an entirely different standard! Say one worker used a tape measure marked off in inches, but the inch marks were actually one and one half inches in length! Another worker used a tape in which the inch marks were really only one half inch apart. The third carpenter had a tape measure which, though marked as inches, actually varied from one inch mark to the next!

No plank or board measured by one tape would fit when cut by someone using a different tape. Nothing would fit unless extra efforts were applied to make it fit. The structure so constructed would be weak, hold up to the elements poorly and fall down under pressure.

A strong, competent mind requires the same dedication to exact and reproducible calculation. Its concepts must be integrated into a non-contradictory matrix standardized to one reference standard, to one *super*concept. To the degree that one, stable, and reliable *super*concept is not adopted, cognition will be compromised. To whatever degree the infant fails to adopt a *super*concept which can eventually be integrated into a conceptual identity, an identity on which successful behavior can be based, is the degree to which the individual's identity will remain in flux. It is around the age critical reason develops, the age we call adolescence, that this process is put to the test. The fundamental development of a person's *super*concept must be well underway by 6 months of age! Conceptual identity slowly develops throughout childhood. How well the two can be integrated into a single, non-conflicted identity will determine whether adolescence is smooth and mentally uneventful or a horrendous psychological meltdown of self-destruction and violence. The *super*concept is a person's standard of truth, it is his epistemological reference for non-contradictory integration. So while the *super*concept is but a portion of a person's **identity**, it is a huge, extremely important, and usually permanent portion.

Spirituality

This powerful requirement that humans have for an epistemological reference, a standard of truth, a *super*concept, for a solid **identity**, is experienced as a ***spiritual need***. The quest for spiritual meaning is experienced as an identity crisis. This is a quest for a better, more cohesive, less contradictory ***identity***. The spiritual quest is the pursuit of a better *super*concept, a better source of ultimate truth. Spirituality is the emotional experience of the importance of this key concept to effective cognition. The emotional quest for spiritual meaning is recognition that cognition will not work properly without an ultimate or supreme reference standard for non-contradictory cognition! For human beings the satisfaction of this spiritual need is as important, and as necessary to survival as air, water, food and shelter! And as we shall see it is in the quest to satisfy this spiritual need that man is able to change his basic psychology, evolve, and adapt to new environments. But it is also this quest which leads to much violence.

We find evidence of man's spiritual quest as far back in antiquity as we find evidence of man. This quest is not about an esoteric or academic truth. This is about the truth required for a person's moment-to-moment survival in the wilds of the forest! Determining what is true and what is false is a matter of life and death. When making decisions in the jungle the consequences of one's decisions are often just moments away! I believe that these conceptual standards are absolutely necessary for all Homo sapiens! I also believe that these *super*concepts are the most persistent and enduring of all man's beliefs.

So the need for a conceptual standard of truth, a *super*concept, a healthy **identity**, is great and virtually every human being will adopt or develop this requirement of proper cognition at a very young age. While a conceptual standard of truth is a survival necessity, it is not fixed by our genes, it must be learned from the environment. And it must be learned very soon after birth.

A person's *super*concept may be configured in any number of ways. In every human language there are words that are used to label this conceptual standard. In English a few of the words used to label this *super*concept are 'reality','nature','creator' or 'God'. Without such a concept, labeled with language or not, the individual will not be able to develop a strong **identity** and will suffer a cognitive disability. Loss of contact with what exists, with reality, with God, results in disorganized, ineffective 'thinking'. The psychiatric diagnosis for this condition is psychosis. Psychosis means loss of contact with what exists. Complete psychosis renders a person incapable of caring for himself, of performing the most basic tasks. Severely psychotic patients will die of starvation, exposure or injury if not carefully cared for and protected. Severe psychosis is incompatible with an independent, self-sustaining life.

As an adult an individual will display psychotic mental processes in those areas of mental functioning not founded on a solid, immutable, and dispassionate *super*concept. So a person must have a strong *super*concept in order to develop a solid **identity** and think effectively.

I have mentioned that this *super*concept may be conceptualized in many different ways, and it may be labeled with language which varies a great deal. Since a child has parents who seem to know how things work, who seem to have an understanding of the world which is superior to that of the child's, humans have a tendency to be anthropomorphic when labeling their *super*concept. So while such concepts may be labeled with such words as 'nature', 'Great Spirit', or 'reality' they are more typically labeled with the word 'God' or given the name of the family's religion or prophet. A person may choose not to label their *super*concept with the word 'God' without any cognitive consequence. However the label used for the *super*concept may have serious social or political ramifications. The name or label given to a belief system may be very important to its adherents, but it is ***the concept behind the label*** that is most important to cognition, and to psychological adaptation.

A person may say for example, that he no longer believes in the supernatural or in the existence of God, that he is an atheist. But if he has not changed his *super*concept, if only the word with which it is labeled has been renounced, then he has accomplished much less then he might think. So, while it is revealing to ask a person if he believes in a God, the more cogent question is: what kind of God does he believe in?

The *super*concept is a crucial component of a person's **Identity.** It is a vitally important requirement of normal cognition. A *super*concept is a survival necessity! It literally holds the individual's mind together. And a person's *super*concept determines to a very great extent what he is able to perceive!

An individual's *super*concept is not held in his mind as a dispassionate intellectual tool. Typically enormous emotion is connected to this belief. It is highly emotional because it is of critical survival importance.

The loss of this critical mental 'compass', even the serious questioning of its authority, can cause severe mental decompensation. While the irrational emotion and violence often associated with allegiance and competition over the **labels** to *super*concepts should be condemned, such commitment is understandable. It is this spiritual need , and man's efforts to satisfy it, which makes him vulnerable to spiritual manipulation. And it is the lack of integration between the *super*concept and conceptual, intellectual beliefs and ideologies to form **identity** which is behind most adolescent angst. It is this competition between religions and socioeconomic philosophies that produces much violence.

It makes an enormous difference just exactly how this standard of truth is conceptualized because this will determine the individual's basic psychology. This standard of truth will effect what the individual can and cannot 'see'. I mean it will make some things easy and other things extremely difficult to perceive. This is not an issue of intelligence. It is an issue of perception like the child in ***The Emperor's New Clothes*[15]**. What a person's standard of truth *is* determines to a very great extent the areas of reality he can best perceive. Problems will force themselves upon our awareness, but the possible solutions may lie outside our perception because of our particular *super*concept of truth.

So exactly what is a person's *super*concept? How can we better understand this very important part of the human mind? A person's *super*concept *is* their concept of God, reality, Nature or Nature's God. While most people have named or labeled their *super*concept with one of these words they seldom have a clear understanding of its explicit meaning. This does not matter, this does not effect its functioning, nor does it hamper acquiring a full **identity** based upon it. But for our purpose here we need to understand that **identity** starts with this base.

A person's *super*concept is an integral part of their **identity** and extremely difficult to alter or change once it is formed. People typically hold on to their *super*concept with a do or die tenacity. But paradoxically it is this aspect of human psychology which provides the possibility of that extremely rare event; psychological modification, evolution and adaptation.

Hindbrain

Now we need to return and finish our discussion of the Hindbrain or Rhombencephalon. Before going on to another Hindbrain function let us complete our discussion of the **self**. It is the **self** that is the 'observer' in the mind. Sometimes referred to as the 'mind's eye', we can think of the **self** as the monitor in our mental computer system. Like a computer monitor our **self** can only 'view' a small portion of our mental content at any given moment. What we bring up on our mental 'monitor' and how we interpret what we 'see' is greatly determined by our **identity**. But unlike the monitor of a computer system it is the **self**, our human 'monitor' which feels pleasure and pain.

And it is here that the Hindbrain has another very important function. Here is located both the hardware and software of the brain's kick-back reward system. Pleasure is mediated in the brain by chemicals or hormones called endorphins[17]. When an animal or human satisfies a biological need, his brain secretes these hormones. These natural chemicals resemble morphine. Secretion of endorphins by the brain evokes the feeling of pleasure. When a person, or an animal, engages in the activities which stimulate the brain to secrete endorphins, they experience pleasure. These endorphins make the individual feel very good.

There is a whole spectrum of endorphins that are produced and secreted by the brain in response to different activities[17]. Physical exercise causes the brain to secret one group of endorphins. Ingesting food, listening to music, sexual activity, and mothering stimulate the secretion of different endorphins. This endorphin kickback system rewards a creature for taking action and satisfying a need. An animal or a human quickly becomes 'addicted' to those actions that cause the brain to secrete these chemicals. Toddlers soon learn to seek the pleasure of these brain chemicals by thinking, solving problems and acquiring the things they need and enjoy. The human mind is ideally designed to succeed at such tasks

When an animal's or a person's needs go unsatisfied, endorphins become in short supply. A lack of endorphins makes a person feel very bad. Low endorphin levels, even for a short time, can lead to depression and despondency. These negative symptoms can prompt endorphin seeking behaviors. This may include the use of artificial exogenous endorphin-like substances. Endorphin starved individuals very easily become addicted to exogenous artificial endorphin-like drugs and chemicals.

Identity Authentication

All living creatures have basic biological needs. A creature will die, or the species will die out, if these basic needs are not satisfied. Humans share a need for air, water, food, shelter and sexual release with most other life forms. But the human, because of our conceptual mind has another innate need. This unique, basic need is the psychological need we have been discussing. This basic and essential psychological requirement for human life is the need for an **identity**. Identity is necessary to **self** definition and self-awareness. Without **identity** there is no **self-awareness**. And without a core **identity**, the *super*concept, there is no effective cognition. The spiritual quest we discussed is the emotional recognition of this necessity.

We experience this need for **identity** as a need to ***authenticate*** our **identity**! This is accomplished through successful behavior taken to reinforce one's identity. Think of it this way. Hunger is a need. But what a person may consider as food differs a great deal from one person to the next, from one culture to the next. The need to authenticate one's identity is a hunger, everyone has this need. Successful action to reinforce one's identity is that which satisfies the need for authentication. The particular action an individual requires to authenticate his **identity** varies from one person to the next and depends upon the nature of his **identity!**

Human beings **will** base their behavior on their **identity**, and they will take action to reinforce their **identity**. Successful feedback from this behavior reinforces commitment to one's **identity**. Successful action reinforces **identity** and thereby satisfies the individual's spiritual need. Satisfying a need provides an endorphin reward. If behavior is unsuccessful, reinforcement is lacking, and endorphin reward becomes scarce. This may cause mental anguish and questioning of the validity of one's **identity**. If the pain and unhappiness continues long enough then a questioning of one's **identity** may occur. This questioning of one's identity is what drives the spiritual quest.

When behaviors based on one's **identity** are successful it reinforces one's **identity**. Sustained reinforcement of one's **identity** is experienced as authentication and is accompanied by a wonderful feeling of endorphin reward. The endorphin reward for the authentication of one's **identity** is very large. *Mental closure on one's **identity** and non-contradictory integration of the mind around one's **identity**, and **the taking of successful action** based on the one's **identity** results in a huge and sustained endorphin reward.* Some have described this feeling as Nirvana.

When an individual suffers mental disintegration it is because of a lack of an adequate **identity**, unsuccessful behaviors and the absence of endorphin reward. The 'psychologically disturbed' adolescent may be undergoing an awful suffering. His built-in Hindbrain reward system propels him toward satisfying this spiritual need, but what if he fails in his quest? For those failing in their efforts to authenticate their **identity** nothing short of anesthesia with drugs or death may quell their misery!

It is in creating an **identity** that this unique spiritual need is addressed. And it is by taking action to reinforce one's identity that this need is satisfied. So the **identity** a person creates in his mind *will* be acted upon! It is not that a person *may* act to reinforce his **identity**; it is that he *will* engage in behaviors that reinforce this vitally important component of his mind! Satisfying this spiritual need is often more important and more powerful than the urge to satisfy any of man's biological needs. The need for **identity** authentication by behavior is often more important than life itself! It is the quest for this reward that motivates much human behavior. Therefore what kind of **identity**, an infant creates in his mind during the earliest months of life is extremely important. It is important because it will affect his cognition and it will affect what kind of **self** he can develop. And since he **must** act to reinforce his **identity**, it will be the motivation behind a great deal of his behavior *throughout his life!*

Summary Of Chapter 2

A baby acquires his *super*concept and his **identity** from the information he receives from his surroundings. This over-arching or umbrella concept begins as a percept which is later conceptualized to some degree. A huge component of an infant's environment is his mother and father. A baby begins the formation of his **identity** utilizing these very early experiences. The need in humans for an **identity** will not end with its formation. A powerful basic need for on-going **identity authentication** remains. Like the needs for air, water, food and sexual release, it is a need that requires repeated and on-going satisfaction. The satisfaction of this powerful need is accomplished through the behavioral reinforcement of **identity**. To satisfy this need for authentication requires **action**. People will respond to this powerful urge and take action to authenticate their **identity**.

Reinforcement of one's **identity** is a must. Human beings have no choice; it is a necessity of life! The kinds of actions they take, the kinds of behaviors they engage in, are determined by the kind of **identity** they have created. When we observe someone engaged in a behavior we may safely assume that one of the goals of the behavior is to reinforce his **identity** and his *super*concept. There are certain rare situations where this would be incorrect, but for our purposes where populations are concerned we can safely make this assumption.

Therefore, if you wish to change a person's basic psychology you must change their core **identity**. If their core **identity,** their *super*concept is changed their behaviors will change! It is easier to change some of the more superficial aspects of **identity**. But this may not affect the individual's basic psychology. But if this change of **identity** includes their *super*concept it will change their basic psychology. This means that evolutionary change usually begins with an infant's environment. To effect an evolutionary psychological change you must begin with newborn infants, or more practically, you must begin by training parents-to-be. Parents then may alter the newborn's surroundings or environment. It is these earliest perceptual experiences of infancy which most profoundly affect the *super*concept and therefore fundamentally shapes **identity,** and thereby forms a person's basic psychology.

Chapter 3

How Human Psychology May be Changed

In every age there are individuals who are unhappy with the way things are. They criticize everything and question everything. Often they are miserable with themselves. In our age they are called 'mentally ill' or 'psychologically disturbed'. In another age they may have been seen as being possessed by 'spirits' or by the 'devil'. In one age these outspoken critics of the status quo might have been revered as a source of great truth. In another they might have been put to death for heresy. In another age they might be locked up and subjected to 'treatment'[16].

These irritating and usually obnoxious individuals are often adolescents or young adults. Perhaps they perceive something better than most of their peers. They are people who for one reason or another have not been able to make a closure on a conceptual standard of truth. Their identity is in flux. While most of their contemporaries may feel no pressing need to change their *super*concept these individuals have an urgent need for change. The *super*concept provided them by their environment, their parents, family and culture is not working. Something about their *super*concept doesn't ring true in the world they now inhabit.

They are seeking an answer to the question: what is the ultimate standard of truth? What or who can I trust and believe in? Maybe there is friction between the perceptual and the conceptual portions of their *super*concept.

The phenomenon which in psychological evolution is similar to genetic mutations in biological evolution, is this adolescent struggle for an integrated, non-contradictory *super*concept! While this quest is often disparaged as 'idealistic' it is actually the most practical of considerations. It is a survival issue as it pertains to the calibration of one's cognition!

The troubled youth, the rebel, the non-conformist, the 'schizophrenic', the 'mad' man are all terms for the psychological angst that many youngsters suffer as they attempt to clarify their spiritual beliefs, put them into practice and authenticate them. They analyze and evaluate what they have learned, what they have been taught, and what they do not know, attempting to resolve contradictions and eliminate discrepancies in their understanding. They seek a better, more reliable standard of truth. They are in quest of a better, less contradictory *super*concept! They seek a stable **identity**.

Some of these young people will become political leaders, some academics, some missionaries for various sects, some will become 'heretics', some will enter the arts, or perhaps become philosophers, journalist, or writers, some will become religious leaders, perhaps prophets of a new ultimate source of truth, the creators of a new ideology or religion. The greater the environmental challenges to the contemporary or popular standard of truth, the greater number of 'disturbed' or 'disgruntled' youth there will be. It is these 'mentally disturbed' individuals that are the 'genetic mutations' of psychological evolution!

Organic species are undergoing frequent, almost continuous genetic mutations. The huge majority of mutations, even those which provide a survival advantage for the individual animal, fail to succeed as a new species. The more dramatically and decisively a mutation solves vexing problems the species is facing, the more likely the change will be selected by the environment to prosper and proliferate.

The same huge difference exist between the many thousands of individuals who are constantly producing basic psychological changes, and the extremely rare occasion when one of their creations is adopted over a few generations by thousands, millions, even billions of people.

Out of the vast panorama of thousands of human beings each shouting their own unique message the environment selects and allows to survive and proliferate only the belief system of the individual with that unique and specific message needed by the masses to deal with some pressing psychological need. All the others may find their offering of some benefit to themselves, perhaps to their families, but many, perhaps most find rejection and anguish to be their fate.

This is the aspect of human psychology that is potentially changeable, or adaptable. It is from such people, many of them very unhappy, most certainly not 'successful' that the evolutionary processes choose and it is from such individuals that the rare epochal changes in human psychology are derived. In every age the vast majority of these individuals fail to ignite an epochal spiritual revolution. One thing or another, usually something unpleasant, happens to most of these youngsters without them accomplishing much of anything. But ever so rarely, extremely rarely, one of these individuals is dubbed shaman, prophet, mystic or savant. His contemporaries crown him with such a label because he provides just the right answer at just the right time. He is considered 'special' and people are willing to follow his lead. What he has which others want is a new or novel **super**concept! They need this *super*concept to complete and solidify their **identity**!

If the new *super*concept offered by one of these malcontents, mystics or prophets is useful at addressing severe environmental challenges it may rather quickly spread as many people, then perhaps most people around the world adopt the new **identity** with its new psychology. Over the span of a few generations a new *super*concept can sweep across the globe igniting a spiritual revolution. These rare events are how mankind adapts human psychology to changes in the environment. It is in the quest to satisfy his spiritual needs that human psychology becomes subjected to evolutionary processes! Out of the many who are spiritually unhappy and seek some new spiritual path, the physical and social environment selects for emulation the rare individual who offers just the right psychological solution at just the right time. In other words to change human psychology you must change **identity** and to do that on a mass scale you must change the popular *super*concept. Change the *super*concept and you change people's basic **identity** and thereby their basic psychology. The fittest *super*concept, that is to say, the fittest psychological adaptation, survives and proliferates.

The Key to Psychological Adaptation

Summary

Man must have a *super*concept of truth in his mind to serve as a reference standard for non-contradictory cognition. The nature of his *super*concept will determine the nature of the individual's **identity**. A person's **identity** determines his motives, goals and decisions; in other words, his psychology. Therefore, for a people to change their basic psychology they must change their core **identity** and to do this they must change their *super*concept. This is a very rare event. I have found only four such changes or adaptations in the existence of our species.

A person's *super*concept may exist in his mind as a conscious explicit belief labeled with a word from his language or it may be unconscious, implicit and unlabeled. Labeled or unlabeled has no effect upon the workings of the *super*concept, but it may affect superficial aspects of a person's identity. A label is superficial; the person's *super*concept is a fundamental part of identity.

If a person's *super*concept is labeled with the name of a God, religion, or ideology, then identification as such a believer may be a socially important component of such an individual's identity. The label given to one's *super*concept may be socially and politically important but it is unimportant to its psychological functioning.

I should mention here that we must not assume that belief in a God or Gods to be a religion. We currently live in a time when religion is our means of psychological adaptation so this is the contemporary assumption. Religion however requires more than spirituality and the belief in a God. We will return to this issue as we draw nearer to our time.

Hunter-Gatherer Psychology

I will now attempt to do the seemingly impossible. I will try to outline the psychology of our ancient hunter-gatherer ancestors! It is an enormous span of time, for I go back to the very beginning of our species! But remember I have access to data acquired by many gifted archaeologists and anthropologists[9, 10, 11]. This includes data from the observation and study of hunter-gatherers of more recent times. I know you are skeptical, as you should be. Let me proceed nonetheless.

Our species lived the hunter-gatherer lifestyle for a longer period of time than we have lived any other. Human beings seem to have been uniquely designed for his kind of existence.

The hunter-gatherer's brain was of course identical to ours. The hardware is the same, but what about the software? What was his standard of truth? Did he have a moral code? Against what did he determine non-contradiction? And when beliefs were found to contradict his truth how did he handle contradictions? More succinctly: what was the hunter-gatherers *super*concept? What kind of **identity** did hunter-gatherers possess?

Hunter-gatherers drew no cave pictures of their God. No temples were built to honor or worship their God. These people lived intimately with their God, not separate and removed. A picture, a temple, an idol or an exclusive spokesman, would have been unthinkable to them. Each individual seems to have had a personal relationship with their God, so no interpreter or go-between was ever considered necessary. Their God was tangible, visible, available and knowable. Ytilaer was not planning on leaving humans and these humans were not planning on leaving Ytilaer. Their God was dependable, sustainable, reliable, concrete, available and responsive. He was not going to end their world or substitute another. The God Ytilaer was a living God. No dead monuments or lifeless images were needed or desired.

The hunter-gatherers God or Gods were **Nature**, or some aspect or aspects of the Natural World. Hunter-gatherers' spirituality was not the unbridled superstition often associated with ancient peoples. Hunter-gatherers of this ancient era appear to have viewed nature, reality, the world or the universe, as God. God was thus tangible, vi sable, knowable, and in constant direct view. There was no supernatural or other-worldly aspect to the hunter-gatherer God. The hunter-gatherers' God, his *super*concept was Nature and it was perceptible and understandable. The hunter-gatherer's **identity** was as a creature of Nature, living in and of Nature.

Hunter-gatherers were apparently respectful with the bodies of their deceased, burying them very carefully. But their views on an afterlife are difficult to decipher. My guess is that their beliefs regarding what happens after death varied greatly from one group to the next. What is obvious is that the hunter-gatherers felt no urge to force any of their beliefs on others. If you didn't like the way your family did things, or their beliefs, no one would prevent you from striking out on your own.

Perhaps from observing their Homo erectus forebears, or from watching animals, hunter-gatherers learned the importance of knowledge of nature. It was a short conceptual leap from studying nature in order to hunt and gather effectively to adopting Nature as their *super*concept of truth. I believe that God and Nature were ***identical concepts*** in the mind of the hunter-gather. We contemporaries generally have at least two concepts where hunter-gatherers had but one. Imagine the concepts of "God" and "reality" merging into one lone concept in your mind. Imagine no contradictions between "reality" and "God" or between "Nature" and "God" in the hunter-gatherer's mind! There was nothing supernatural about the hunter-gatherer God and there was no conflict between Gods realm and man's! There were no contradictions between the **perceptual** part of an infant hunter-gatherer's *super*concept and his later learned **conceptual** part of his **identity**. His *super*concept was integrated, and it was non-contradictory. And furthermore it worked! When the hunter-gatherer took action to authenticate his **identity** the behavior was generally successful! His *super*concept was internally integrated both perceptually and conceptually, ***and*** his **identity** provided the basis for successful action in the natural world! Behaviors based on his *super*concept of Nature and his **identity** as a creature of Nature resulted in two rewards! First there was the reward of the successful action taken in the natural environment, perhaps the successful catching of a fish. This would provide food for himself and his family to enjoy. The second reward was the endorphins he experienced from the authentication of his **identity!** (See! The universe really is the way I believe it to be! The beliefs I have, pointed me to this fish and provided a way for me to catch him!)This spiritual or mental reward was added to the physical reward of enjoying the meal of fish. (Actually all rewards are mental in the sense that all are mediated through endorphins, but there are different endorphins for different activities.)

This *super*concept of Nature also influenced the way hunter-gatherers thought. Thinking, the hunter-gatherer type of thinking, where concept building starts with Nature favored percepts. When thinking begins with percepts and is integrated without contradiction from the ground up, it is called percept-driven or 'empirical'. This type of thinking is often called 'common sense.' I believe this 'empirical' type of thinking to be characteristic of hunter-gatherers. Such thinking is 'bottom-up', from perception to conception. Such thinking begins at the ground level. An empirical thinker does not allow concepts to overrule perception. So imagination and conceptualization is kept in check by a diligent, non-contradictory cognition. A cognition that is kept subservience to perception, objective reality, and experience, or as the hunter-gatherer might explain, to God. The hunter-gatherer is not 'up in his head' rather he is focused on God, Nature, or objective reality. He is looking out at the world, out at the environment, out at an objective and Natural God.

The hunter-gatherer was looking for 'signs' from God, for God to 'talk' directly to him. God 'talks' to the hunter-gatherer by 'what works'. He is trying to understand God's laws, Nature's laws. 'What works' tells him he is closer to an understanding of

God's laws. In this sense the hunter-gatherer was a natural scientist. He was curious about his world, his Natural God.

The hunter-gatherers unifying concept of truth was derived and vindicated through perception. The hunter-gatherer God was an objective Nature, God Ytilaer, or reality. This God, this *super*concept, was stable, benign, predictable, dependable and non-interventionist. Nothing in biology, nature, or reality was in conflict with this God. Some *super*concepts favor perception, while others favor conceptualization. Nature or Nature's God as a source of truth favored a perceptual or empiric cognition. By strict worship of this God man was protected from conceptual manipulation and trickery.

For hunter-gatherers then the conceptual portion of his *super*concept was Nature's God. This conceptual portion did not conflict with the perceptual portion forming a coherent *super*concept. This all-encompassing idea of God as reality or Nature drew children to focus on their social and physical environment. The worship of and study of their natural world formed their **identity** and it was both a spiritual and academic enterprise. This equating of their spirituality with their material existence prevented a schism between perception and conception, between mind and body.

I doubt that hunter-gatherers considered ethics as a separate question from understanding Nature. Their dependence upon nature and their **identification** with nature would have lead them to an implicit natural morality. Such a 'morality' is implicit in the behavior of animals. Few animals routinely kill members of their own species. Fight and perhaps injure yes, but murder, no. There is competition with other members of one's species, but it is seldom lethal. We might even go so far as to say there is an implicit recognition of individual sovereignty. Animals of the same species generally live and let live!

Animals are largely disinterested in others of their same species except as satisfaction of sexual needs are concerned. Among animals satisfaction of sexual needs is generally accomplished through mutual cooperation. Exploitation of members of one's own species is rare. In the animal world each individual is doing that which satisfies its own needs. I find no examples of animal behavior that are not selfish. This comes very close to the ethic of rational self-interest. I believe this natural or animal morality is exactly the type of ethic to which our hunter-gatherer ancestors would have subscribed.

When we ask how the hunter-gatherer handled contradictions we must conclude that he tended to use the mechanism of rejection. Rejection requires little discipline or will-power. It is the easiest, most natural way for the human mind to handle concepts that do not make sense. It is therefore reasonable to assume that rejection was the hunter-gatherer's principle method of dealing with contradiction. That which did not make sense was simply spit-out of the mind and ignored.

So to summarize I believe the hunter-gatherer's *super*concept was **Nature.** He developed an **identity** as a natural creature at home in his natural world. From this natural epistemological standard I believe his cognition to have been empirical and his ethic to have been rational self-interest.

Hunter-Gatherer Social Structure

The social organization which grew out of the hunter-gatherer's belief in Nature as his God, his priority of perception, and his selfish efforts to satisfy his needs was the biological family. Sometimes these groups grew in size by including members of the extended family to form a clan. But just like Homo erectus, and many other primates extinct and extant, Homo sapiens found the family organization to be the best way to survive, live and meet their needs.

Their concept of Nature, or their God Ytilaer, provided hunter-gatherers the reference standard of truth they needed. Utilizing this concept as their *super*concept for non-contradiction, our ancestors could make sense of their world, survive and flourish. They undoubtedly saw themselves as beings that understood Nature and by so doing were able to survive and prosper. Their identity was as 'Nature understanders' or those who know and worship Nature. Each new generation of children were exposed immediately to the *super*concept of Nature, which then guided them to revere and respect nature and acquire knowledge of the environment. This belief system also created a certain attitude in parents toward their offspring.

When an infant is viewed as a natural part of life he is valued by his parents as a natural aspect of God or Nature. In this view he is not either 'good' or 'evil.' I think hunter-gatherers viewed the pursuit of the satisfaction of one's needs as natural. When a child is valued by his family he will easily learn to value his **self**. Under such circumstances no explicit instruction is required for an infant to acquire a positive identity including a positive moral orientation. An infant learns to love his self from the love he is shown by those around him. Unless undermined in some way the simple act of feeding an infant and sustaining his life will be interpreted by an infant as proof of the value of his **self**, as a positive value placed on his life. This will create the positive moral orientation of self-love. The child will then develop a conscience that places a high value on self-sustaining, self-benefiting, and self-attaining behaviors. The hunter-gatherer's conscience and his needs were in alignment.

I believe that the hunter-gatherer child drew a sharp distinction around his **self** as a consequence of his morality of rational self-interest because a clear and precise **self** is necessary in forming the concepts and making the decisions needed to survive in the natural world. In evaluating his survival skills for example, the hunter-gatherer had to make a judgment of his individual prowess. When hunting, stalking a prey and getting up close and personal for the kill, the hunter-gatherer had to know his capabilities and his limitations. Assuming too much knowledge, strength or skill would result in injury or death. In order to survive it was very important to know one's capabilities and weaknesses. One had to evaluate their **self** as objectively as possible. I believe the hunter-gatherer conceived of his **self** as an ***individual*** human being.

The principal disadvantage to this psychology that I can suggest is that it may have engendered a kind of smug certainty. Hunter-gatherers were highly competent at dealing with their world. Information from the few interviews of hunter-gatherers by anthropologist's available[11], reveal them to be extremely self-confident of their view of how things worked. This hard-headedness, stubbornness and closed-mindedness may have promoted a sense of finality. Hunter-gatherers may have believed their world to be perfect, unchangeable and in no need of improvement. From our prospective the hunter-gatherer's biggest failure was the inability to establish permanent settlements.

Perhaps they believed that they were powerless to effect any changes or improvements. Consequently they made few discoveries or inventions. Now I could be completely wrong in this assessment. Perhaps our hunter-gatherer ancestors discovered and invented much more than we can now perceive and appreciate. Perhaps I am too psychologically naive to perceive all that they knew, understood and accomplished. They seem to have enjoyed their lives, and over enormous periods of time seem to have made few changes in the way they lived.

In sum I think we should give our first ancestors credit for what they accomplished. They managed a sustainable and rewarding life-style largely free of famine, warfare, murder, slavery, and child abuse. Humans over this huge span of time appear to have behaved ethically and morally, in other words these ancestors behaved as do all the other creatures in the earthly kingdom.

Hunter-Gatherers
Contemporary Perspective

Pause for a few moments and think back to the little story about the Hollow Tree and the missing fishing poles. Is this story naive? If you think so, you have the reaction many others have had. If you think this story is too childish or too simplistic you are not alone. The story is surely all of that. But maybe some of your criticism of the story as unsophisticated arises because it does not contain a sufficient amount of violence! You will remember the statement about hunter-gatherers that *"many generations of people may not have witnessed a single episode of aggressive, violent, murderous behavior!"*

Perhaps we all now equate sophistication with the acceptance of a certain amount of aggression. A story without bloodshed and killing may be considered a children's story. We must consider the biases of our perception.

What a person perceives is to a great extent determined by what his mind has been programmed to receive. That which forms the 'noise' around you is very much of your own making. If for example you have lived for some time among the hustle and bustle of a modern industrial city you probably tune out much of the noise and racket, ignore the vibration and focus on your concerns. Someone just arrived from a quiet home in the country might find the loud noise and clamor of your city disconcerting.

If a city dweller went alone deep into the forest and remained alone in an isolated location, changes in his perception might slowly occur. Alone, with plenty of provisions but without a cell phone or wireless computer, or any other communication device, the environment might seem very quiet. For a while he might be very interested in his environment. Fear and concern regarding his safety might stimulate him to explore the area and influence his choice of where to camp. But after a while the lack of modern sensory input would probably lead to boredom.

In spite of the boredom let's say that our subject remains deep in the woods without any sound of humans or any signs of human life. Very slowly and subtly his hearing acuity may improve. The stream he camped near so water would be handy seems to be slowly becoming louder. Yes the rippling of the fast moving water definitely has grown louder!

Also as time passes he notices that the chirping and singing of the birds is clearer, crisper. Even the buzzing of insects now has a greater intensity. He is becoming much more aware of his surroundings!

His perception, his hearing is becoming more sensitive. His hearing had adapted to the heavy load of sound he had to contend with in the city. Now out in the wilderness his sensitivity to sound is returning.

This compensation of our senses to the environment is not limited to hearing. A heavy or loud sensory load impacts all of our senses in this way rendering us less sensitive to the subtler, quieter or 'softer' sensations around us. This ability of people to numb themselves to an overload of stimuli is not limited to the senses; it is also a psychological phenomenon.

Consider drama for a moment. Television, radio, magazines, movies and books of our contemporary era compete for our attention and our money. Each medium ratchets up the violence and bloodshed as it attempts to gain and maintain our attention. The background 'noise' of violence goes up. This 'noise' of violence seems to be constantly increasing. We become more and more numbed by the incessant violence and killing. The American Psychiatric Association says that the average youngster in the United States of today has by the age of 18, through television, and the movies, witnessed over 200,000 episodes of aggression including over 18,000 simulated murders[24]! Inundation can overwhelm our senses. Perception can alter a person's view of things, including violence. Drama it seems cannot exist for us without aggression, violence, bloodshed and murder.

If you look around our contemporary world you will find that this high level of violence is everywhere, not just in drama or entertainment. Every news source reports daily on the violence, rape, torture and killing of human beings at home and abroad. This violence is not just among individuals, for many nations, regions, ethnic and religious groups around the world are busily engaged in destruction, aggression, violence, torture and murder.

We might reasonably refer to this constant and excessive violence as the 'noise of evil'. This heavy and constant exposure to evil is pervasive. It is nearly universal. Evil is so prevalent that we have nearly lost out ability to perceive it. Evil is everywhere. We are inundated with it. But we have accommodated to it. We now pay little serious attention to evil. Evil has become our entertainment, our trade and occupation and our national anthem. Yet we have become largely oblivious to its existence. This saturation with evil affects us all. But we are largely numb to its affects.

I recall coming home late one evening. Tired from a long day I sat down in my easy chair in the living room to catch my breath. My wife asked if I wanted something to drink. I answered in the affirmative. As she went to leave the room she asked if I would like the television turned on, and if so to which channel. I asked her to turn on the television and tune it to the 'shoot'em up' channel. As she clicked the remote I added, 'I just need to relax for a few minutes'. When I heard myself say those words they struck me as discordant in some way. Watching people shoot and kill each other for entertainment and relaxation!

Perhaps we all, or at least many contemporaries, have come to expect a certain amount of aggression, violence, bloodshed and murder as 'normal'. Then any story or theatrical production which does not portray at least a minimal level of such evil is experienced as juvenile or childlike. What I am suggesting is that we live so immersed and surrounded by evil that we have become addicted to it. We have not only adapted psychologically to a huge diet of evil we have come to embrace it as normal and 'realistic.' Worse, we have become complacent with evil, content to live with it, even accord it respect!

Here is another *News Flash!*

A military airplane is flying at 15,000 feet. It is a pilot-less drone armed with explosive missiles. It flies quietly for a period of time. The drone then enters the airspace of a sovereign nation. This nation has declared war against no one. The pilot, remotely controlling the airplane from 10,000 miles away, receives his target instructions. The coordinates of the target is punched into the computer and the missiles are armed. Moments later the missiles are fired.

Less than a minute later a private home, in the mist of enjoying a wedding celebration is obliterated along with the lives of dozens of family members and friends. The target was **one *suspected* terrorists.** The deaths of the newlyweds, children, proud parents and grandparents, and dozens of guest are considered 'collateral damage'. Millions of people around the world 'understand' the necessity of this kind of killing.

How did we get from the hunter-gatherer to where we are now? There may be more to our accommodation and embrace of evil than simply setting our perception lower. Our ignoring of the loud decibels of city noise may be an innocent physiologic reaction to sensory overload. Our relationship to the massive overload of evil may not be so innocent.

Section I
Summary

1. For psychological evolution to be a viable theory the human mind must be capable of fundamental change.

2. The mind can only change in a fundamental way if there is the capacity for such change in the human brain.

3. The anatomy and functions of different areas of the human brain is explained.

4. The importance of identity and how it is incorporated by the infant is explained.

5. The importance of a standard of truth to epistemological identity and how it is incorporated by the infant into his *super*concept is explained.

6. Endorphins and the Hindbrain kick-back reward system are explained.

7. The huge endorphin reward for action taken on the basis of one's *super*concept is explained. This is the process of authentication.

8. For man to change his psychology he must change his *super*concept. This is extremely difficult and only very rarely occurs.

9. Homo sapiens do not automatically choose force as their first option in dealing with other humans and are not innately violent. Homo sapiens are innately about as aggressive as other primates. Humans, like all other living creatures, are about satisfying their needs.

10. Homo sapiens appear to have a biological need for family just like many other animals. This is also about satisfying needs.

11. The first psychological adaptation Homo sapiens made was the Hunter-gatherer's belief in a natural spirit, Nature or God. This *super*concept holds that the universe is fixed by natural laws enforced by a supremely powerful and benevolent Spirit or God.

12. The hunter-gatherer's constant effort to know and study his God Reality held his mind in intimate contact with the perceived natural world which gave him a special mental strength protecting him from easy conceptual manipulation and deception.

13. Hunter-Gatherers, because of their God Nature, had a tendency toward empiric, perceptual, rational thinking and behavior. This is the type of thinking that gives the best results, the best chance for survival.

14. Hunter-Gatherers had a very strong tendency to develop a positive self-regard. This type of self-evaluation provides the best chance for survival.

15. Because the hunter-gatherer's spiritual quest led him to select Nature as his *super*concept, as his God, his cognition was effective in solving his problems of survival for over 180 millennia. The hunter-gatherer's **identity** was that of a person who knows Nature and who is able to survive and prosper in the Natural World. Man's first psychological adaptation was a great success.

16. The biggest disadvantage in this psychological adaptation I can suggest is that it was too successful. The hunter-gatherers were too comfortable and secure, experiencing few pressing needs. Consequently they discovered little and invented little. Their life-style remained virtually unchanged for 180 millenniums.

17. Our hunter-gatherer ancestor's biggest failure was their inability to find a way to create a sustainable permanent settlement. Therefore they were never able to give up their trekking after game and other food.

18. This first psychological adaptation is characterized by a *super*concept of Nature, an ethic of rational self-interest, and empiric cognition.

SECTION II

Tribalism

Mankind's Second
Psychological Adaptation
20,000 BC- 2100AD?

Chapter 4
Problems in Paradise

As idyllic and Utopian as the hunter-gatherer lifestyle may seem, it had its harsh side. The hunter-gatherers nomadic life was a strenuous one. Certain hard facts had to be faced and endured. Those who could not walk and keep up with the family had to be left behind[11]. Those left behind almost always died very soon. The old, sick, injured and weak fell behind and were left. Perhaps for a time they found food, but without the strong and healthy to assist them their survival was probably short. Deformed and handicapped infants could only be carried by their parents for a while. Once they reached a certain weight, if they could not walk and keep up, they too had to be left behind to certain death. The hunter-gatherer had no choice. This was simply a fact of existence. They had to deal with it. We might consider them stoic or uncaring.

Another accusation we might level at our hunter-gatherer ancestors is this. It is a cliche' that 'necessity is the mother of invention'. If that be true then the hunter-gatherers must have experienced few pressing needs, for they came up with few inventions! Perhaps they were too comfortable, or maybe they were having too much fun, to concern themselves with improving the world. Perhaps they just could not conceive that the world needed improving, or that doing so was in their power. They did not invent the wheel[10] (4000B.C.) or domesticate farm animals[11] (8000B.C.). But they did invent the bow and arrow[11] (40,000B.C.)

Hunter-gatherer women of child-bearing age delivered an offspring on average every 4 years[11]. About 20 millenniums ago, the fertility rate of child-bearing aged women suddenly doubled to one child every 2 years[11]. Rather quickly the total worldwide population of humans doubled. As the human biomass doubled from one million to two million people the populations around key, exceptionally lush and fertile areas of the earth may have become quite dense. Competition for these fertile areas could have developed.

Hunter-gatherers were physically very lean[11]. The constant walking prevented the accumulation of much body fat. Body fat produces estrogen, the female hormone. So as the women gained a little more body fat this could have increased their fertility[11]. But this theory assumes an improving diet, or a decrease in trekking, before the advance of agriculture. Another theory is that the bow and arrow made a bigger difference in people's diet than scientists generally assume. Maybe more red meat added more fat to the body and that triggered the population increase[11].

Perhaps both the increased fertility and the adoption of agriculture began as local phenomenon with such areas as Fertile Crescent of the Near East by providing the hunter-gatherers abundant food without as much strenuous searching.

Then perhaps it wasn't the pressures of increasing population at all. Perhaps out of sympathy for the old, sick, or handicapped, or from laziness, someone saw a solution to the constant traveling in pursuit of food. Hunter-gatherers had long known about seed and planting[11] but it was just a curiosity, they never seriously engaged in agriculture. As one hunter-gatherer told an anthropologist, "why would we spend time and effort planting a crop, when so much food just rots on the forest floor?"[11] But agriculture would allow permanent settlements and stop the constant nomadic search for food.

In fact agriculture is practically a prerequisite to the establishment of permanent settlements. Some fishing villages became permanent or semi-permanent, but few of these became very large. Agriculture is the essential requirement for permanent settlements with large populations. But the agricultural enterprise requires a great deal of energy. And at this time in pre-history the only source of energy that could be applied to agriculture was man power. The ox and the horse had not yet been domesticated. Even after animal power was harnessed 8000 years ago agriculture remained an extremely labor intensive enterprise. Only with the availability of steam and petroleum powered machinery would the great potential of farming be realized. 20 millenniums ago the only available power was human muscle. But the hunter-gatherer found it too easy to live off the land to devote himself to the monotonous drudgery of working the soil. Therefore the only way to effectively harness human effort, the only way to make people work the soil, was through the invention of slavery.

To fully appreciate why agriculture required slavery we must again reflect on the fact that agriculture is an enterprise which requires a huge amount of power. Power is needed to plow the fields, dig irrigation ditches, plant seed, cultivate and de-weed, harvest the crop and transport the harvest to storage. Before the invention of the wheel even more power was required. Before the domestication of the ox and horse, the only power available was manpower. Agriculture is just so very labor intensive. Even with the mule, iron plow and wheeled equipment available in the 19th century it remained an arduous undertaking. At agriculture's beginning 20,000 years ago, humans had only humans for this work.

Permanent settlements would be a lot easier on the old, weak, sick and handicapped. For those too lazy to walk and hunt, permanent settlements could mean a less strenuous life. Settlements would allow more easy communication and cooperation among people and could lead to the cross-fertilization of ideas and the faster advancement of knowledge. And for those who did not want to work, but wished to gain power over others, agriculture offered a possible means. Or perhaps the motivation for permanent settlements was fear! But whatever the motivation for permanent settlements may have been, agriculture was required. And agriculture on a scale needed for a permanent settlement at this point in time required slaves, lots of slaves.

So if the establishment of permanent settlements was the highest priority then agriculture was a necessity. If agriculture was a crucial need then the only means of exploiting it sufficiently at this early point was through the use of slaves. But to turn hunter-gatherers into slaves would be quite a challenge. It would require a new kind of psychology. A new psychology, a slave psychology, would be needed and that would require a new and very different *super*percept!

About 20,000 years ago this was finally accomplished. After 180,000 years of successful hunter-gatherer adaptation mankind changed human psychology in order to exploit agriculture and create permanent settlements. It was quite an accomplishment.

Shama

Very early one morning, 20,000 years ago, when Shama was only 6 years old, his father, Nadar took him deer hunting. As it turned out this would be the first and only hunt of his life. Deep in the forest, while it was still very early, he and his father spied a large buck. Slowly, very slowly and quietly they stalked the animal trying to get closer for a better target. They were both so intensely focused on getting to the deer that neither noticed that a huge cougar was stalking little Shama!

Just as Nadar closed in on the deer and drew back the arrow on his bow the cougar pounced on Shama's back! The boy screamed out as he fell to the ground under the big cat's charge. Instantly the archer's target was changed and with deadly accuracy the arrow found the cougar's heart! The big cat fell dead to the boy's side. But Shama was mauled badly. He was bleeding and cried out in excruciating pain. And he was very, very frightened.

Nadar carried his son back to the camp in great haste. Shama's mother, Lu Mon and other family members immediately ministered to Shama and tried to relieve his pain. They administered a special juice and had Shama chew on a special bark. His family worried that he might not live.

Hunter-gatherers were acutely aware of the need to avoid injury. Much of their hunting strategy involved carefully avoiding getting hurt. The biggest benefit of the bow and arrow was that it allowed hunters to kill game at a safe distance. The club and even the spear required intimate contact with the game. Because of the invention of the bow and arrow the danger of hunting had greatly decreased. The bow and arrow had also saved Shama from immediate death, but he could still die of his injuries.

Even after arrival in camp, with his mother at his side, Shama was unable to calm down. In spite of everyone's efforts to comfort him, Shama remained intensely frightened! Lu Mon devoted herself completely to her sons care. He remained in constant need of his mother for a very long time. Slowly, very slowly, over the course of many weeks he began to improve. It took a very long time but he did recover physically.

Even after Shama's physical wounds were well healed he continued to cry out for his mother whenever she left his side. He still could not relax. He could not get over the panic that he now lived with constantly. At great inconvenience to his entire family Shama remained in camp and at his mother's side. Everyone in the clan had to make adjustments to their hunting and foraging so that Lu Mon could stay at Shama's side. He was so fearful when she was out of his sight that Lu Mon had to devote herself almost exclusively to him. Her life nearly ceased except for her caring for her boy.

Shama suffered with nightmares constantly. Every night the boy awakened frantic with fear. With screams and tears he recovered only though the consolation of his mother. Weeks turned into months with little change. As Shama grew older he remained extremely fearful of the forest and would not leave the camp to go hunting. People came to think of Shama as 'strange' and 'special'. Shama became very knowledgeable of medicinal plants. He was soon an expert on pain reducing potions and mind-altering extracts. Shama used these substances daily. His friends brought him the plants he needed from the forest.

Shama's father Nadar could not accept the way he lived. One day he left Shama and Lu Mon and returned to the hunting life. He was never seen by them again. This did not seem to bother Shama that much, but it added another burden to Lu Mon's life.

Shama had a great talent for making-up stories and could tell them with spell-binding drama. His stories were always about monsters and demons that lurked in the forest and attacked and killed those who went there. These monsters and demons had sharp claws and long teeth with which they caused great pain and death to all they caught in the forest. Shama did not wish to travel through the woods. He wanted the camp site to be permanent. He never wished to leave where he was and he never again wished to go trekking after game. He always looked for ways to avoid the search for food and wanted others to bring food and the magic plants to him.

Shama found that others would bring him meat, fish, fruits, nuts and medicinal plants in exchange for a good story. People especially liked stories that explained why God had chosen to have things happen as they did. People were particularly interested in why people had died. Why did grandpa die in his sleep? Why did one of the boys in their clan drown while swimming in the river?

Shama enjoyed his skill at making his listeners laugh, cry, or become very excited and frightened by his stories. He took to dressing in a strange fashion. He wore brightly colored bird feathers in his hair and painted his face with red, yellow and blue dyes. He looked fierce. He was especially pleased by his ability to create great fear in the imaginations of his young listeners. He learned to beat the drums and to throw powders into the fire that created clouds of colored smoke. He had a great talent at creating a spectacle.

As Shama grew into manhood he accumulated a following of young people. A few orphaned children were always around him and they catered to his every need. Other youngsters left their families to stay near Shama, hear his stories and enjoy his potions. Shama established a permanent settlement at the bend of a great river. Here the vegetation was so lush and the wildlife so plentiful a permanent village could be maintained almost without agriculture. Shama was able to maintain a permanent encampment by sending out members of his group to procure food and bring it back to the settlement. His group of followers began to plant, cultivate and harvest crops. These followers stayed with Shama, worked the fields, hunted and cooked. They obediently carried out his commands and listened attentively to his every word. Shama told even more fantastic tales of monsters in the forest and of the supernatural powers of these monsters. Shama believed his own stories and often scared himself as much as his listeners with his tales.

Shama's mother, Lu Mon died suddenly. She had been extremely frail and pale. She had also been very depressed since Nadar left. She may have taken her own life. Her death shocked Shama profoundly. For days He muttered "why, why, why"? Shama was unable to eat and lost all interest in his surroundings. He had trouble sleeping and when he did doze off his nightmares quickly awakened him. These nightmares were severe. He dreamed about demons and he dreamed that these demons were immortal. He believed that these monsters were Gods and that they inhabited another dimension. Shama came to believe that the Gods had punished him by taking his mother from him. How had he displeased them?

Shama 'saw' his mother often. He 'saw' her walking in the village. She appeared to him as he sat in his hut alone at night. He would often see her among other women along the path. But when he ran to her screaming 'mother, mother', she disappeared! Shama came to believe in the existence of another world, a realm only a few, special men, like him, could discern. His mother and all the departed resided in this supernatural realm. In Shama's imagination the Gods grew even greater in their viciousness and fierceness.

Shama was now taking even greater amounts of his mind-altering brew every day! It permitted him to 'see' what others could not! Finally Shama concluded that the Gods had to be appeased or they would punish him again. He concluded that the Gods would punish all mankind if they weren't satisfied. If one expected to go into the forest to gather food or hunt game and come out alive, the Gods would first have to be appeased.

This realization by Shama, that the Gods must be appeased, brought him back in touch with his followers. His appetite and sleep improved. He returned to his preaching and storytelling with renewed vigor. Now he preached the need to appease the Gods. He preached the importance of making offerings to the Gods. He preached a new concept which he had conceived. This new idea or belief he called EVIL! Evil; he explained was the worst of the bad. Evil was a built-in desire and tendency to do terrible, awful deeds and commit great harm. And he explained, mankind was the source of all evil. Every child, he explained was born evil! Every human was evil at his very core. This is why the Gods did not shower man with blessings. Because man was the essence of the bad, the awful, the worse of the worse, in other words, man was evil. The monsters and the animals of the forest killed and devoured men because they were evil. Evil man should not enter the woods.

When asked how he could know of Gods which others did not apprehend, he explained that one had to force oneself to believe that which could not be seen. It took a strong mind; weak minded individuals would never see the messages of the Gods. This type of force, this mental force or self-discipline he called faith. He had faith in his beliefs and his followers must also have faith.

Shama started carving wooden figures of the Gods he saw in his dreams. After a while he had many Gods. Some of the Gods looked like deformed animals. Other Gods looked like monsters no one else had ever seen, while other figurines of Gods were non- descript globs of painted wood. He carved his figures larger and larger. Soon some of Shama's God figures were as large as a log and took almost an entire tree to construct. He dug holes and stood the figures up as if they were holding a meeting. After a while he had a special tent erected over his Gods.

Shama set up a system of accepting offerings on behalf of the Gods to ameliorate their tempers. He encouraged food offerings mostly, but ornaments and precious shells, colorful bird feathers, bears teeth and sparkling rocks were accepted as offerings as well. This tent was devoted solely to these Gods and it became a place to accept offerings to them. As Shama explained to the villagers often, only offerings to the Gods calmed their tempers and made it safe to go into the forest to hunt. Only offerings to the Gods made it possible to survive, kill game and return to the village alive.

A question many followers had of Shama was this: why are the Gods so mean and angry with us humans? Shama's answer was: "because we humans are so evil!" But why was man so evil they asked? Shama taught his followers that mankind had brought upon themselves the wrath of the Gods, that humans deserved the way the Gods were treating them! In what way were humans' so disappointing to the Gods? What disappointed the Gods and incurred their wrath was man's selfishness! If a person wished to find acceptance in the eyes of the Gods they must renounce their concern for their self and sacrifice their life to the service of the Gods. Nothing short of an offering of their life and effort, which the Gods would use as they saw fit, could prove their suitability in the eyes of the Gods.

According to Shama human beings were born evil because they were selfish and only suitable as offerings to the Gods. Only selfless obedience to the Gods could partly redeem a human being.

Shama provided an altar in his special tent where more offerings could be placed. Boer's tusk and bear claws as well as birds and fish were placed on the altar as offerings for the Gods. These food offerings provided sustenance for Shama and his increasing number of helpers. Offerings of medicinal plants were especially appreciated. Shama would wear boar's tusk and bear claws in a necklace around his neck in the fashion of a hunter. But he never ventured into the forest; these trophies were of a different type. He obtained them from the offerings left by real hunters on his altar.

In time people asked Shama to witness them kill a captured animal as an offering to the Gods. They would place the bound animal on the altar and killed it there. Shama on such occasions, wearing his special regalia, would perform a chant, play the drum, burn incense, and recite prayers to the Gods. He was rewarded for his performance by an offering of food, shells or shinny rocks to the Gods.

Shama was stunned when, one night, Hadim brought in his beautiful little 5 year old daughter and offered her as a gift to the

Gods. Hadim had lost his mate in childbirth when little Meele was born. Hadim had listened carefully to Shama's sermons and had learned to blame himself for his woman's death. Maybe he blamed Meele as well. He now knew what Shama knew, that bad things happened because the Gods were displeased with you. So mostly he blamed himself.

Before Shama could respond to Hadim's offer he sat the child on the altar and with the skill of a well experienced hunter he drove a sharp flint tipped knife into her little heart. Little Meele muttered only a tiny whimper before collapsing onto the altar. Right before Shama's eyes and before he could utter a word she was dead! Her limp body lay there on his altar. Her father Hadim cut out Melee's heart and slapped the beating organ into Shama's hand! Shama looked at the heart, he looked at his hand gripping the beating heart, and he looked at the blood dripping from his hand, he was enthralled! He tried to drop the heart but he couldn't let go! In panic he tried to sling the bloody organ from his hand but it would not dislodge! In desperation he pressed it to his chest over his own heart. Hadim looked at Shama in awe! The father of the dead child came over to Sasha, looked him in the face, and then fell to his knees. After a moment of meditation Hadim kissed Shama's feet!

Shama was shaken to his core by this experience. He could not sleep for many nights. Then in a frightening nightmare the Gods spoke to him. They explained to him that he had been given a great gift, the ability to bring his followers to the Altar of the Gods where, at last, they could find salvation. He took this vision as his cause, and overnight he lost all his fear. For the first time since 6 years of age, he was no longer afraid! Now he felt supremely courageous, confident of his competence and self-assured. He knew his life's mission and his goal and he faced his destiny with no further doubt or indecision.

Shama realized that what the Gods had been telling him all along was that they wanted human blood, human life, in exchange for their help and approval. He didn't know why it had taken him so long to see this. But now that the Gods had spoken clearly, now that he truly understood, he would cease to doubt, he would cease to hesitate, and he would preach the true words of the Gods!

Shama immediately became even more eloquent. He mesmerized his followers with even more spell-binding stories. Offerings of all types increased. He graded his oratory by its results. More and more people turned to Shama to tell them what the Gods wanted of them. More and more villagers' offered their children to the Gods. Shama could see that the suffering of the sacrificial victims was short, while the feelings of purity and holiness, the feelings of security and serenity that was wrought upon the family, nay the entire community, was great and lasted a for a long time. The Gods were good to the people when the sacrifices were plentiful. So as the settlement grew, as the crops were bountiful, as the weather was temperate, as the fish and game were abundant, many children were sacrificed and abundance was great.

Abundance and prosperity was great for a long time. Then a great pestilence struck! Many villagers became very ill and many died. No one had any idea what caused such a calamity, or what to do about it. No one understood what was happening except for Shama. Shama preached a great sermon. He explained to the people that the sickness was a punishment from the Gods. He told the followers that the sickness and the deaths was vengeance of the Gods against them for their false faith. He explained to the people that they had not been sincere in their sacrifices. They had grown arrogant and self-important in their own eyes. The Gods had seen into their hearts and saw their wicked selfishness still hiding there. The sickness and the deaths would continue unless a solution were found, a way to appease the Gods! There was really only one solution. The tribesmen should cease using the sacrificial offerings to rid themselves of recalcitrant or deformed children. It was time for the faithful to be honest with themselves and with the Gods! Shama explained that a true sacrifice demands giving up, giving to the Gods, that which one prizes the most! A sacrifice must hurt he emphasized. Only by sacrificing their most loved and highly prized children could the Gods be appeased and the community saved! He demanded that the villagers sacrifice their most precious, loving, and beautiful children.

Shama oversaw the sacrifice of many beautiful little girls and dozens of promising little boys. The blood flowed almost continuously for two solid weeks. Then low and behold, the pestilence began to abate. Within a few days the sickness slowly began to subside. After more days the pestilence went away and the sick recovered. The sun once again shone bright, the leaves on the trees came out, and the earth was renewed.

Shama was hailed as a great prophet. He was known as 'the one who talks with the Gods'. Whatever Shama now said had the effect of absolute law. His wishes were obeyed immediately and with absolute fidelity. Anyone that opposed Shama or even appeared less than enthusiastic at following his orders was quickly sacrificed to the Gods. Shama was the absolute dictator not only of behavior but of thought! No one dared oppose him.

Shama ordered the people to build a great temple with monuments carved from logs as tributes to the Gods. He established a number of Holy Days and laid out the plans for how they were to be worshiped. He planned the many sacrifices and dances. Shama prescribed the occasions for the use of mind-altering plants. He chose the music, chants, and songs to be used in all the rituals. And he decreed when sex was allowed and encouraged in the ceremonies. Shama's next order was that the women work the fields, plant and harvest the crops while the men hunt game and fish the Great River. All were to bring their bounty to the Great Temple where the Gods had decreed, Shama would have first choice of the bounty.

Shama established all of the annual rituals of his village, and required changes be made in the daily life of the settlement. It became mandatory that all the children over 4 years of age be brought to the Temple each morning, except for Temple Day which was held weekly. Shama's Temple helpers, now called Priest, taught the children from morning till evening nearly every day.

In these very first schools children were taught by the Priest that their sole purpose for existence was to serve the Gods. The Gods demanded that everyone live to serve the Gods and to do what the Gods demanded. The children were taught that every person's job was to serve others before they served themselves. Yes, the highest moral duty and purpose of a human being was to serve the Gods by sacrificing their lives in service to the Gods. The Gods commanded that everyone sacrifice themselves for their community, for the tribe, and for its chief, Shama!

So all the children, and in time, all the people came to understand that their purpose in life was one of sacrificing their self in the service of the tribe and its leader, Shama. This is the way it was to be because that is what the Gods wanted and demanded. Self-sacrifice was the highest virtue and must be promoted regardless of the time required or the cost. Nothing was more important to anyone or everyone than self-sacrifice!

Shama explained that death was nothing for the sacrificial victim to fear. This life on earth was a mere illusion, a test the Gods had established to determine who were worthy enough to enter the real world of the afterlife! Nothing visible and tangible in Nature would last. The wise and faithful would sacrifice their selfish happiness on this fleeting and temporary earth for an eternal happiness with the Gods in their world of luxury and abundance in a life after death.

During the time Shama was establishing his permanent settlement at the bend of Great River, another settlement was developing some miles away on a cool spring near the base of a great mountain. This village had no recognized leader or tribal structure. There was a market area, a common kitchen area, and a designated latrine. But there was little else in the way of organization. There people were still hunter-gatherers, they had no tribal identity, no tribal religion, no temple or religious rituals. They had no chief and had received no schooling in force, supernaturalism or self-sacrifice.

One day a group of men from the Great Mountain encampment met a group of men from Shama's Great River Tribe. The

Great River tribesmen told the men from Great Mountain that they worshiped a great Chief, Chief Shama. They told the fishermen that Shama was the one who talked to the Gods and demanded human sacrifices. They explained that they were fishing for their village, Shama's village.

The men from Great Mountain did not understand what the tribesmen were saying. They laughed and poked fun at the idea of a supernatural realm. They voiced concern that the tribesmen had been fooled. They ridiculed the idea of the Gods talking exclusively to Shama. They told the tribesmen they thought it was very wrong to kill human beings and could not understand such vile and hateful behavior as killing children. They told the tribesmen that the true God Ytilaer was undoubtedly very angry with them! And when the men of Great Mountain caught many big fish and Shama's tribesmen caught only a few tiny ones they laughed and said the true Great Spirit was obviously very angry with them.

When the tribesmen returned to their village they immediately went to the Great Temple and met with Shama. The tribesmen told Shama about their encounter with the fishermen from Great Mountain. They told Shama all that had happened. He instructed the men to go into the Temple and pray, he would seek counsel with the Gods.

It was not until morning that Shama emerged from the Temple Sanctuary to address the crowd of men who had assembled to hear his proclamation. He spoke in a slow quiet voice but his fury could not be camouflaged. "These men of the Great Mountain encampment are savages of the worse sort. They are lazy. They hunt, fish, and live off the land. They wander the forest and have no loyalty to any land and they know nothing of agriculture. The Gods have not chosen them and they remain ignorant of the Gods and their demands. These barbarians are selfish, they live only for themselves making no sacrificial offerings, and they have no temple and no priest or chief. But their most contemptible behavior is their blasphemy of the Gods! They make fun of sacrifice. They ridicule the Gods and dare to call the offerings of our most cherished children to the Gods an affront to their Great Spirit! In the name of the Gods I will lead you to this encampment and we shall force these heathens to bow down to the Gods and make sacrifices to them!"

All the able bodied men of the village took up their bows and arrows, flint tipped knives, bone tipped spears and stone axes and followed Shama. Out of the village compound they walked headed for the Great Mountain encampment. After more than three hours they arrived in sight of the Great Mountain village. They formed a circle to discuss strategy. They would enter the camp just as if it were a herd of deer.

Two groups of tribesmen were formed. One group circled around to the far side of the encampment and then the tribesmen made their move. Shama wearing his most frightening costume of skins, bird feathers and paint, walked boldly directly into the middle of Great Mountain camp. His drummers and chanters loudly announced his arrival. A few men came out of their huts to see what the commotion was about. Shama asked them were they prepared to bow and kiss his feet and acknowledge their total obedience to the Gods. When one of the bewildered men laughed a shower of arrows killed him instantly.

More men came out into the clearing, as they ran to assist their fallen friend they too were slain. The tribesmen went through the village killing all the men and everyone who offered resistance. The women and young men who were not killed and the children were tied up. Then crying and afraid they were herded back to the Village of Great River.

As Shama and the tribesmen returned to their village people came out to see what was causing all the crying, shouting and noise making. The people encircled the hostages. Shama stood on a platform in front of the Great Temple. Shama spoke "These people have blasphemed the Gods and the Gods now bring their wrath down upon them. The men have been slain as a sacrifice to the Gods. Others will be sacrificed on our Altar to our Gods in due time. These children will be taken by our Priest, raised in the Temple School, and trained to serve the Gods. These prisoners are to be tied up and guarded so they cannot escape the vengeance of the Gods".

The next morning a number of the tribe members went back to the desolate Great Mountain camp to search for signs of blasphemy which they planned to destroy. Instead they found a large supply of animal skins, pelts, baskets full of nuts and fruit, dried fish, and a large quantity of very fine flints. A few dogs were also present. The fishermen's huts were burned down. All the loot was taken back to Shama who shared some of it with the villagers.

Over the next few days the older men and women captives were sacrificed to the Gods. As time went by the older children who could not or would not learn self- sacrifice were also sacrificed on the altar of the Gods. The children who learned their lessons of self-sacrifice were turned into slaves. They became slaves to the tribe and its chief, Shama.

Shama immediately saw the benefits of conquering other tribes. More slaves and more sacrificial victims can always be used. The bounty taken from conquered tribes enhanced the wealth and power of the tribe. And the tribesmen enjoyed the glory of the easy kill and the valuable trophies warfare brings.

Shama's Invention

Shama was one of those 'mentally disturbed' youngsters that finds nothing to his liking and wants to change everything. Completely breaking with his hunter-gatherer tradition, he hated and feared the forest. Like his peers of every era he knew everything and wanted to straighten-out his world.

Shama wanted a permanent settlement and he wanted to never have to enter the forest again. He knew that agriculture was the key to accomplishing this. What Shama needed was the power to do a lot of farming to produce a lot of food on a regular dependable basis. But this was a time before any farm or draft animals had been domesticated. The only power available for anything was manpower. He needed men who would follow his orders and raise crops!

Shama intuitively realized that to accomplish this he would first have to create in people's minds a *super*concept and an **identity** which was very different from that of hunter-gatherers. So long as his people believed their own perceptions, their own understanding of Nature, so long as they believed Nature to represent the Truth, he would not be able to harness them. He knew he had to create a power much greater and much more powerful than Nature. He knew he had to create a new 'reality' for people, a 'reality' that **he** controlled. Shama realized that only by controlling people's minds would he be able to control them and make them work for him. He had to find a way to sever people from their natural tendency to rely upon their own perception. He had to find a way to sabotage people's tendency to 'see' God in Nature. He had to find a way to substitute people's reference of truth, their *super*concept of Nature with a *super*concept of himself as Chief! He, rather than Nature, had to become people's standard of truth! He had to create a new **identity** for his followers!

Let's look at the situation from Shama's perspective. Shama wanted a permanent settlement. He was fearful of the forest. He knew that a permanent settlement would require farming. For farming to produce enough food to sustain a permanent settlement Shama knew that many people would have to work the soil. His challenge was how to get a bunch of independent, laid-back, easy-going, self-sufficient hunter-gatherers, to become dedicated farm workers and till the soil? Hunter-gatherers loved the outdoors and preferred to hunt and fish, they didn't want to work the soil. They saw no need for a permanent settlement.

Shama needed people who would take orders and do what they were told to do. He needed slaves. The ancients had learned long ago that people learned best when taught as young children. Possibly from the experience of raising and training dogs, they learned that obedience was something that had to be taught and that the very young learned their lessons best. For a belief system to really stick in a child's mind and really alter his psychology it needs to be taught during that very early period in a child's life when imprint programming is still in progress. Shama established the first organization to teach his program and produce slaves! The purpose of this organization is to teach the slave psychology.

You see slavery is first and foremost a psychology. And the psychology of the slave is very different from the psychology of a hunter-gatherer. Slavery may or may not involve brutal physical treatment, but it does require acceptance. A people can be subjugated by force, but only sophisticated psychological seduction can induce someone to accept slavery. Until an individual acquiesces psychologically to his subjugation he may be a prisoner, even a prison worker, but not a slave. But once a person resigns himself to a fate that he be ruled by others, he is then truly a slave. So to enslave a person we must enslave his mind. His mind must be transformed. He must believe that he truly needs a ruler. He must be psychologically altered to adopt the slave mentality. The institution which Shama created to accomplish this is known as the Pagan Religion.

To produce good slaves they must be properly indoctrinated. Pagan religious indoctrination was not optional. Every child and every tribesman was forced to endure the litany. The slave is forced to learn the Pagan theology. Shama's religion, Paganism, may have initially offered to his adult followers only weekly Temple Services to reinforce their Pagan beliefs. But in order to produce lots of slaves on a continuous, assembly-line basis, schools were needed. Pagan religious schools were organized to be the factories needed to provide the on-going production or manufacture of slaves!

What is this indoctrination expected to do? What are the attributes of a good slave? What do we want taught to children to transform them into slaves? What are the attributes of this 'slave' or 'religious' mentality?

The first and most important slave trait is **obedience**. Yes, we can't get much benefit from our slave if he is not obedient. If a slave insists on doing what he wants to do, rather than doing what we want him to do, in other words if he is selfish, then he's not much of a slave, and not of much use to us. We want our slaves to be selfless and live and work for us, not for themselves!

The next attribute you desire in a good slave is that he be **obsequious**. We don't want a pouty, out-of-sorts, ill-natured slave. We want him to be cheerful and willing to serve us. In other words he should 'know his place'. Quiet, meek, timid, humble, dependent and appreciative are very desirable traits to have in ones slaves.

Next a good slave should have the **morality of self-sacrifice**. He should believe in the virtue of sacrificing his *self* for the benefit of others. His conscience should punish him with anxiety, panic, shame and depression whenever he notes any infraction of this slave ethic. Gaining the approval and avoiding the disapproval of others should be a slave's highest goal. To do for others, to share, to give freely of his life, time, energy and possessions should be his code of conduct.

If you capture a child that you intend to make your slave you would have the problem of him running away. It takes a lot of guards and prison walls to control the unwilling. It takes a lot of guards to operate a prison today. In the very low, or no tech world we are discussing it would have taken even more guards. And of course where would you find the guards?

With hunter-gatherers you have an additional problem. Here you have a people well equipped to escape into the forest and survive quite nicely. You want the slaves to work the fields. Fields are located very close to the forest making escape very easy. You simply can't watch every slave every minute. You need the slaves to guard themselves! You need each slave to mentally watch and guard himself!

So the final attribute required of a good slave and demanded by Shama is **automatic self-policing**. This is very important. We can't always be watching to see that our slave doesn't steal and that he obeys his orders. Therefore we must train him to monitor himself, take note of infractions and punish himself for any disobedience. We wish to inculcate in him a proper slave conscience that will force him to toe the line. This will work better than any guard or police.

People who have these attributes have been indoctrinated and programmed with the 'slave or religious mentality'. If children are taught these lessons during their period of imprint programming it will 'take' particularly well. They will develop a slave **identity**. Properly indoctrinated and trained slaves are obsessed with pleasing others and terrified of disapproval by 'authority'. A well trained slave is obsessed with a need to conform, obey, serve, and win the approval of their 'betters.' Slaves will even commit suicide if they believe they have irreparably violated their slave ethic of self-sacrifice. A properly trained slave will never question his need for a master. If a master fails to materialize from the environment the slave will exercise his imagination and creativity to 'produce' one!

An adequately programmed slave will believe in a supernatural God or Gods with human-like features and behaviors. He will believe that his ruler is endowed with a miraculous ability to communicate with these Gods. He will believe that the universe, nature, reality can and is changed by the Gods often at their rulers request. He may even believe that under special circumstances he himself might influence the Gods to alter reality in his favor. These beliefs give the slave a *super*concept of truth that is people or human centered. Truth will be what the tribal authorities say truth is and gaining the approval, and avoiding the disapproval of the rulers of the collective, i.e. tribal authority is the epistemological reference! The tribesman's mind does not function all that well. It requires a collective of men as its cognitive authority for his mind to function at all!

The rulers of the hunter-gatherer family, if you can call them that, would have been those with a superior knowledge of hunting and gathering. The attributes of a good hunter are many. A few of those attributes are knowledge of various animal species, their habits and habitats. Knowledge of animal behaviors relative to the seasons and weather, and specific details of hunting each species under various conditions are part of the hunter's repertoire. Then of course knowledge pertaining to field dressing various species and how to transport the carcass back to camp is essential.

The gatherer, who might be the same individual, needs another very comprehensive body of knowledge. Which berries, nuts and vegetables are safe to eat? Which potential foods can be safely ingested raw versus which need some processing or cooking before they are safe to ingest? Which foods are safe under some conditions, season perhaps, and dangerous to eat at other times. What foods have medicinal uses?

The hunter-gatherer elder is likely to have accumulated the greater knowledge of hunting and gathering. So the most likely 'leader' of a hunter-gatherer family would be one of the elder members. This leader is the 'leader' because of his superior knowledge of Nature. His skill-set are those that improve his chances of success at solving survival problems in the natural world.

The skill-set required for 'survival' in the tribal collective is entirely different. The skills of the members of the ruling class are those of a politician. Oratory, charisma, celebrity, showmanship, acting ability, and psychological manipulative talent, are the skills valued by tribal peoples. These are the skills you need to impress and placate both the Gods and everyone up the chain of command. As a slave the ability to project submissiveness, adoration, obsequiousness, awe, subservience, dependency and thankfulness are assets.

When you live in a world of Nature and your standard of truth is Nature, the skills which are valued are those needed to understand and relate to Nature. When you live in a world of people and your standard of truth is the collective of people, then the skills valued are those needed to understand and relate to people, and the rulers of people. In the first case what counts is what you know, in the second it's who you know!

So the slave, or the tribesman, lives in a world of people, in a world of politics! Everything, and I do mean everything, is determined by the rulers! If the rulers say the sun revolves around the earth-then for everyone in the tribe-the sun revolves around the earth. If the rulers say that disease and death come from the anger of the Gods-then angry Gods cause disease and death. And anyone that might cite evidence in refutation of the tribe's beliefs runs the risk of condemnation and death by sacrifice. I repeat: reality is what the **collective** says it is. The slave mentality is the collectivist mentality.

Shama invented a world famous and highly effective process and curriculum for producing slaves. Without Shama's invention a permanent settlement at this point in time would not have been possible. He provided just the right *super*concept needed to create the slaves required for agriculture. He created a new psychology that could more effectively adapt to a **collective** societal environment. This new collectivist psychology relied upon several key concepts. These concepts or beliefs are taught through Shama's invention: **religion.**

Religion* is *an* institution** which promotes the belief in a supernatural God or Gods ***combined with a narrative, ideology or theology. If the theology supports more than one God the religion subscribes to polytheism. If the theology promotes just one God the religion advocates monotheism. **Religion** is thus an **institution** which **teaches and supports a narrative, an ideology or theology**. The institution of religion has an agenda. The agenda behind the religious curriculum is psychological modification. The purpose and mission of religion is to produce the collectivist or slave psychology.

Over a relatively short period of time Paganism and the tribal societal structure spread around the world and developed into thousands of varieties. Each tribe created its own unique theology, or narrative portion of religion. This theological portion of religion was customized by each tribe to their own specifications which included tribe specific myths, the worship of a particular set of Gods, its history and a group of chiefs. Each tribe also had their own particular rituals and taboos. Superficially tribes appear very different. But all tribes retain in their religion the same fundamental, generic bundle of religious concepts. This group of highly effective concepts forms the basis of **all** religions. These concepts form a curriculum that is called **theism**.

Chapter 5

THEISM

The Religious Curriculum
The *First* Stage of Theist Indoctrination—Creating a Prejudice

Lesson # 1 Innate Human Evil

This first stage of indoctrination of children in theism does not begin with children! It begins with their parents! Parents are taught by their religion that the first 'truth' revealed by the Gods to humans is that humans, all humans are born evil! This explained why the Gods so dislike humans and why humans found it so difficult to understand and please the Gods. Once parents understood that their infants, children, adults, themselves, everyone, all humans were innately evil, then their punishment by the Gods was understandable.

Religion further explains to everyone including parents, and parents-to-be, that there is no way that mortals can fully understand evil. It is enough for humans to know that evil is the worst thing a person could possibly be. ***A human is innately the worse***

thing there could possibly be, and there was nothing that a person by himself could do about it.

A newborn infant does not wait until he knows his name to begin 'thinking'. He does not wait until he knows the meanings of certain words. And he doesn't wait to begin 'thinking' until he can talk! No, a baby begins to 'think' as soon as he is born if not before[21]! Now, granted, his 'thinking' is not yet conceptual and he is years from being able to think critically, but he is born already much smarter than most animals are at their maturity. So what kind of 'thinking' is the baby capable of?

All animals 'think' perceptually and the newborn human can 'think' in this way with ease and agility. Dogs and a few other animals can do a pretty good job of determining our mood or emotion from observing and listening to us. Infants are hundreds of times better at this. So infants are ready for training the moment they are born and whether we know it or not they are 'learning' a tremendous amount about us every single day! They are learning what kind of world they inhabit, what the people are like who are around him, and he is learning who he is! He 'learns' all this and thousands of other things **perceptually** long before he learns the words with which to label these ideas and can **conceive** them. Perceptual 'learning' is not inferior to conceptual learning; they both have their assets and liabilities. These two ways of thinking do work differently and both are needed if a human infant is to mature into an intelligent adult.

So, initially, this Original Sin concept that humans are born evil is not taught to infants as a **concept** at all! Innate human evil is taught to infants from the moment of birth, before they are capable of creating concepts, as a non-verbal **percept**. The purpose of this earliest training is to thoroughly instill into the infant's *self* a prejudice against his *self*. An infant can **perceive** that he is considered defective and evil long before he can **conceive** it!

This prejudice against infants is communicated by parents and others verbally and ***non-verbally***. Innate self evil is taught with the efficiency and thoroughness as only a prejudice can be taught. Innate human evil is 'taught' to infants as a *feeling* by an attitude which permeates the family and the tribe. The newborn's parents, family, community and tribe all communicate this same anti-self attitude. The entire social milieu of a tribe is one which is prejudiced against the *self.* Often the newborn infant is up against **everyone** in his world! This is character assassination on a community wide scale!

By the time an infant can *conceive* that he is evil, he will have long before incorporated the *percept* of an **evil self** into his mind and will have developed the beginnings of his *super*concept based upon such an identity! By the age at which he can begin to conceive that he is evil it will *feel* to him that such an assessment is true and correct! The concept of innate human evil will be thoroughly conceptualized by the child later. It will then validate this self-appraisal and he will adopt it for life. The other concepts required to complete the child's indoctrination in theism will follow and will be added later to this initial belief in Original Sin or innate human evil. This prejudice against the self, this percept of **self evil** thereby becomes part of the infant's early perceptual portion of his *super*concept. Self evil becomes an integral part of his self **identity.**

We can be sure that an infant picks up on any negative attitudes that his mother and others around have towards him. An infant's innocent and naive trust of adults is easily influenced. This is because infants rely on how lovable and cute they are to ensure their survival. The infant 'believes' his 'goodness' to be crucial to his survival! Infants act as though they believe that their survival depends upon their parent's recognition of this. Infants and children are very suggestible. They want to please. They seek affirmation. They want their goodness to be recognized. They expect the 'lovable-ness' of their **self** to be appreciated and admired. The desire to be valued and loved is very strong in infants. An infant's notion of his value and self-worth may not be held as a conscious and explicit concept but it is still felt and it is extremely important to him. Infants are very sensitive to the ambiance around them. Fear, panic, apprehension, and pain are the consequences to an infant of the feeling that he is evil and therefore defective. Parents that are prejudiced against the **self**, their own and their infants, cause their infant to feel unworthy and unlovable. Parents cannot fake a feeling of **self** love. Their negative feelings toward their **self** and their infant's **self** will be communicated to their child. Even if such communication is limited to non- verbal means its power will be sufficient for an infant to get the message.

When an infant's needs are met and he begins receiving proper training from the day he is born he will embrace the guidance provided. Proper training of very young infants is identical to the proper training of pets and other animals[25]. It is done calmly, and in a firm and loving manner without hysterics and with the smallest amount of negative reinforcement and a ton of positive reinforcement. Observe how animals train their offspring. Long before the child can conceptualize such training has made him a delightful addition to the family. In time such children mature into thoughtful, respectful, opinionated, and self-sufficient adults.

But this is not the way infants are treated in the tribal theist family. No positive proactive training or guidance is given to the young infant. Parents provide for their infants basic needs and provide affection but mostly they wait and watch. While the infant is busy trying to discover how to survive he is forced to come up with his own plan. The plan improvised by most infants left to their own devices is usually the one which works best among tribesmen. Infants quickly learn to manipulate their parents in order to get the things that they need and want. Crying, throwing tantrums, biting and kicking are usually the ways to get attention and get what you want. This is not much different from the way the adults behave in a tribe! It becomes a battle of wills to see whose sadism is the most powerful, parents or child! And this plays right into the theist playbook.

Tribal peoples have been trained to look for and to spot the signs of evil in their newborn child. The tell-tale sign of innate evil is **selfishness.** It is selfishness which parents look for in themselves and their infants and children. So when tribal women become mother's they carry out the training this Pagan doctrine has taught them in regards to their infants. So when the child who is receiving no positive training begins to act-up there is plenty to interpret. Parents then 'see' what they have been trained to look for! No newborn initially behaves 'selfish', he is just trying to satisfy his needs and survive! But he can behave nasty, mean, obstinate and cry and scream incessantly especially after being trained to understand that this works! So tribal parents stand by, do nothing to guide or train their offspring, they let him flounder around searching for a method of getting the attention of his family, and dependably he resorts to tantrums and terror tactics. With such behavior underway parents have no problem in spotting evil right off! The parent's observations are vindicated as they observe evil selfish behaviors in their baby! The parent's behavior, how they now react to their infant, will reflect their attitude toward their babies **self**. So instead of taking proactive action to calmly train their infant so that he can learn how to behave and have his needs satisfied, they sit back, watch and criticize as their infant makes mistakes. Then they typically take one of two equally destructive paths.

By far the most common reaction of tribal parents to child-rearing challenges is to punish the infant and child severely for his evil. When an infant feels stigmatized, spiritually unclean, and deficient he will not react to it happily. When he meets with sadistic discipline and severe physical punishment instead of training and guidance he will learn the lesson of sadistic bullying. This treatment of little babies as if the evil in them must be beaten out of them scares them profoundly. This fear is not only profound, it is existential! That is, it is a fear for one's very existence, for one's survival! If the non-verbal message is that you are evil, and you are repeatedly whipped where will this lead? If your **self** is evil can you expect to be loved and nurtured! Verbal accusations, screaming, condemnation, beatings and attempts to instill guilt only teach a child to copy such behavior himself for later to use on others! Watching for sin and then applying punishment makes no sense as a child rearing technique. It is just training in violence. We might call this the sadistic approach to child rearing.

The opposite extreme in raising children, the technique favored by the ruling elite, is total permissiveness. Their child-centered, ultra-permissive approach to child-rearing is also the result in the belief in innate human evil. It amounts to a resignation and acceptance of it! Rewarding, even paying children to behave, is very good training if the purpose is to produce a sadistic coercive ruler. Since the parents must bear the pain of such poorly behaved children let us refer to this as the masochistic approach to raising children.

Both the punitive, severe physical punishment approach and the overly permissive, let-him-get-away-with-everything approach to child rearing grow out of the same prejudice against the **self**. The excessive permissiveness and the sadistic punitive approaches to child rearing emanate from tribal sadomasochism and teach this orientation to children. So child rearing technique introduces and reinforces for tribal theist children the essential elements of tribalism. Neither approach is rational but rational behavior seldom comes from prejudice. If you wish to understand the foundations of any culture study the fundamentals of its most common child rearing beliefs and practices.

The corporal punishment of children was seldom needed in the natural world. The little pain inflicted by hunter-gatherer parents in the training of their infants carried no morbid consequence. But the corporal punishment of infants, in combination with an existential threat to their life implicit in the prejudice against the **self** makes a deep and permanent scar on the emerging infant mind. When a very young child begins to exhibit shyness we can assume he has learned this lesson.

Corporal punishment to inflict physical pain is a frequent tool used in the indoctrination of infants and toddlers with this Original Sin concept of theism. Whippings, beatings and psychological intimidation are the basic tools of theist child-rearing. The Pagan Religion and its religious schools are nearly synonymous with child-abuse. The existential fear and depression which results from these early experiences remain with most theists all their lives! Many theists are so frightened and terrorized by their treatment as children that they *never* get over it!

Sit quietly by a mother as she feeds her young child and listen to her 'baby talk'. Observe and listen as a barely walking infant is scolded. He is constantly asked to 'be good' with the attitude that no one really believes that that is possible! Soon the incessant 'be goods' are joined by 'don't be selfish' and 'share'. Such accusations increase when the infant behaves badly. To this is added the pain and fear of corporal punishment designed to inculcate in the child's mind a strong fear of tribal authority! All very good lessons for a slave-in-training!

Mother and father slave are not being mean. They are being realistic! They are engaged in the psychological mechanism of projection. Projecting onto their infant their prejudice against him creates a self-fulfilling prophesy. An infant's motive is to satisfy his needs. There is nothing evil about that! But theist parents project onto their child's behavior motives he does not have, but which they assume he has. In time as he grows older, because of the way he is treated he will develop the very motives his parents originally accused him of having as an infant. Slave parents believe that their child was born evil. Evil, they believe, is revealed in their infant and child by his not cheerfully and totally devoting himself to fitting in with their life and serving their needs! Slave parents believe that childhood should be a rehearsal for later serving the elites! But infants and children apparently didn't receive the memo and think their parents will take care of them! Slave parents are looking for the little slave to serve them for awhile. But their selfish little baby has other ideas!

This 'Original Sin' doctrine is what the parents have been taught by their religion. Every religion teaches this in some way or fashion. Disapproving slave parents therefore see evil motives in much that their child does or does not do. This reaches a crescendo with their teenagers. Parents are constantly point out to their child what they observe, along with their interpretation of what the behavior means! So when the 11 month old pours his milk out on the floor, the interpretation is that he wishes to cause Mom more pain. As now Mom must bend over again and clean up the floor! As she wipes up the milk she says in her most haughty voice "I know you think it's funny, me getting down here on the floor and cleaning up this mess. You make more work for me on purpose, you selfish little brat!" The parents projection of 'selfish', 'evil' motives onto their child's behavior assures that in time this prophesy for their child will come true. This prejudice against children appears ubiquitous wherever theism is predominant. The more precocious the infant the earlier he assimilates the prejudice his theist family, and culture have against his **self.**

Dutiful slave parents will not miss any opportunity to interpret their child's behavior as evidence of his evil **self.** The child will learn from his parents that the meaning of his feelings and behavior is that he is basically evil and must work mightily to be worthy of serving his masters. So from mother to infant, from parent to child, for the past thousand generations, this ancient Pagan evil we call theism has been passed unchanged and unmodified in its wickedness! And if the parents should drop the ball and do not sufficiently teach the lesson of innate evil, then grandparents, other family members, neighbors and teachers stand-by ready to fill-in and teach the awful prejudice in their stead.

Sometimes, occasionally, something much more terrible happens. The theist infants I have been describing are loved! They are perceived by their slave parents as innately evil *as a class* which includes everyone in the family. But most slave infants are loved and cherished *as an individual.* The cultural prejudice against his **self** is awful but it is limited. Relatively speaking, especially among his kin, he is of value. But if his parents, especially his mother, perceives him to be evil not only as a member of a class, i.e. human infants, but also *as an individual* then he is not only evil but maybe unloved as well! A theist parent, because of her belief in Original Sin, may conclude that her infant is responsible for the trials and tribulations of her own life. She is unhappy and her child's evil self is to blame! Under these circumstances theism can produce human beings capable of the greatest evils known to man.

Innate human evil or the Doctrine of Original Sin is the most important concept in the theist repertoire! The use of this prejudice to inculcate the slave psychology is Shama's most important discovery for it provided the essential ingredient to his invention of religion. Innate human evil is the most important concept in the theist package of concepts because it creates a gateway through which the other theist concepts can pass. Once this first or lead theist concept of evil is incorporated into the infants mind, the other theist concepts easily follow! If the theist message of his innate evil can be introduced into an infant's mind early enough, before the end of the period of imprint programming, then the individual will probably be mentality enslaved for life. This is the concept responsible for religion's tenacity across generations. And it is this concept which is behind masochism or self-hate and a great deal of evil in the world. This concept is so powerful that once incorporated into a child's mind he will **seek** a religion or ideology which dis-values his **self**, even where one is not offered by his parents or community.

By the time a theist child has reached the age of understanding he will have incorporated the percept or feeling of his innate evil into his developing *super*concept. This prejudice against one's **self**, this feeling of one's **self** evil is incorporated into the theist's **identity**! This will adversely affect his cognition. At the time the percept of one's **self** evil is incorporated into the infant's mind it causes a cognitive error. This logical mistake may become the infant's cognitive methodology and remain with him all his life! Later we will examine this issue in more detail.

Individuals who, for one reason or another, do not during infancy incorporate sufficient perception of the evil of their *self* into their **self** identity will usually find in later life that they are unable to truly and fully believe in religion. Although such persons may try mightily to 'believe' they generally know in their hearts that they do not. Such people may truly wish to be 'True Believers'[18] but ultimately must accept the fact that they do not possess the appropriate psychology to do this. Many people profess belief in a religion that they really do not believe.

Shama's invention, religion and the religious or slave psychology marks the second type of psychological adaptation invented by Homo sapiens. It provided man the **collective** *super*concept necessary at this early point in time to create the slaves necessary for the development of permanent settlements. This collective *super*concept creates a collectivist psychology. The religious psychology, the slave and tribal psychology, and the collectivist psychology are all the same. All these terms refer to the psychology that is required to support tribalism.

A Personal Story

I was born in 1939. So during the 1940's I aged years 1 through 10. During the last years of that decade I had one of the most profound learning experiences of my life. Yet I was well past middle age before I fully understood its significance.

1939 was when I was born. This is important. Of equal importance is where I was born. I was born in the United States, in the Southern State of Louisiana and raised in the small town of Wisner. Although Negros had been emancipated by President Lincoln in 1863 the order of the day when I was a child was 'equal but separate' facilities. The South existed in Apartheid with the Black and White races separated by law and custom. It would not be until the 1960's and Martin Luther King, Jr. that any degree of real equality would come for Black Americans. During the 1940's, the decade of my childhood there was not the slightest hint in Wisner that things as they stood would not remain as they were forever. All seemed stable and fixed and accepted by all. White children went to white schools and black children went to black schools.

I was 5 years old when World War II ended. It is one of my earliest memories. I was pedaling my tricycle down the blacktop road in front of our little red brick house. Mother came out on the front porch and shouted, "The War is over! The War is over!" At that moment an airplane flew overhead. I pointed up to the aircraft and shouted back at Mother, "The boys are home! The boys are home!" (Thinking back on this in later years I realized that the aircraft I observed that day was probably a crop duster.) I referred to the military personnel as "The boys" because the news reels at the movies of that time referred to U.S. Soldiers as "our boys overseas."

Shortly after that memory I met Odell. I know that we met and became friends before starting school. So I think we were both around 5 years old. Odell was a Negro boy the same age as I who lived over in 'Nigger Town' on the other side of the railroad tracks. One day he came to our house to play with me, possibly at the invitation of my mother. We hit it off immediately. We liked the same things, anything with wheels, cowboys, horses, airplanes and everything military. We liked to play 'Indians' and we loved swimming in Turkey Creek.

Odell and I were almost perfectly matched physically and mentally. He was a slightly faster runner and swimmer that I and he could climb a tree or a rope slightly faster. I was a little better with horses and bow and arrow. Behind our house was a large forest where we spent every free hour available to us.

Practically every day, for most of several years this forest was our real home. My parents were both working long hours then and Odell seemed to have no restrictions, so we spent our lives in the woods playing, swimming and riding horses with almost no adult supervision and it was wonderful! Often we had other children join us. We played Indians, cowboys, Tarzan and army from sun up till after dark and never tired of shouting "Bang! Bang! You're dead!"

During the summers we spent our entire day playing together. Even when the schools were in session we spent every afternoon and weekend playing in our woods. Odell of course, went to the black school and I went to the white school. Once school let out Odell would run all the way from school to my house. He always arrived at our kitchen door bright eyed and happy with a big smile and some intriguing story from across the tracks. Occasionally I would meet him at his house if we planned to go somewhere near his home. The junk yard was closer to his home and we occasionally went to Mr. Jackson's salvage yard. There we made important purchases of wheels, frames, and assorted screws and bolts for our tree house, fort and other building projects.

I can still remember Odell's laughter, such a sound of freedom and complete abandonment to the enjoyment of running free and unrestricted. For many years my Mother had a picture of Odell and me sitting nude in a number #2 washing tub while she sprayed us with the water hose. The looks on our faces said it all. No actor can put-on well enough to pose and make a picture like that!

And so it was that for half a decade we lived this way. School and doing our chores were just diversions from the real joy of life, our playing in the woods! Then one day it came to an abrupt and complete end.

Odell came over to my house that afternoon after school as usual, but I could tell from the way he walked that there was nothing usual about him. His head was hanging low. I didn't know he was capable of such a posture. He came up to me in a very formal way and said in a very formal sounding voice, "Master Gene, My Daddy has told me that I must learn to know my place! I am not supposed to hang around you and your house anymore!"

I had never heard such talk before and nobody had ever called me 'Master'. I asked him what his Daddy was angry about. He said his father wasn't angry. He said his father just wanted him to "grow up". I asked Odell if he could put off growing up and ride the horses today. He said his Dad told him to do whatever I told him to do. I didn't like telling Odell what to do. It was something new to me and I couldn't get used to it.

We rode the horses, but it wasn't the same anymore. Something had changed inside Odell and after a while something changed inside of me. We ceased meeting after school and the forest behind our house lost its magic. After a while my family moved to the country. I saw Odell in town a couple of times but he had new friends and had no time for me. I went my way, he went his.

It is almost impossible to escape the corrosive effects of prejudice when it is as widespread in a culture as it was in the South of the 1940's. When prejudice is the law, the custom and the mindset of everyone, all those who are the objects of prejudice are victims. The Black children were as completely brain-washed to racial prejudice as the White children.

Such prejudice can be subtle, often non-verbal and so pervasive that it enters the human mind as if by osmosis. A person has to do nothing active to acquire such a wide-spread belief system; it just seems to stick to your mind like glue. As a white citizen of the South I never received a lecture on the value of segregation or attended a course on the behavior or the mental attributes of Negros. Prejudice does not require formal didactic teaching. It is conveyed very effectively non-verbally by the turn of one's lip or the raising of an eye lid. The undercurrent of racism is amply communicated by the choice of phrase or the context of one's story. The racism that I later learned was taught, not by my parents or teachers, but by a few of my schoolmates and peers. If you are brought up in this kind of milieu, whether you are black or white, it is almost impossible to escape its debilitating consequences. Black parents were as thoroughly indoctrinated in the belief in innate evil as white parents. Only if you think hard and work constantly to keep prejudice from harming your mind do you partially escape its mind-damaging effects.

It is the same with the prejudice that contemporary theist culture has toward all children, particularly toward the child's **self**. The purpose and goal of theism is to create this prejudice against the self. This prejudice is everywhere. It is several times more powerful and profound than racial prejudice. You simply cannot avoid it. It is everywhere. You cannot escape it. It will affect you. Only by thinking about it can you recognize it and diminish its awful effects upon you.

Now we will return to the exposition of the theist curriculum. The Doctrine of Original Sin we have been discussing is the parental concept which forms all of *stage one* of religious indoctrination and it opens the gateway for the other theist concepts. Now we go to *stage two* and the other theist concepts required of slave indoctrination.

The *Second* Stage of Theist Indoctrination
School Aged Children

Theist Lesson # 2 Revelation

Revelation is the very first theist **concept** the child must learn and accept. The infant has been prepared for this belief; he already has the **percept** of his innate evil, so he has little resistance to it. Especially when the first example of divine revelation taught to him is the **concept** of innate human evil. Something about this idea rings true to him. He already feels his inferior evil condition, so who is he to buck all the adults, all the authorities in his family and culture? This first revelation confirms some of the earliest beginnings of his *super*concept. The percept of innate evil incorporated during infancy prepared the soil for the concepts of revelation and innate evil. Now this prejudice helps these concepts take root in the young mind and grow. The theist child now learns all about Shama and has no reason not to believe all that he hears. So as the theist child learns all about Shama his *super*concept, his collectivist **identity** matures and becomes firmly established in his mind.

    Shama held, as the child is told, that he and he alone was able to communicate with the frightening Gods which inhabited the supernatural world and control the entire universe! Shama held that only very special people like himself could apprehend the vast supernatural realm and communicate with its rulers. Only he could understand their seemingly contradictory demands. Only he and his clerics could comprehend the divine mind! And it was in these communications with the Gods that they **revealed** to Shama their demands of human beings!

    The first revelation taught to the child is familiar; it is the word from the Gods that he, all humans, are born evil. He hears this again just in case he missed it as an infant. And then the idea that evil is just too awful for mortals to understand is revealed. There is just no definition that a human being could possibly understand. Evil is simply the most awful thing a human could possibly be. This should be sufficient. A human being is simply the most awful thing there could possibly be! This is why humans are so scorned by the Gods.

The next lesson the Pagan clerics teach is the importance of revelation itself. A few very special humans like Shama have the miraculous ability to understand and communicate with the Gods. Regular people are too defective and sinful to understand or to know the Gods. How fortunate for humans that this is so, for now through Shama and special people like him we can receive the word from the Gods and pass it down to others!

The Supernatural Gods revealed to Shama that they were far more powerful than anyone or anything that existed in the physical world. They controlled and commanded Nature. Nature was but the tiny, visible and temporary portion of their vast universe of both the natural and the supernatural worlds. These Gods also revealed to man through Shama that a truth based upon Nature was flawed, imperfect, incomplete and inferior to the truth provided them through Shama.

Truth based on the standards which Shama's Gods provided was very different from the truth hunter-gatherers had known. Shama's truth was malleable, unreliable, changing, and revengeful. Only with Shama's help could any mortal expect to successfully run the gauntlet of life on earth. Only the professional expertise of Shama and his clerics offered any safety in guiding a person through such a dangerous 'reality'. Shama's truth was what Shama said the Gods demanded and such a truth was inconsistent from one time to the next.

With Nature's truth discredited tribal man was left to the whims of chiefs, rulers and other humans. Now human's, not each and every human, but the leaders of a human collective, the rulers of men, personified in the person and office of the chief, became the authority over the entire universe, including nature.

Now it was not Nature, not reality, not Ytilaer that was the authority over all things, it was the ruling elite. Now it was a few men who ruled over nature, man, and through association with the Gods, ruled over the entire universe!

Lesson # 3 Force

Shama embraced force. He used force and he taught the righteousness of force. Force and ***might make right***. There was no place for weakness in the Pagan Religion or in tribal life. Pagan religious schools begin teaching the use of force right away with each new group of students beginning with the compulsory nature of school itself. On the first day of school the child must learn of the absolute power of the tribe. The compulsory requirement that he attend school says this loud and clear. He is under the rule of his slave parents, but they and he are under a greater power: the Chief. He learns now, if his parents have failed to teach it to him earlier, that force is the method used by the tribe to make him comply with the psychological modification he is receiving. Force and fear become associated in his mind with tribal rule! The concept of force is of central importance. The physical violence of corporal punishment metered out by his Pagan parents, teachers and clerics accentuates this show of force. The child was forced to leave his family, forced to sit in silence for long periods of time, and forced to accept many unpleasant ideas into his mind. And then he is forced to retain them in his mind and recite them on command. Child abuse and pedophilic exploitation are core attributes of religion and this is very clear when you examine the treatment of children in many religious schools.

Next the child is taught to use force on himself. Children are easily intimated by over-bearing aggressive clerics and Pagan teachers. Physical punishment is used to instill discipline and force the child to force himself. Bullying begets bullying. Strict discipline teaches strict self-discipline. Self-discipline is intrapsychic force. It is the use of this psychic or mental force to enforce compliance that characterizes self-discipline. The teacher insists, and wants the child himself to insist, that he learn the concepts being taught and that he develop the attributes of a good slave. The child is taught to use self-discipline to force himself to be a good slave. He learns to force concepts into his mind whether they make sense or not. This teaching of the use of force is more than just teaching physical bullying. It is also training in the use of force as an epistemological tool!

Force is used by the slave's conscience to punish his **self** for infraction of tribal rules. The slave judges his thoughts and behaviors by the criterion of his collective abstraction or tribal religion. When he finds that he has not complied properly or completely to the tribal rules he punishes himself inside his mind. Depression is anger and hatred turned inward on the **self** and is punishment metered out by the conscience for being a poor slave. This punishment, or the threat of such punishment, is generally sufficient to keep the slave in line. Depression is a consequence of tribalism. Next force is taught as the theist notion of morals.

When 'evil' is combined with 'force' it results in the development of the concept of morality. Morality is a set of rules or commandments the fundamentally evil individual must adhere to if he wishes to mitigate, as much as possible, against the effects and fall-out from his evil **self**. As Shama outlined to his villagers the concept of evil he also informed them of the steps they must take to avoid the Gods' wrath. Don't be selfish, make donations to the Gods, sacrifice your children to the Gods, pray, have faith, participate in tribal rituals, and most of all sacrifice yourself! These actions or behaviors constitute the first set of morals given to man and they are the epitome of evil marketed as virtue! Depression is the result of applying intrapsychic force to one's *self*. It is as evil a form of coercion as any other. A healthy human, uneducated in the evils of theism, no more needs an artificial conceptual morality for his mind than he needs an electronic pacemaker for his heart!

Lesson # 4 Supernatural Gods

Shama was acutely aware of the fact that hunter-gatherers had a high regard for Nature. They revered the very same Nature that he so feared and hated. Hunter-gatherers saw Nature as powerful, benevolent and bountiful. Shama wished to destroy that view. He wanted all to see Nature as not so powerful, as evil and of danger to man. And he needed people to view the world as one of scarcity not abundance. If they didn't think this way they wouldn't work strenuously and put in long hours in the fields. He needed Gods that were much more powerful than Nature and who would change man's world view to what he needed!

But hunter-gatherers are 'hard-headed' and were a difficult group to convince. They knew that Nature was far more powerful than they. They knew what limited powers they and their families possessed. Up against the power of a flowing river or the wind of a storm, personal or family power was very, very puny. Hunter-gatherers would not forsake their awe of Nature to worship the puny power of a group of people or their chief! Even when Shama spoke of the power of many families, unified as a settlement and promised that they would be very powerful, hunter-gatherers were skeptical. So when the idea of many families living in the same permanent village was touted as very powerful, hunter-gatherers didn't buy it. How could many puny individual families add up to a power that could rival Nature? How could a collection of families become more powerful than Nature? How could a collective of human beings morph into a power greater than Nature's God?

Shama intuitively realized that to accomplish what he wanted he would first have to create in people's minds a *super*concept which was very different from that of hunter-gatherers. He knew he had to create a power much greater and much more powerful than Nature. He knew he had to create a new 'reality' for people, a 'reality' that **he** controlled. Only through controlling people's 'reality' would he be able to control them. So long as his subjects believed their own perceptions, their own understanding of Nature, so long as they believed Nature to represent the Truth, he would not be able to harness them. He had to find a way to sever people from their natural tendency to rely on their own perception. He had to find a way to sabotage people's tendency to 'see' God in Nature. He had to find a way to substitute man's reference standard of truth, his *super*concept of Nature, with a *super*concept of himself as Chief! He, rather than Nature, had to become people's ultimate source of truth!

Shama began the indoctrination of his followers by describing an elaborate and detailed fantasy. He was a great story teller; he could spin a dramatic and emotional tale. He created in his listener's imaginations a **supernatural** universe. He populated this imaginary realm with many animal-like and human-like supernatural creatures. These creatures possessed omnipotent powers and he referred to them as 'Gods'. Shama discovered that he could use people's ability to conceptualize and their imaginations to wrench them from the objective natural world. He learned that he could get people to abandon reliance upon perception by substituting this supernatural world which they could not 'see' except in their imaginations! He realized that he could use the fact that people had trouble seeing his supernatural Gods to his advantage, as evidence that humans were defective. He created a way to substitute imagination for perception as the source of information about 'reality'. He invented a technique for substituting the *super*concept of Chief, or ruler of the collective, for the *super*concept of Nature.

The supernatural Gods which Shama created in his subjects' imaginations were supremely powerful, much more powerful than nature and thus they ruled over Nature! They ruled over a realm much larger and greater than the perceivable world. In this imaginary world the material world was but a puny portion. These supernatural Gods were the rulers of this vast combined universe of both supernatural and natural worlds and were much more powerful than Nature.

Shama's Gods were also capricious, vindictive, vicious, merciless, and barbaric. In addition to their mindless cruelty they were unpredictable. They intervened and meddled in people's lives when unwanted but were seldom of help when needed. Shama's Gods were indeed very frightening Gods.

Each new class of slave children was taught about Shama's Gods. Nature or reality was belittled as these supernatural personages or monsters were endowed with omnipotent powers. Nature and Natures God became unimportant to man as man turned his mind away from the earth and towards the heavens.

Theist Lesson # 5 Morality of Redemption Through Human Sacrifice

I want to engage you in a thought experiment. What is the most awful and evil concept you can think of? What is the worst idea or belief you can imagine? Yes, what belief do you think is the most awful, despicable, and evil concept there is? What is your vote for the most terrible, most evil idea ever conceived by a human mind?

Perhaps you would choose genocide. Genocide is the idea that you annihilate every member of a race of people because you don't like some of them. I agree with you that this is right up there on the short list of the worst ideas ever conceived. And it is something the Pagan Gods regularly endorsed. But I think there is still worse.

Perhaps you would select the concept of slavery itself. This is the idea that it is acceptable that some people force others to do work for them with little or no compensation. This denies the 'slave' any choice in the matter. The slave is required to live not for himself, but for his 'master.' A slave is not allowed to have a life of his own. I agree with you again that this idea is right up there near the top of the list. It is another concept universally applauded by the Pagan Gods. But I still don't think it's the absolute most evil idea of all time. Let's look some more.

The most evil idea of all time may be the acceptance of the practice of cannibalism. The idea is that it is morally proper for people to kill and eat other people. This is pretty repugnant as well as about as evil as you can get. I certainly agree that this selection goes to the top of our list. This is also an idea promoted by many Pagan Gods, but I think we can find a belief that is still just a little worse than this, the most evil idea ever conceived!

My vote for the all-time most evil, most awful, most diabolical, idea ever conceived of by the human mind has to go to the concept of human sacrifice. The picture in my imagination is of those 5 and 6 year old little girls walking up the long flight of stone steps of the Mayan Temple to the altar. There they were murdered by having their hearts cut out by the clerics. As a cleric held the child's heart high in the air, still beating in his hand, the little girl's life was offered as a sacrifice to the Pagan God Chaac. Reportedly over 5000 children brutally killed as offerings in a single day! The Aztec also killed thousands of children to incur the favor of their God Huitzilopochtli. Sacrifices such as these were not rare for many millennia! Quite the contrary, such religious murder rituals were standard practice in nearly all tribes and not rare at all. The idea of killing innocent people to appease or bribe some God seems particularly evil and repulsive to me. The fact that it was practically a growth industry for thousands of years, to me is sickening! Human sacrifice has my vote as the most evil idea ever conceived!

Wouldn't you know that it is this evil concept, the belief in the virtue of human sacrifice, which Shama and theism choose as its greatest value! This is one of the pivotal concepts chosen by Shama and the other inventors of religion. Yes, it is human sacrifice, self-sacrifice that religion has chosen as its highest ideal! It is this concept of human sacrifice, this epitome of the most evil concept imaginable, that religion seeks to teach children as its supreme moral principle and the ultimate virtue!

What this ideal really teaches a child is that his intrinsic value is extremely low, he is almost worthless. This rings true to the theist-prepared mind. The purpose of this lesson is to further undermine the child's sense of self-worth and value. It will further undermine self-confidence and help with obedience. The child is taught that only through tribal rituals, and anointment by tribal leaders, can he over-come his intrinsic evil. He is taught that only religion can bestow value upon him. This means that any value he might obtain can only be granted by the **collective**, through tribe or religion.

Self-sacrifice is if anything, even worse than being the victim of sacrifice at the hands of others. With self-sacrifice the individual is expected to willingly harm his **self!** He does this supposedly for the benefit of the collective as a means of moral cleansing. The concept of human sacrifice or self-sacrifice is taught to slave children to prepare them to accept exploitation by the ruling elite. Theism is among other things training in masochism. Theism damages the mind rendering it vulnerable to enslavement, vulnerable to victimization. The fact that the victim goes willingly to his end does not mitigate the evil of the perpetrators. To take advantage of others, those unable to properly defend themselves, as with children, is the very essence of evil. The concept of self-sacrifice is every bit as evil as sacrificing others. When you admonish someone 'do not be selfish' you are commanding them to prostrate themselves before you. Why not simply tell them to heel! When you comment that someone is 'selfish', you are complaining that a slave does not know his place! And when you teach people that it is a virtue to sacrifice themselves such teaching inevitably produces people more than willing to sacrifice others!

The concept of human sacrifice is used to inculcate the utter worthlessness of the individual-absent his utility to the tribe, State or other collective. The victim of sacrifice has been taught that he is born evil. He is taught that he is selfish, loathsome and undesirable. He is schooled in the view that concern for one's own self-interest is bad, concern should be for 'others' i.e. the ruling elite. He is taught that he should turn against his **self** and 'live for others'. We don't want a slave living for himself! He should be concerned only for his masters' desires. He is pressured to embrace an anti-self ideology. This is what religion is! Theism is an anti-self ideology which has the effect of turning the

Forebrain against the Hindbrain. It creates self-destructive masochism and names it virtue! Can anything be more diabolical and evil than that?

Shama then received another revelation from the Gods which offered a ***solution*** to the sin of innate evil. This revelation confirmed that the individual could not redeem his evil **self** by himself. But through a collective of men, or the ruler of a collective of men, **redemption** might be provided. It was possible to be forgiven and made worthy if the collective chose to bestow its blessings. Forgiveness and redemption were possessed by the collective. It was for the rulers of the collective to decide who was forgiven and how redemption could be earned. Morality was no longer an individual issue; it was expropriated and now a collective possession. Through appropriate institutions applying the correct rituals a collective of men, or the rulers of a collective of men, could improve a person, anoint him with forgiveness, and make him less evil and more acceptable to the Gods. Redemption provided hope for man who could now take solace in the promise of forgiveness of his evil through collectivist rituals.

So the purpose of public rituals of human sacrifice is to demonstrate to the entire tribe, but particularly to children, the utter worthlessness of human life, absent its utility to the Gods and to the tribe. ***This reinforces the belief in the innate evil and the near worthlessness of children!*** Only human blood, the death of sacrificial victims, was of any small importance to the Gods. It is man's death that the Gods demand in exchange for some small, temporary forgiveness of his innate evil nature! Thereby man's innate evil is countered by sacrifice. Human sacrifice was then the currency that every member of the tribe had to acquire in order to buy forgiveness of their Original Sin.

When a child is taught that human sacrifice, the killing of human beings is the highest virtue one can obtain, the psychological message is that it is Holy and virtuous to sacrifice oneself and others. This is a clear and unmistakable message advocating **sadomasochism**. The masochistic sacrifice of oneself and the sadistic sacrifice of others are equally admired tribal behaviors.

Either or both types of behaviors glorify the Gods and earn accolades from the tribesmen. Both the sadistic killing of tribal enemies and the masochistic killing of one's **self** in combat are Holy actions which earn blessing from the tribal elite and the tribal Gods. These are the type of actions which reinforce a **theist identity** and authenticate one's collective *super*concept. Sadomasochism is the theme underlying all relationships within and between tribes. Sadomasochism is the world view of tribal peoples and they simply cannot see around or through this attitude. Sadomasochism is the ethic of the collectivist *super*concept!

The hierarchical structure of the tribal leadership reflects this commitment to sadomasochism. The societal or political principal of ***might makes right*** is reflected in the psychology of sadomasochism. The Gods supposedly sadistically command the chief and his ruling entourage. They in turn are sadistic toward all tribesmen and slaves below them in the chain-of-command. The slaves all have those above them, who exercise sadistic license toward them, while they masochistically prostate themselves before those above them in the power structure. All relationships are this way child-parent relationships, student-teacher relationships, and warrior to leader relationships.

This sadomasochistic collectivistic ethic also applied to sexual relationships. The concept of romantic love is a rather new, modern development which closely followed the adoption of such ideas as individualism and privacy. Even so, hunter-gatherers probably chose their sex partners on the basis of preference and availability, probably not too differently from modern people. But the collectivist mentality changed this individualistic choice for tribal peoples profoundly. The sadomasochistic theist ethic of 'from each according to their ability-to each according to their need' combined with a *self* identity inseparable from the tribal collective, meant that promiscuity was promoted and encouraged. Women generally smaller and weaker were the losers. The sadomasochistic ethic was applied to sexual relationships just as it was to everything else. Every woman's sexual need was every man's responsibility to satisfy and every man's sexual need was every woman's responsibility to allay. Such promiscuity or 'free love' undermined the biological family and discouraged a return to the hunter-gatherer lifestyle. But over time, because tribal leadership was usually inherited, the separation and distinction of the families of the ruling elite required an accounting of parentage. The offspring destined for leadership had to be identified correctly. The need for designated mates and restrictions on sexual partners developed. Such restrictions applied primarily to women of the ruling class as their offspring were the most valuable and destined to rule. Then as the various slave classes more fully developed and the ownership of lower slaves by individual members of the ruling elite became common, there was a need to know which offspring by different slave women belonged to which member of the ruling classes. So, just as with their domesticated animals, tribal rulers wanted their female slaves to be bred to specific males. So the rudimentary beginnings of the institution of marriage developed. This tribal 'marital' relationship was not expected to be monogamous for males. The tribal marriage was a way to know the parentage of the offspring so that his or her place in the tribe could be accurately assigned. The tribal sexual ethic for men remained the right and responsibility to satisfy their sexual needs with as many women as possible. This tribal sadistic sexual exploitation of women is technically not rape since there is general tribal acceptance of the sadomasochistic ethic. No one questions sadistic or sexual exploitation from their superiors and no one hesitates to exercise their prerogative of using their subordinates in any sadistic manner they wish! No slave questions the masochistic requirements of his or her station. **Theism** teaches the slaves that they *exist for their use* **by the elites who rule the collective.** This means everyone above you in the chain of command.

Theism also teaches that to save the **collective** from some of the effects of innate evil the **self** must to be sacrificed. The **self** must be sacrificed to save the collective. The tribal slave must sacrifice his children, his family and his **self** for the benefit of the tribe! Thereby man's **moral identity** was changed. No longer was the purpose of ethics the long-term success and happiness of the individual. It was now required that the individual be sacrificed to insure the survival and happiness of the collective! Rational self-interest of the hunter-gatherer was replaced as man's morality by belief in the virtue of human sacrifice. Rational self-interest was replaced by sadomasochism!

This concept: ***redemption through the collective***, gave man hope of forgiveness of his innate evil but it created additional cognitive problems for him. For now to improve his **self** man had to incorporate into his *super*concept, into his **identity**, the concept of the **collective**. He could no longer hope for redemption on his own account! To be forgiven of his evil, to be deemed worthy, he would have to form a conceptual alliance or partnership with a collective, he would have to piggy-back his *self* onto a tribe, an Empire, a State, a gang, a group, a collective of some kind! He could not expect to overcome his evil without joining a collective of some kind and having it at his side. A partnership with a collective of people, a syntheses with a collective, was necessary in order to be worthy of forgiveness and anointment! This expanded the person's **self** greatly for now it included not just his individual self, but the collective as well. While it gave a person a way to obtain forgiveness of his innate evil it also made him responsible for the welfare, survival and moral condition of the entire collective, and for maintaining its ruling elite. This was because the tribe or some other collective was *now part of his* **self!** A child raised in theism and indoctrinated in its principal concepts will then have a need for an abstraction, conceptualization or collective. He will ***require*** this mental construct to replace objective reality or Nature in the functioning of his mind. ***He will need a collectivist abstraction for his*** **super*concept!*** At this early stage in man's pre-history this need was met by the tribe and its ruling elite. Tribal man now depended upon a merger of his **self** with a collective, with the tribe. Man's **identity** now required a tribe. His identity was now a **self/collective** identity! A tribal *super*concept required a **self/collective** identity! With the proper merger of the **self** into the collective the **self** could actually disappear completely. The tribal identity could become pure **collective**! Now Shama no longer had to fear people leaving him alone in his newly created village and returning to a hunter-gatherer life-style. He no longer had to fear this because now the village, the tribe, was a *psychological necessity* for the slaves! Now people had to have the collective to complete their *self!*

The tribesman with his collectivist *self* cannot conceive of private property let alone the concept of private property *rights.* In the tribe there is no private property except as the control over some objects or persons may result from superior force. Might makes right for the tribesman, indeed might makes *rights!*

The **collectivist** type of *super*concept places the tribal theist in an untenable situation. He must have the collective to complete his identity. But the collective morality of self-sacrifice requires that he be masochistic! To satisfy his need for *super*concept authentication, and obtain that endorphin reward, he must take self-destructive action! Yet self-denial and self-sacrifice will thwart endorphin rewards from the satisfaction of his other basic needs. He cannot win. Whatever he does causes pain. If he upholds his Pagan morality he suffers the pain of his masochistic behaviors. If he acts against self-sacrifice and obtains some selfish endorphin reward he feels guilty and sinful. He can only enjoy himself if he intoxicates himself with alcohol or other mind-altering substances. Still guilt and depression await him when he sobers-up. Shama's invention religion, by producing the collectivist *super*concept produced the slaves needed to establish permanent settlements. But the collectivist identity does not come without a price.

Shama's religion was institutionalized as the Pagan Church and it was very successful at creating slaves and tribes. Because of strategic military competitiveness it was copied and very rapidly the tribal structure spread around the globe. The concept of a supernatural realm populated with omnipotent Gods was created and used to provide a substitute for Nature or objective reality in man's cognition. By this means the **collective** of men or the rulers of a collective of men, through revelation and exclusive contact with these supernatural Gods, could be substituted in man's *super*concept for Nature, or objective reality. This left tribal man with the **collective**, rather than reality as his epistemological reference of truth, his *super*concept. And it left him in dire need of the tribe and totally enslaved to tribal leadership!

Lesson # 6 Punitive Conscience

The day that Shama created a conceptual morality of human sacrifice he created the punitive religious conscience. A conscience is formed because of the way the human mind is constructed. Conscience is a mental function created in the imagination designed to monitor ones thoughts, feelings and behaviors and judge them as to whether or not they comply with one's **identity**. Those thoughts and actions then judged by the conscience to violate one's ethic elicit mental punishment in the form of guilt, shame, anxiety, panic and depression! The conscience is very powerful at causing pain and commanding obedience to one's moral code. Slaves will sometimes murder their **self**, commit suicide, as punishment for major violations of their religious conscience.

Once a punitive religious conscience was established in the minds of all Shama's congregation the refusal to do as Shama commanded would unleash upon them, inside their own minds, their self-punishing conscience. And Shama's number one command was not to be selfish! He taught that evil was untribal behavior like not sharing your booty with your tribe! The religious conscience, by creating guilt, shame, anxiety, fear and depression keeps the tribesman obedient to his chief and his Gods.

This religious conscience operates as a system of self-policing. By self-grading, self- judgment and self-punishment the slave is kept in harness. Such a system must be created in the slaves mind to keep him in line. This 'social' or 'tribal' conscience must be established to police the slave when he is out of sight of his master. This tribal or religious conscience punishes the individual when he makes decisions that are good for his **self** and rewards him when he sacrifices his interest for the tribe. This tribal conscience is not directed at keeping the individual true to the principles of self-survival it is now an instrument for enforcing compliance with the collective. It is now a collectivist conscience!

This system is enhanced by teaching the child a mythical narrative that outlines a system of reward and punishment. This sadomasochistic belief system must promise a great reward and threaten a great punishment if it is to function effectively. The slave must experience mental pain, guilt, depression, and shame, when he perceives that his obedience has been less than perfect. By teaching the child how to develop a punitive religious conscience an automatic self-indoctrinating, self-punishing control system is established in the slaves mind. Now we no longer need as many guards to oversee our slaves. Now Shama had a congregation of obedient hard-working slaves who emotionally need a tribe and chief, and who guard and punish themselves!

Theism Summary

It is the purpose of religion to take children and inculcate the slave mentality. Much of the apparatus and bureaucracy of the tribe have but one goal, the abuse of children processing them into collectivists. Theist concepts will create the slave psychology needed for a person to accept his role in life as a slave. An infant's religious parents aided by the tribal culture, inculcate in him the feeling of **self** evil. Then religious schools follow with further theist indoctrination to produce a near perfect slave. I say near perfect because slave indoctrination tends to have a relatively short half-life. The indoctrination needs regular weekly repetition and reinforcement to remain strong in the slaves mind.

The hunter-gatherer way of thinking is the easy way for Homo sapiens to think. Non-contradictory self-interest requires no force, no will-power; it is the easy, natural way to think. The slave mentality requires ongoing indoctrination and reinforcement. So regular, weekly meetings with rituals of human sacrifice, stories of transgression and punishment, examples of sacrifice with rewards after death, must be held to reinforce the slave training. The frequent meetings which are held for the slave population also help these frightened, confused, dependent and obsequious individuals find comfort and consolation amongst one another.

Theism then consists of the concepts of **evil, revelation, force, supernaturalism, human sacrifice** and **collectivist conscience**. These concepts are taught to children to create the collectivist mentality they need to be obedient, hard working slaves. It is this unique combination of concepts, taught as early in an infant's life as possible that forms the psychological lure and entrapment that captures and holds the individual in bondage.

Millions of slaves were needed to provide the power needed to gain mastery over agriculture, create permanent settlements and maintain the tribe. Religion and religious schools are the institutions created to teach this ideology. Millions of Pagan religions were invented by millions of Shama's around the globe. **Theism** is the adhesive that creates **slavery** and creates the **tribe**, and then holds it all together. This three-some is called **tribalism**.

Theism is an attack on children, on human life, on women because they bring forth human life, on life in all its dimensions. Theism is anti-freedom, anti-earth, anti-money and anti-business and commerce. Theism is for the tribe, the ruling elite, slavery, obedience, other worldliness, and evil. Theism is the worship of death. Theism is the worship of sadomasochism. Theism is the worship of coercive authority. The most astonishing phenomenon I have witnessed in my lifetime is the fact that most theists are oblivious to the real meanings of their beliefs.

Theism can therefore be seen as a psychological technique uniquely designed to enslave the human mind. It is the result of over 20,000 years of constant research and development. It has been modified, changed, altered and buffed into near perfection as a mind enslaving program. It is so efficient that no aspect of human brain function has been overlooked. It is a mind puzzler or 'computer virus' capable of shutting down the cognitive processes of the mind by riddling it with conflicts and short-circuiting its electrical and hormonal mechanisms. It is so good at entrapment that the smarter you are the better brain you possess, the more profoundly you are ensnared. Once caught within the theistic maze ordinary reason and logic only deepen your entrapment.

Slaves Indoctrinate Slaves

Just as horses are bred to replace themselves as they wear out and die, so human slaves were required to reproduce and replace themselves. Fertility in one's slaves is as desirable as fertility in other livestock, and it was encouraged and highly prized. Thus religion encourages unbridled reproduction, the more children the better. The ruling elite, the religious establishment, the tribe does not care how many people are destined to die of famine or war, the more slaves the better. As slaves came to make up a larger and larger portion of the population of these permanent settlements, the slaves themselves were given the job of teaching all the tribal children. With more and more slaves it became impossible to segregate the indoctrination of slave children from the schooling of the elite's children. Soon all were receiving the same theistic indoctrination in tribal schools.

After a time everyone in the tribe was thoroughly indoctrinated with the collective theist mentality. Most tribes continued to have a base of designated slaves. These people had almost no choices or freedoms. People of this class could be sold or sacrificed with little recourse. Above this base caste of designated slaves were one or more levels of slaves with more privileges. Then above these groups is the ruling class. Some of these upper castes may even have conceived of themselves as 'freemen'. These people were not free by any modern definition of freedom, but in relative terms they did have more choices. But regardless of the relative freedoms or the lack of it for any caste, all the classes within the tribe in time shared the same slave psychology. Thus all tribesmen are enslaved by their psychology to the tribe. All tribesmen, designated slaves and 'freemen' alike are enslaved to the collective. Each caste is enslaved by its collective mentality, and thus by its commitment to the tribe's expectations. Everyone in the tribe is enslaved by their minds which enslave them because they have adopted the ruling elite as their *super*concept of truth!

How then do members of the ruling elite distinguish themselves and exempt themselves from this collective mentality? The answer is they don't. Leadership is the mirror principle to slavery. Sadism is the twin to masochism. Sacrificing others is the other side of self-sacrifice. Leaders who have been indoctrinated with the virtue of self-sacrifice will eagerly sacrifice others once given the excuse. Grown men and women do not need a leader. A leader is a Judas Goat leading the slaves to do the tribe's bidding. How the leaders are chosen varies. In some tribes it is inherited. In others the role is won in combat. In a few tribes it was bestowed by the vote.

However one receives his anointment as a member of the ruling elite the individual **will consider the appointment as a leader to be sanctioned by the Gods**. He will then believe it his right and duty to order and coerce the lower slaves and have them do as he and the tribe demand. A collectivist, willing to sacrifice himself is usually more than willing to sacrifice other people to a Holy tribal cause! Psychologically the relationship between ruling elite and tribesmen, or leader and slaves, is one of co-dependent sado-masochism. Anointment with a leadership position is often taken as a sign from the Gods to shift more fully from the masochist role to that of the sadist!

Chapter 6

Tribe

Shama's invention included the hierarchical social structure we call the tribe. This was the permanent settlement he longed for. Utilizing a pyramid-type chain-of-command with Shama at the apex of power and his lieutenants and helper clerics just below, he held absolute power over all the slaves below him. The tribal structure would endure to modern times.

The tribe was designed to be, and is, a totalitarian societal structure. For millennia tribes exercised absolute and total control over all thought and action. As polytheism was institutionalized into religion and slavery, this over-arching support institution, the tribe was established. The tribe was invented to provide support to the agricultural and military enterprises, to better exploit slavery, maintain Pagan religious schools, and promote the religious institution.

But the tribe was not a separate invention. The collective identity which Shama created to produce slaves created the need for a collective, a tribe, which had to be created to complete the slave psychology. The slave/tribe relationship is a psychological interdependency. Slavery and the tribe were created together at the same time, they came into existence together. One is necessary for the other. If you remove the tribal slave from his tribe he will experience severe mental problems. Destroy a slave's tribal culture and you will have psychologically destroyed him.

Over the past 20 millenniums thousands of permanent, agricultural based settlements were established. The social structure invariably adopted by these settlements was Shama's societal structure. At the top of the ruling class was a chief (king, emperor, czar,). Right under him was his lieutenants (generals, priest). Next were the other tribesmen who are members of the ruling elite. Then there were the common or regular tribesmen. At the bottom of the pyramid-like organizational structure were the military slaves, the administrative slaves, the domestic slaves and the lowest of all, the plantation or farm slaves.

By tribe however, I mean more than this physical arrangement of the leadership structure. We must look past any superficial physical dissimilarity of social organizations to understand the deeper psychological meaning of the tribe for its members. On a more fundamental psychological basis a tribe is a group of people who are emotionally and mentally interdependent, all sharing a common belief system grounded in theism.

The love of one's tribe, meaning the people, their customs and lifestyle, their property or geographical area, all the various concretes which make up living among a group of people, is just a portion of what I mean by the psychological commitment tribal peoples have toward tribalism. There is more than this involved in the love and trust tribal peoples have for their belief system. Tribal people's love and trust of tribalism also exist in the abstract! By this I mean they are in love with the idea or their belief in tribalism! They have a kind of super patriotism toward their chief, their ruling elite, all the layers of slaves, the tribe's history and mythology which goes well beyond loyalty to the present chief or tribal members.

I know you have experienced this phenomenon. You enter a theater or sit down to watch a television movie. Within a few minutes of watching you have thoroughly identified with the star or main character. Now his problems are your problems and you are emotionally busy trying to protect his interest and find solutions to his struggles! I just came into the building or sat down in front of the television and suddenly I am transported off to another 'world', off to the world created by the writers, actors, director, musicians and set designers of the 'show' I am watching! I become truly emotional. I am angry that the leading character is being mistreated and jubilant when he is triumphant at the end!

What happens to us when we have an experience like this is that our imagination has been used to create a fantasy world that we inhabit while we are in the theater and under the spell of the drama production. In my case this fantasy tends to linger for a while after the show is over as my emotional involvement slowly subsides and I come back into focus on my real life.

For the tribal slave most of his life is one of living in such a fantasy world. Tribalism is such a fantasy. At night, sitting alone in his hut with his mate, if he is allowed a mate and such privacy, he may enjoy a few moments when he can separate himself from the emotional furor and tribal 'noise' and experience some real moments. But most of tribal life is lived in public; there is little or no privacy.

So a tribe is more than a social structure, it is a ***fantasy***. I say this because tribal slaves imbue the collective, the tribe and its chief or rulers, with a supernatural and spiritual quality. Tribal peoples have as their conceptual standard of truth a collective made up of the tribe, its Gods and its clerics and rulers. So the tribe, as an abstraction, is worshiped along with the abstract Pagan Gods and its idealized, often deified, tribal rulers. So the tribe is in a way, another set of Gods. The tribal collective with its ruling elite almost replicate the situation with the supernatural Gods as they also supposedly exist in a hierarchical structure in their alternative realm or Heaven. There may be little separation in the slave's mind between the tribal ruling elite in their palace and the supernatural ruling elite in their heaven! Perhaps the earthly authority is a subdivision of the Heavenly authority. It is the earthly elite who issue the laws and extract the offerings, so the earthly division is the more visible and tangible. The ruling tribal elite, whether or not they are deified, are the representatives and interpreters of the Gods. So if they are worshiped less it is not much less.

Thus tribal members feel a strong duty and obligation and a loyalty to their tribe not only in the concrete but in the abstract or conceptual as well! This feeling of love and patriotism goes beyond all the individuals and concretes in the tribe and is a devotion to the abstract notion of the tribal **collective.**

All tribal peoples have this strong love and trust of their tribe, as a collective. They are 'True Believers[18]', they believe that their tribe has some type of spiritual essence which exists above the common laws of nature and they furthermore believe that in the moral sphere their tribe can do no wrong. Tribal peoples are furiously loyal and patriotic to the belief in the moral superiority of their tribe. They are super patriots that place the survival of their tribe above the importance of their own survival.

During the first millenniums that followed the invention of the tribe, when there were still lots of hunter-gatherers around, the tension between the two was very great. Hunter-Gatherers did not like nor trust tribalism. Many hunter-gatherers simply couldn't buy into the tribe-worship that tribal peoples demanded. Tribal peoples considered hunter-gatherers to be selfish lazy savages only concerned with their own families, always hunting and fishing or taking it easy and enjoying life. Tribesmen viewed hunter-gatherers as a sub-human species. And the tribes nearly hunted the hunter-gatherer families into extinction. This genocide of hunter-gatherers continued well into the twentieth century. As recently as 1936 in South Africa you could purchase a hunting permit that allowed you to legally go out into the jungle and shoot yourself a hunter-gatherer for a hunting trophy[11]!

It is important to point out the fanatic and psychotic devotion to tribalism, its delusions and myths that are required to sustain a tribe. In psychiatric lexicon 'folie a deux' means a delusional or psychotic belief system held in common between two people. The term 'folie a society' has been used to denote the lies, beliefs and delusions held as common belief by most members of a group or population. The tribe is a delusional belief system, a 'folie a society,' which is necessary to maintain a tribal collective.

Just as the Witch-Doctor Shama needed fear, myth, lies, force, intimidation, terror and murder to start and maintain his tribe, all governing collectives require these same elements to continue in existence. Tribal people's need the tribe for the same reasons Shama needed the tribe: to save them from hunting, and gathering, to save them from having to go into the forest and to provide a justification and rational for their killing and stealing from other tribes. The tribe, its rulers, and the polytheist mythology that all the tribal members share, provide the rational for tribal slavery, murder, war and theft. People are murdered as sacrifices to the Gods. Slaves die in combat killing the people of other tribes. Tribesmen steal the lives and energy of other slaves. Tribal peoples steal the products of other tribes destroyed in warfare. By adherence and devotion to the tribal polytheist belief system each tribal member is guaranteed a share of the booty obtained through this exploitation of slaves in agriculture and warfare. The modern term 'gang' might convey some of the feeling of a tribe. The sadomasochistic aspects of all this stealing and killing should be noted. All socioeconomic philosophies which support tribal collectivism are rationalizations for stealing and killing. In addition to the delusional concepts of theism, what other beliefs are required for the creation and maintenance of the tribe?

The **first** delusion required for a tribe to remain in existence, is the belief by a majority of the people that the tribe is needed and necessary. The tribes' existence reflects the fact that a majority of the tribesmen cannot imagine surviving very well without slavery, stealing and killing. This is the typical feeling of people amputated of their objective orientation to Nature and imprinted on a theistic collective or tribal standard of truth. Having a *super*-concept based on the collective or the worship of a chief practically guarantees theft and murder. This is because the usefulness of slaves to the ruling elite is limited to agricultural and military purposes. A few females can be used as concubines, but the masses had to produce or steal. Stealing seems invariably to lead to murder.

In a certain sense this first delusion is not a delusion at all. As explained earlier the slave really does need the tribe. The slave has a psychological need for the other two components of tribalism. He needs the tribe and his Pagan Religion to function mentally! The tribe is his *super*concept, his epistemological reference point. And his Pagan Religion with its clerics is his authority and experts at the interpretation of his tribal 'reality'. The typical tribesmen cannot leave his tribe and move to another any easier than a person can change their *super*concept.

Likewise there can be only one brand or variety of Pagan Religion per tribe. Any breach of loyalty or minor change in belief does not simply lead to a competing sect as may be the case in a modern democracy. A slight change in belief is a huge fracture in the tribal structure and cannot be tolerated. It will lead to a fracture in the tribe! So the loyalty a tribesman has to his tribe must be toward its political leadership and toward its single Pagan Religion. These two authorities are unified and inseparable in the tribe, and in the minds of the tribesmen. Separating the ruling elite from the religious elite or splitting the religion into more than the one exclusive brand, threatens to split up the tribe.

The **second** delusion required of tribesmen is the unquestioning belief in the supremacy of force. The force and violence polytheists utilize on themselves inside their own minds causes them to project a similar need for force and violence in their dealings with all aspects of reality. This psychodynamic sadomasochism is no less obvious in their dealings with their social reality, in dealing with people. Indeed in polytheist societies there truly is a need for force! It is required to secure order!

In tribal communities impulse control must come from outside the individual. Paganism sets up a system of rules, regulations and expectations in the tribe. The tribesman must then depend upon his will-power to force his adherence to this legalistic structure. He will then project onto his tribe a need for societal rules and regulations which he demands his rulers create and enforce. Since the tribesmen have no true understanding of the need for social rules and are operating solely out of will-power, which is weak and unreliable, social order is always precarious and problematic. For tribal peoples force is a much more reliable means of maintaining order. It is indoctrination with polytheism that creates the cognitive and emotional acceptance of intra-psychic force which in turn creates the need for force and a preference for force. Thus force and violence are the standard of practice throughout tribal society. This reinforces another rationale for the existence of the tribe: to maintain social order.

The **third** delusion required of tribal members is the unquestioning belief in the righteousness of the tribe's ruling elite and their actions. The tribesmen are thereby relieved of any guilt they should have for participating in slavery, war, killing and pillage. Like all co-dependent relationships both the ruling elite and the common tribesmen acquire emotional support for their evil by belonging to the tribe. The tribal member's deep devotion to his tribe, and his delusional belief in his tribe's moral superiority, makes him blind to any atrocities the tribe's elite may order and which the tribe perpetrates. So regardless of how extreme the tribe's evil, a little oratory will make it alright with members of the tribe.

Tribal peoples must also accept massive deception and delusion in order to believe that slavery, murder and theft are holy and noble. They must believe that anything and everything sanctioned by the tribal collective, as communicated to them by the tribal elite, is acceptable, necessary and noble. Polytheism is the pure essence of evil, and the tribe shares this same attribute. But the tribesmen do not recognize this, cannot recognize this evil. All of this slavery, stealing and killing is viewed as grand, glorious, and noble.

The tribe is a very formidable machine and the term 'machine' is accurate here because everyone in the tribe is expected to 'play his role'. Every member of a tribe is intimated by the expectations of every other member in the tribe, especially those of higher 'rank'. Everyone in the tribe, from top to bottom is required to be a 'cog in the wheel' that the tribe needs and demands. Force and violence permeate tribal society. In a very real sense everyone in a tribe is a slave. There are just different classes of slaves, but from chief to the lowest of the low, everyone is expected to serve the tribe. Anyone who fails to live up to the tribe's expectations could find himself ostracized or worse. Both kings and the lowest of the low have lost their heads.

Each tribe's belief system, indoctrinated into the slaves when they are children, is a special version of a generic polytheism. Tribe specific taboos and ruler worship dogma is easily piggybacked onto this generic polytheism to suit the needs and specifics of the ruling tribal elite. Once polytheism is inculcated into a child's brain it is relatively easy to modify or to change the details and particulars of the tribe's special dogma. The tribe may change its name, geographic boundaries or ruling families, but the fundamental delusions regarding human sacrifice, murder, slavery, theft and the nobility of the tribe remain intact and persist.

The ruling elite cannot allow the slaves to know how wicked and evil they and the tribe really are! They do not want to know this themselves! So the ruling class works hard to convince themselves and the masses that the tribe and polytheism is essential to everyone's survival. Therefore anything that reduces or negates dependency of the tribesmen on the tribe threatens to expose this truth and could potentially dissolve the tribe. The ruling elite must therefore eliminate anything which encourages independence because it is a threat to the tribe's survival!

You might think that the tribe is very vulnerable to the truth. But the Emperors' nakedness is almost always interpreted as fine clothes[15] because of the tribesmen's profound dependency. Their need to believe is so great that they see what they are told to see! The booty obtained through the stealing, killing and enslaving is divided up among tribal members. The acceptance and enjoyment of one's share of this loot demands that one not examine the ethics of the tribe's behavior too closely. The tribe's expectations and the tribesmen's 'conscience' control everyone's thinking and behavior at all levels within the tribe.

Initially the purpose of the Pagan religion was to produce large numbers of slaves in order to harness agriculture. The institution of religion with its temples, schools, priest and teachers were charged with turning out large numbers of slaves on an assembly-line basis. Utilizing young children as their raw material these mind modification or slave producing factories perfected their techniques and developed a high degree of efficiency. Soon, where farm slaves had been the sole goal this was expanded to include military slaves as well. Domestic and administrative slaves were soon required. Soon the larger tribes required not only many plantation or farm slaves, domestic slaves for the ruling class, and full time warrior slaves, but a full administrative bureaucracy to manage the religious institution, agricultural enterprise and the military as well.

The tribe was always at war or preparing for war. When you believe humans are born evil and that only the correct tribal or religious rituals can make people a little acceptable, then the members of other tribes, worshiping different Gods and rulers and doing things differently, cannot be doing things correctly. The members of other tribes must be fully and completely evil! The Gods would certainly sanction and approve eliminating them!

War and military conquest is therefore as important from the psychological standpoint as it is as a source of sacrificial victims, spoils, territory and slaves. Constant warfare, or the constant threat of it, serves to reinforce, for all the tribal members, the necessity of sticking together for their mutual defense and survival. Neighboring tribes are always planning to attack, so every tribe has to stay vigilant and ready to defend itself. The psychological dependency inculcated into the slave mentality is thereby reinforced by the reality of tribal life. Polytheist will always work things around in their tribe to reflect their inner fears. So what starts out as a fear in their religious imagination comes to be created in their lives. Societal organization recapitulates psychological organization.

The **forth** delusion required of tribal peoples is the belief that only the tribal leadership can be trusted with storage and control of the tribe's food! The ruling class knows that the key to its rule over the masses is the control of food. While slave labor produces the food that allows the existence of permanent settlements and thus the very existence of the tribe, and its ruling class, it is the ruling class that commandeers the food at harvest time. They place the harvest in storage under armed guard and then dole it out to the slaves in subsistence amounts. Just as surely as domesticated animals can be made to comply and obey by control of their food, a slave must do likewise or starve. Those that refuse to work or resort to criminal activity give the tribal leaders another reason to exist. Capturing and executing criminals and 'trouble makers' is an example to the obedient slaves and encourages continued obedience. This forth tribal delusion holds that the ruling elite are the best, most appropriate entity to hold control of survival necessities, most importantly food. The tribe cannot tolerate a tribesman, or group of tribesmen, having an independent source of food! An independent source of food could easily lead to psychological independence, and that could easily lead to the recognition of the unnecessary and evil nature of the tribe. Sedition and secession might follow.

These four tribal delusions all work to create the illusion that the tribal collective is 'real' and to reinforce dependency upon it and its leadership. What then works against dependency? Anything which reduces or negates this delusional tribal fantasy threatens to expose the truth and could reduce psychological dependency. Weakening any of these tribal delusions could potentially expose the illusion and dissolve the tribe. The ruling elite must therefore eliminate anything which encourages independence because it is a threat to the tribe's survival! What must the tribe fight against because it tends to undermine the slaves dependency on the ruling elite?

Psychological insight into the evil of slavery, human sacrifice, murder or theft would certainly threaten the survival of the tribe. Atheism, deism, anti-religious belief, anarchy, slave mutiny, pacifism and refusal to work or serve in the military are obviously very serious tribal offenses and must be dealt with swiftly and harshly. Few tribesmen are mentally equipped for heresy like this and in most situations it would amount to suicide.

The tribe is a collective. The tribal ruling elite are a collective. The Pagan Gods in their Heaven are a collective. These abstractions mingle, merge and interact within the slave's mind as the collective portion of his **self/collective** *super*concept. A tribal member feels as responsible for the welfare of these collectives as he does for his self, perhaps more so. Actually he is unable to differentiate one from the other! These collectives live within his mind and are an integral part of his **self**. To threaten the tribe, its rulers, its Gods are the same as threatening the slave or his family. He has no self that is independent from and separate from these collectives. His self *is* these collectives! There is no difference between **self** and collective. It is all one and the same. Now the **self** is just self/**collective**.

To the extent that tribal 'noise' interferes with or prevents the individual from perceiving objective reality, nature or Nature's God the person is, by definition, psychotic. Tribal peoples are psychotic or near psychotic much of the time. A minimal appreciation of reality is necessary, at least on the part of the tribal leadership, just to successfully manage the agricultural and military operations. But many in the tribe are mostly psychotic most of the time.

Ruling Elite

Who are the ruling elite? What are the motives of the ruling elite? Every tribe has a group of people who control the apparatus of the tribe and profit there from. That profit is among other things, material goods. The ruling elite profit materially in a greater degree than their time and effort would justify. The chief for example might take 60 to 80 % of all harvest, spoils, loot and domestic production for himself and his entourage. While profiting in this great amount he and his elites might contribute less than 1% to the total time and physical effort of the agricultural and military enterprises. Material gain is a very real and substantial lure for elites.

But another even more powerful motive for elites is, as it was for Shama, the **control** of the tribe. This allows them to avoid their fear of the forest. This fear of the forest can be translated as fear of reality, or in hunter-gatherer terms, fear of God. This fear comes from a feeling of underlying incompetence at dealing with God. This is an existential fear that emanates from their existential incompetence. This is caused by a change in *super*concept from nature to rulers. Now relying on humans, on a collective of people, rather than nature creates a distrust of nature and a need to control people. Thus the ruling elite's motivation is the fear of the forest. They would rather have their slaves kill and steal from others than to be forced to survive on their own. They would simply rather murder than work.

Going into the forest to hunt and gather is so fearful for some that they seek to force or trick others into making the journey for them. They seek power over those who are not afraid to go into the forest. They wish to control those who are not afraid to confront God face-to-face. No amount of manipulation, destruction, deception, or killing is too great to use in promoting this agenda.

Theism is the belief in a theology. This theology has an agenda. The agenda behind polytheism is the survival and maintenance of the tribe and its ruling elite by slavery, stealing and killing.

From the earliest tribes down to those of the Tigris and Euphrates Valleys, to Sparta and Egypt, the Empires of China and Japan, the Empires of the Mayan and Aztec, to the Nazis, and the Soviets, all have depended upon theism, some variant of Pagan religious belief, and on some degree of terror to control its citizens to enforce submission and obedience.

The tribe is therefore the prototypical totalitarian organization. The level of terror, violence and absolute control that the tribe holds over its people is difficult for moderns to appreciate. Huxley and Orwell painted a mild version of this ancient reality. People brought-up and living in a liberal modern country may have trouble imagining this level of absolute brute power and savage force. The words total and absolute control of everybody and everything comes close. I would add the designation of absolute and unremitting evil.

Chapter 7

The Reification Error

The cognitive error which the theist prejudice of **self** evil institutionalizes in the infant's mind is called reification! Reification[19] is the type of cognitive error caused when an abstraction is mistaken for a concrete, real event, or physical entity. In other words, it is the error of treating as a "real thing" something which is not a real thing, but merely an idea or a fantasy. Reification is generally accepted in literature where reified abstractions are understood to be intended metaphorically, whereas the use of reification in logic is a cognitive mistake.

A cognitive reification error is often committed by confusing a behavior, feeling, belief or fantasy with a real entity! It is confusing an action with the actor. A person may mistake behavior for the entity exhibiting the behavior. It is an extremely common cognitive fallacy among theists. This type of cognitive mistake is also more common among theist when they are dealing with the areas of reality we call the 'social' or 'behavioral' sciences. This is caused by theist's inability to differentiate the **self** from the collective.

Such errors can often be identified by a shift from verb to noun. Take for example the concept 'alcoholic'. This is a fallacious concept. There is no such thing as an 'alcoholic.' There are human beings who drink too much alcohol. But 'drink' is a behavior, a verb! Or consider the word 'homosexual.' There are human beings who prefer sex with members of their own gender, but that is a behavior. There is no such thing as a homosexual. Many such errors of thinking begin as metaphors. Then when the metaphorical nature of the concept is lost or forgotten the concept continues as a fallacious one.

My profession, psychiatry, is based on such a fallacious concept. 'Mental Disease' is a metaphor[16], no such thing as mental disease actually exist! Disease refers to damaged tissue not to incorrect ideas. There are diseases of nerves but there is no such thing as 'mental disease'. 'Mental' refers to how one thinks, behaves and feels. It makes sense like referring to a shoe repairman as a 'shoe doctor' makes sense. It may be a cute metaphor but it is a logical error. Almost all the so-called 'psychiatric diagnoses' are fallacious. There is no such thing as a 'schizophrenic.' 'Depression' does not exist as a noun! It may be logically correct to say "Jim is depressed." But what is meant by "Jim suffers from a depression."? As a psychiatrist I might be able to assist you in changing the verb which accurately labels your behavior but even a surgeon would have difficulty changing you as a noun!

This kind of fallacious thinking is ubiquitous in the theist world and it causes all manner of problems. It also makes solutions to many kinds of problems almost impossible for theist to see. I suppose that we should be thankful that at least in the 'hard' sciences most theists can think clearly.

The cause of cognitive confusion on a scale as massive as this begins in infancy with incorporation of an evil self identity. The concept of 'evil self' is itself a reification error, as are all prejudices. The acceptance of stereotypes is a reification error. Stereotypes and prejudices like 'evil self' are no more real than Green Gremlins. That of course does not prevent people from believing that they actually exist. And infants are almost completely defenseless against a culture-wide prejudice like the notion of innate human evil.

So the initial and prototypical reification error occurs when a child accepts his identity as evil. This belief in innate evil is first experienced by an infant as an attitude, a bias or prejudice against his **self**. Then during the development of their *super*concept the theist child incorporates the percept of an evil **self**. This prototypical reification error is incorporated into a child's mind at a point when the individual has not yet reached the age of critical reasoning. It becomes his fallacious cognitive methodology before he has any means to resist it! The acceptance of **self** evil, creating this fallacious cognitive methodology, then facilitates the incorporation of the other theist concepts.

This reified type of 'thinking' is referred to in the psychiatric literature as 'magical thinking.' Utilizing this same kind of 'thinking' the other theist concepts are easily incorporated into the child's mind to produce the **collectivist *super*concept.** This self/collective now effects how the theist thinks about his **self**. This expanded **self**, this amalgamation of **self** and collective, created a new kind of **self** and it reinforced this new reified way of thinking. Now when a slave thought about issues evolving his **self**, he thought differently. He has incorporated into his mind an evil **self** and institutionalized the use of reification as his method of 'thinking!'

Tribal man was then tied to the use of this cognitive error. Now it was his way of thinking in all issues involving his *self*. And since there was no separation between the **self** and the tribe this kind of thinking applied to all things tribal as well! But the 'tribe' itself is reification! Fallacious thinking now came not only to morality but to cognition as well! The theist child's **self** is then both evil and merged with a collective. The theist has a fallacious **self** inhabiting a fallacious universe! He cannot renounce and discard the one without renouncing and discarding the other! Shama was now secure that his slaves could not leave him for now they had no other place to go!

So a **tribal identity**, the collective **self**, the **self/collective** is a phony conceptual **model** of a **self**, it is a false, fantasy **self**. It is a **self**, a type of **identity** which is not totally real, it is mostly floating abstraction! This causes the tribesman to 'think' in this particular, peculiar, fallacious way. Tribal 'magical thinking' is excessively conceptual and abstract. The slave's 'thinking' has a tenuous attachment to perceptual reality. The slave 'sees' many abstractions which he believes are real. He 'sees' reality where there is none and is easily influenced by narrative and drama. Children indoctrinated with such a *super*concept develop a cognition which lacks a stable, objective, impartial reality by which to gauge contradiction. The tribesman suffers from *egocentric, over-inclusive categorization*!

Now conceptual models, abstractions such as theological narratives and tribal legends and myths became more 'real' than perception, more 'real' than objective nature. The human mind now functioned differently. Now truth was by these abstractions and conceptual models rather than by a natural perceptual truth. Now men were swayed not by their own perception and personal experience, but by their imaginations stimulated by rhetoric and ritual. Oratory has a magical effect on the minds and emotions of theist! He is highly suggestible and gullible. He easily believes in the reality of such abstractions as 'tribe,' 'supernatural,' 'Gods,' 'heaven,' 'demon,' 'monster.' Revelation is more real than sensation, perception and experience. And he must have an abstraction of some collective to complete, nay **create,** his **self**.

While the creation of this cognitive disability in children is perhaps the worst thing the belief in innate evil does to a child, it is by no means all the damage. The Doctrine of Original Sin damages more than just cognition. It also destroys man's natural morality.

The concept of innate evil undercuts the very purpose of a moral code of conduct. The purpose of a morality should be to guide decisions and thereby assist in successfully solving the problems of survival. The purpose of effective moral principles is to sustain one's life. The concept of innate evil removes all control and choice from man's powers. There is nothing a person can do himself, as an individual, to improve or better his **self**! He is powerless at effecting and improving his moral stature, his worth or his chances of survival.

The concept of Original Sin is therefore very destructive to morality. It means that there is no direct connection, no payoff or profit in one choice of behavior over another. Individual choice does not amount to anything. It means that sustaining one's life and the lives of one's family is not very important, not a moral accomplishment. This is because their life and his life are so unimportant! It means that for humans, life is not much better that death. Life is not morally superior to death; humans are evil in either case. The belief that one is innately evil can be a basis for nothing other than masochism and destruction, death and suicide.

By these means Shama changed collective man's **super**con**cept**. Now the epistemological reference was not an objective, fixed and inelastic truth of Nature, but a truth based on the ever changing whims of the ruling elite. Now the measure of 'good' was that which benefited the collective. Therefore human sacrifice became anointed as the highest virtue! Changing man's epistemological and moral identity changed his *super*concept. Changing a person's *super*concept changes their psychology. Tribal man's *super*concept became the **collective** of man.

Because a collective *self* makes it difficult to comprehend and respect private property rights, conflict and violence are assured. The two great laws of civilation[26] (1) do not trespass and (2) do all that you agree to do, both require an understanding and commitment to private property only possible to an individual with a *self*. It is a commitment that is not possible for the tribal collectivist. Thus the morality of self-sacrifice makes human relationships very difficult in many ways.

The Origin Of Evil

The correct definition of evil is: ***coercing human beings***. Evil is behavior which emanates from the conviction that one has the right to coercively exploit other people for one's satisfaction. It is the belief that one human being may forcibly use another, that people are chattel. Evil is the initiation of force against another person. Trickery, deception or manipulation does not alter the fundamentals. Cannibal, slave master or tax feeder, the principle is the same. The theist is taught that evil is not evil when done in the name of his tribe.

Shama, in the process of inventing collectivism invented evil. Shama by inventing a way to enslave people created evil. For slavery is a system based on the ***belief that other human beings may be manipulated and exploited for my benefit and satisfaction.*** We now know why Shama refused to define evil, for it would have revealed his plan! So with the creation of theism and religion, evil was created. Religion, not man was 'born' evil!

If man is not born evil then from where does evil emanate? Evil is taught. Evil is learned behavior. But evil is not necessarily taught directly and explicitly. Evil is the result of indoctrination and not necessarily an intentional outcome, or result. But regardless of intent, evil is the result of theist indoctrination.

Evil is always required in the production of slaves. The human mind is constructed in such a way that it only works properly when it is free! If you want to create slaves, you must accept the fact that that much evil will be necessary to their production! The forced and reified indoctrination of children in theism in order to turn them into good dutiful obedient slaves runs into a peculiar problem.

Any parent can probably guess what this peculiar problem is. And certainly anyone who has done any counseling such as clinical psychologists and psychiatrists knows where I am going.

Prison wardens and criminologists perhaps know what I am about to discuss best of all. I wish to discuss the psychological blow-back, the so-called 'acting out' or rebound that occurs from heavy discipline. The discipline applied to infants and children to indoctrinate them in theism and produce obedience, obsequiousness, and automatic self-policing inevitably leads to rebellious and sadomasochistic behaviors. This back-lash of rebellion and sadomasochism can be very violent and destructive. We see this consistently in clinical practice. A child or an adult may be 'brainwashed' or indoctrinated into obediently following commands. He appears totally docile, compliant, masochistic, generous and altruistic. Then when you least expect it he rebels. And his rebellion is often much more violent and aggressive than an evaluation of his past treatment would seem to warrant. Indoctrination is after all, **force-feed** instruction! The human spirit refuses to remain subservient indefinitely! Dependency often breeds a violent rebellion. It is this trait which is responsible for most of the evil in the world but it is also this trait which has saved the human race from totalitarian subjugation on more than one occasion.

This back-lash of sadomasochism and rebellion against indoctrination with theism and the effort to teach the desirable slave attributes amounts to the creation of a second set of slave attributes. So, we end up with **two** sets of slave attributes or traits. The slave is taught the 'good' slave attributes, those masochistic attitudes and behaviors which the authoritarian elites desire. And then there are those 'bad', evil, rebellious, violent, sadistic and destructive slave traits that develop from the push-back against indoctrination, force, discipline and coercion.

Tribalism is based on human sacrifice. Tribes sacrifice members of their own tribe as offerings to their Gods. Tribal peoples sacrifice themselves to their Gods. It is usually considered a great honor to be chosen as a sacrificial victim. Giving generously of one's life, time, energy, food and personal possessions to other tribal members is a tribally approved behavior. A tribesman is expected to offer his abilities for use by members of his collective before being asked to do so. This shows a proper attitude towards one's duty and loyalty to one's tribe. Joining in and participating in work projects and military raids is applauded, especially when done in an exemplary and heroic manner. So the traits of self-sacrifice, obedience, obsequiousness and self-policing are very desirable in one's slaves. These 'good' slave attributes are the foundation of slavery and crucial to the existence of the tribe. The code of the tribe is ***'from each according to their ability-to each according to their need'***.

Note that all these 'good' slave attributes are good for the tribal collective, but they are not necessarily good for the individual. These 'good' slave traits are actually the very essence of masochism! **Masochism** is defined as obtaining gratification from pain, deprivation, humiliation, degradation, and suffering, ***inflicted on oneself***. It includes the condition in which sexual gratification depends upon suffering, physical pain, self-denial, submissiveness and humiliation. The act of taking from oneself and giving to others can be an act of masochism. Masochism also includes the act of turning one's self-destructive tendencies inward or upon oneself. Self-flagellation and 'cutting' behaviors are masochistic rituals. Pain is a signal from the body that something is wrong. Pain signals events which could lead to or result in death. Pain is the indicator of approaching death. Pain is the closest a person can come to death and yet not go over the edge permanently. So it is that suffering is the theist *silver* coin and death the *gold* coin with which one's redemption from the sin of innate evil is paid.

When observing another person's behavior we might note an act which we believe to be one of extreme generosity. Or we might label this act one of altruism. If we could get inside the mind of the person who acted so generously we might find that he acted as he did because it was painful for him to do so. In other words his generosity could have been motivated by masochism or self-sacrifice. An act of masochism may be observed by others, approved and applauded as an act of altruism, when psychologically it is also an act of self-sacrifice. Masochistic slave traits may therefore be beneficial to the tribal collective, especially its ruling class and may be viewed by others as virtues and labeled altruistic. So if we are to understand human motivation we must look deeper than behavior. Are those who consider self-sacrifice to be a virtue encouraging masochistic behavior?

Masochistic behavior is often considered altruistic by the recipients or the benefactors of such sacrifice. So the assessment of motivation by an observer of such behavior, especially a benefactor of said behavior, may be very incorrect. If someone is the recipient of someone else's generosity, they seldom question the deeper motives of the giver. They simply thank the giver and then focus on their good fortune. The ruling elite certainly do not question the motivation of the slaves as the slaves do as they are required. The ruling elite consider it their *right* to profit from the slave's masochism!

Here we are addressing the slave's true motivation, as difficult as that may be to determine, and not how an observer might evaluate a behavior. In the final analysis only the individual themselves can answer the question of the true nature of their motivation. From the psychological perspective masochism and self-sacrifice are almost synonymous.

We can be certain that theism, religion, as a belief system approves and promotes masochism. Theists expect their lives to be one of suffering and self-sacrifice and as dedicated masochist this is the kind of life they accept or create. A tribe is a collective of masochist and exists in conditions as you might expect from such an orientation.

Now to discuss the second set of slave attributes the **sadistic** traits. Sadism is also about pain, the inflicting of it on others. It is making others pay-up for their innate evil. The first sadistic trait I wish to mention is the slaves' hatred of other slaves that display poor slave traits! The theist is so thoroughly indoctrinated in the virtues of self-sacrifice that he abhors and angrily rebukes any sign of selfish individualism. He especially condemns any sign among his fellow tribesmen, or in his own motivation, of selfish, self-serving behaviors. For the tribal slave no accusation is worse than that he has placed his own interest above that of the tribal collective. Such behaviors are universally condemned and every tribe punishes such behavior. The punishment metered out for 'selfishness' in most tribes is extremely harsh, often lethal.

Next I wish to point out how intolerant tribesmen are of deviant thought. Because the theist has a *super*concept based upon a collective of people rather than nature, his standard of truth is often very demanding of the moral performance of his fellow slaves. Tribal purest are often critical of the adherence of their fellow tribal members to the official tribal doctrine. Therefore ***any*** variation of belief from the official tribal consensus is dealt with harshly, usually by execution.

These are just a couple of the sadistic behaviors tribal slaves relish. There is no end to the human imagination so there is no end to the ways tribal peoples have created to inflict pain. Such evil or **sadistic** traits of slaves are the psychological attributes which are usually an ***indirect*** result of theist indoctrination. These indirect results of theist indoctrination of children are more or less side-effects. Nonetheless some of these side-effects are very useful to the tribal leaders.

Sadistic tribal attributes are useful to the rulers of the collective so long as they can be controlled. Warfare allows the collective to vent its sadistic traits and acquire prisoners for sacrifice and other valuable spoils. Murderous vengeance and blood sacrifices of one's enemies are desirable tribal attributes. As long as the rulers of the collective can keep these feelings directed at tribal enemies there is relative peace within the tribe. People who display selfishness towards the tribe and its members, those who do not recognize the importance of one's tribe and its beliefs, those who denigrate one's tribe, are some of the groups who are the objects of rage, hate, torture and death . No degree of torture and no number of deaths is too great for the enemies of one's collective. The sadistic portion of a collective identity will unleash its homicidal rage when it is summoned by the ruling elite and directed at tribal enemies.

As I mentioned above the existence of the sadistic part of the collective identity does not result necessarily from the *direct* indoctrination of children with violence and killing. While these sadistic traits are often taught to tribal children they will exist without any help from direct indoctrination! The **sadistic** attributes of the slave would not disappear simply because religions or theist peoples ceased to explicitly promote them. ***The major, most powerful cause of evil results from the way the human mind works!***

The **masochistic** tribal traits are based on *self-sacrifice*. The **sadistic** traits of theism are directed at the *sacrifice of others*. Yes, humans are not born innately evil, but indoctrination with the theist masochistic beliefs produces the evil sadistic attributes as a rebound or side-effect. ***The evil, sadistic* traits of a collective identity grow directly out of the effects of the *masochistic* indoctrination of children!**

The reified collective identity is always present as this duality. The **masochistic** traits of a collective identity always travel with the **sadistic** attributes. Sadism is always present as the other side of the coin from masochism. Self-sacrifice is the other side of the concept of sacrificing other humans. Psychologically it is not possible to have one side of these tribal traits without the other. They are the two faces of Janus. As mentioned before: a person willing to sacrifice himself is usually more than willing to sacrifice other people! This is a very important principal of human psychology. I want to be sure you understand how it works. The cause of this is a psychological phenomenon I have labeled the **Dialectic Effect**.

The Dialectic Effect

A fully integrated mind such as that of a hunter-gatherer will be constructed of four-dimensionally integrated concepts, which means that they are integrated back and down, over time, with associations that tie them, through perceptions, to external reality. In the healthy mind, no matter what level of conceptualization may be reached, the associations can be traced back in time and down through each level of abstraction to a perceptual underpinning. The hunter-gatherer's *super*concept was formed through the direct perception of nature. The hunter-gatherer's *super*concept will contain a self **identity** that is an individual perception rather than a collective reification!

A collectivist *super*concept develops out of a child's indoctrination in theism, the development of reified 'magical thinking' and the Pagan morality of human sacrifice. This collectivist *super*concept will contain a **collective/self** identity. A collectivist identity is not completely real! It is not fully integrated into real percepts! It is **not** fully attached to perception! Indeed perception is useless at providing information about a person's theism, a better understanding of their collective reifications or their supernatural Gods! The collectivist *super*concept is created by the processes of conceptualization and reification **absent a perceptual base**. It is a floating abstraction with no attachment to perception and thereby to objective reality. So when the theist refers to his source of truth, the human or tribal collective, he is referring to this abstract reification. For information about his reification he turns to another of his reifications! Such is the delusional fantasy world inhabited by tribal peoples!

So the collectivist identity is **not** based upon perception, objective reality or Nature. The associations connected within a theist's collectivist mind may be interconnected with each other, but they do not connect to perceptions. Such an identity, the **theist**

identity differs in a very fundamental way from an identity that develops from a belief and commitment to objective nature and perception. This difference between a **percept** based identity and one based on floating fallacious **conceptual** abstractions is very important.

One of the most profound effects of a floating, unattached **collectivist identity** is that such mental constructs *cannot be limited to a single entity!* Human cognition works in a dialectic manner. When a concept is difficult to integrate, the mind works on it from every angle. It tries breaking it down into smaller concepts and it tries combining it with other concepts to make a wider one. Then, if a concept cannot be integrated into one's knowledge, a healthy mind tosses it out. But if the individual fears spitting a concept out of his mind, if we are dealing with **a tribal delusion that the individual is afraid to reject,** then he must allow the mental construct to stay in his mind! This principle applies not just to identity; it applies to all floating and unattached tribal concepts. It applies to all tribal concepts not attached by associations back and down to perception. Beliefs, sins, commandments and taboos also exhibit this dialectic effect.

The slave has many reasons to fear throwing out of his mind any tribal belief. He has no way to determine whether or not the belief is true. The ruling elite are his 'experts', those he turns to for evaluation of truth. Without an independent and objective standard of truth how can he make such a judgment? The tribal elite are for him the source of truth so how can he question beliefs the tribe sanctions? The tribesman's need to **obey, conform, and believe** the required tribal belief is much more powerful for him than is the need to think and understand the beliefs. He would not be so impudent as to question tribal dogma or to think it over for himself. Under this circumstance the dialectic aspect of cognition causes the Dialectic Effect.

Whenever one floating, unintegrated, mental construct or belief is accepted into and maintained within a mind, it creates an opposite or mirror image of itself -- another mental construct or belief. An action causes an opposite and equal reaction. This is true in outer space where there is no gravity. Because of the dialectic of human cognition, this is also true in the mind. Abstractions that are *unconnected to reality*, without links to percepts, floating loose and unattached create their opposites the more they are reinforced. This is the consequence of forcing unintegratable or unintegrated concepts into the mind! This is the Dialectic Effect.

A body of beliefs, a matrix of concepts, integrated four-dimensionally, founded on a connection through perception to external reality, or nature, does not rebound, and does not create a mirror matrix, is immune to the Dialectic Effect. An identity dedicated and faithful to objective reality or Nature does not trigger the Dialectic Effect.

For the tribesman whatever religious fantasies, floating unattached beliefs, unintegrated rules and commandments are accepted into his mind will be inversely replicated! Therefore whatever the child is taught by his religion, whatever has been decreed to be sin or taboo, whatever theist virtues the slave child has been forced to accept, whatever the indoctrination has been, **will automatically create their opposites in his mind!** The theist virtue of masochism will automatically generate the evil, sadistic motivation! The mind of the theist worshiper will create a sadistic, rebellious identity to break the rules and oppose those decrees created in the 'good', obedient, masochistic 'mind.'! Actions motivated by beliefs that originate in the tribally approved obedient area of the collective mentality will reinforce not only that desired aspect of tribal belief and behavior, but will also concomitantly reinforce the mirror or ghost 'mind' containing the sadistic and rebellious beliefs as well!

The sadistic, rebellious aspect of the collective mentality will generate selfish or heretical fantasies! These rebellious or heretical fantasies of a collective identity may initiate selfish, sadistic, hedonistic, narcissistic and aggressive thoughts and behaviors. This is often the cause of the self-fulfilling prophesy. These rebellious, sadistic or heretical feelings and beliefs violate the slave's moral code of sacrifice and represent sin and transgression. This will frighten the slave and have him question his faith and the sincerity of his devotion to his supernatural Gods. He may wonder that perhaps the Devil is trying to possess his mind and cause him problems with his theist Gods! Thus the more the slave prays and tries to do 'good works', *the more he reinforces his obedient, self-sacrificing, and masochistic identity, the more he reinforces his sinful, sadistic, evil and rebellious fantasies as well, and vice versa.* **Thus where floating abstractions are concerned the more one obsesses over adhering to 'the straight and narrow' rules and commandments, the more one compulsively tries to behave in a righteous way, the more fantasies of breaking of the rules will be generated.**

Most tribes do in fact teach a great deal of evil, but such teaching is not necessary for evil to be created. Tribal peoples also demonstrate by example sadomasochistic behaviors at all levels within the tribe. But these object lesions are not required to ensure the propagation of evil. All that is needed to create evil is to **force children to believe in the virtue of human sacrifice!** Yes, all that you have to do to produce evil is to indoctrinate children, especially those afflicted with the reification cognitive deficit, with abstract, anti-self, self-sacrificing, masochistic notions of virtue! These self-sacrificing, masochistic floating abstractions will generate their evil opposites in the person's mind without any need for didactic instruction.

Thus the theist individual who tries the hardest to live a perfect and 'Holy' life, completely devoted to the tribal collective and its ruling elite will have the greatest difficulty of experiencing 'sinful', 'selfish', "evil", 'sadistic' fantasies that he must combat with will-power ! If a theist tries very hard to lead an altruistic, obedient, tribally acceptable life he may then find himself thinking about things that will improve or enhance his own, personal, selfish individual life. This will cause him to feel great guilt. This violates his collective moral code of self-sacrifice. He is engaging in selfish thought. He becomes angry with himself. He may then turn his anger inward on himself, punishing himself inside his mind. He may 'preach' to himself inside his own mind, accusing himself of awful things, calling his **self** horrible names and condemning his **self** to a sadistic punishment. This will cause him to become depressed. He experiences anxiety when he realizes he is violating his collective morality of sacrifice and that this may anger the Gods and the tribal rulers. Panic may overwhelm him at those times when he realizes that he may be caught being selfish inside his mind. It is his tribal conscience; the judgmental portion of his collective identity which persecutes him.

The **dialectic effect** of the theist mind will therefore create the selfish, rebellious, evil, sadistic traits as long as the obedient, self-sacrificing and masochistic traits are taught and reinforced! The slave will therefore develop fear of his own thoughts and feelings. His own imagination threatens him with heretical notions and with images of punishment and retribution from his tribe. Fear and dependency keep him in line. The 'slave mentality' is the same as the tribal identity. They are all the same and perhaps best labeled the **'theist identity.'** They develop from the change in *super*concept from Nature to a ruling collective or Chief.

We might ask: why does the dialectic effect exist? Is it not proof of man's innate evil? The dialectic effect is a normal phenomenon. Children and creative people use it in their fantasies without any undesirable consequences. The dialectic effect is a result of the human being's great freedom of conceptualization. It causes them no harm because they are not forced to believe the concepts they are playing with. The dialectic of cognition **is** man's free will! And it is this attribute of the human mind which necessitates a *super*concept. If man did not have a dialectical cognition he could not make this error, he would then not need reality or Nature's God as a concept and he would not have free will! This is the state of most animals. Only man is free enough to be capable of evil. A capability does not insure inevitability. If evil were innate then man would have no choice and if he had no choice he could not be evil.

As a result of the invention of evil and its use to sever man's *super*concept from perception and Nature and substitute imaginary conceptual Gods, the collective **theist identity** was created. Tribal peoples incorporate into their minds a *super*concept of truth which holds that truth is whatever the tribal Gods say it is! This means that truth is whatever the tribal elites' decree. There are many Gods so there are many reference standards. It is difficult to make closure on one *super*concept. This means that tribal people developed a greatly compromised cognition. This 'fuzzy thinking' could result in total psychoses. But typically some portion of the slave's mind is tied to perception and thereby to objective reality. Just to farm, or to engage in warfare requires a minimal degree of rationality.

Therefore the adult tribesman, because he believes in multiple Pagan Gods has much of his mind standardized to multiple measures of reality. When he is engaged in some mundane manual task related to his work he might demonstrate rational mental operations and appropriate behaviors. But in all other mental operations this tribal person might be, by definition, psychotic. When praying to or obeying God X he would think as he has been taught to think in regard to God X. When praying to or obeying God Y he would think and behave according to what he has been taught of the requirements of God Y. He may have a conceptual matrix or personality or 'mind' for each of his 'Gods'. A tribesman's mind would have to do this to the extent that a particular God's concept of reality varied from that of the next. A tribesman might even have another 'mind' for relating to the chief and perhaps another for his relationship with the tribal witch-doctor. This would also depend upon the edicts and requirements of the particular tribe.

This meant that only a small portion of the tribesman's mind was concerned with what we moderns call reality. The major portion of the tribesman's mental capacity would have been devoted to constructing and maintaining these elaborate mental matrixes, these numerous 'minds', one for each important figure, real and imaginary, in his panoply of Gods and rulers which he had to obey!

A person like this would appear strange to us. He would display some competence as we observed him go through his daily routine. But if we could converse with him we would find his explanations of the world, what worked and why, to be almost incomprehensible. If we were very patient and listened very carefully we might learn that he explained the occurrence of some event or phenomenon with several different scenarios of cause and effect and would explain that it depended upon whether this God or that one was behind the event. He would further explain that it would require the witch-doctor or the chief to determine exactly what was going on! You can see that such a psychology requires most people to defer to the ruling elite as the experts necessary to answer questions of truth.

The human mind is designed for **one** *super*concept. This is the individual's default belief system which serves as his ultimate standard of truth. The tribal type of psychology, where there is a belief in multiple Gods creating multiple reference standards may prevent exclusive closure on one *super*concept. This can lead to both visual and auditory hallucinations. The tribesman will typically interpret these mental aberrations as visits from the Gods or Devils. These hallucinatory events as well as the general state of mind created by many reference standards create a lot of anxiety and fear in these people over their thoughts and feelings.

The next psychological result of the Dialectic Effect I wish to mention is a predilection toward even more extensive cognitive error. Human cognition is far from infallible in the best of circumstances. The best modern minds are susceptible to errors in cognition. The cognitive error upon which theism relies, reification was central of course, but the other cognitive errors which have been identified were also rampant. These tribesmen made almost every error of thinking that is possible and in the most gross and flagrant manner[20]. They lacked any objective standard of truth with which to anchor their thoughts. Tribal peoples are conceptualist; they are easily moved by a narrative or story. They easily ignore their own perception in favor of a conceptual model. What an 'expert' tells them is more 'real' to them than their own sensations and their own experience!

As much as the ruling elite search for, identify, and punish slaves for 'selfishness', it is the conscientious spying by their fellow slaves which brings most selfish tribesmen to 'justice'. This should not be a surprise. If slaves can punish themselves so harshly as to suffer depression, even suicide, then 'outing' their fellow slaves for conduct that is selfish and un-tribal is a small inconvenience. Much of the indoctrination in schools is devoted to the teaching of this peer pressure technique of enforcing compliance to tribal rules.

The manufacture of this new theist psychology on an assembly-line basis required the development of the Pagan religion and Pagan schools. This religious institution then created the slaves needed to support a permanent settlement, the tribe, its ruling elite and its farming and military operations. The human race then entered into the darkest period of its existence. Once the human race was taught to hold the Pagan religion and its consummate evil as virtue there was no stop or end to the awful behavior that commenced. Twenty thousand years ago a huge, black theist blanket of evil fell upon the peoples of the earth. Man now suffocated under this religious blanket which snuffed out the fresh air of freedom. Instead of pleasure and happiness man now sought pain and suffering. Not life but death became his goal. Honesty and decency almost became extinct as man now valued violence and the sadistic domination of others.

For the next 15 millenniums, almost everywhere on earth wherever there were more than a few scattered people, they were organized into tribes. While these thousands of tribes differ in many specific and detailed ways, they are uniform in their dictatorial and collectivistic nature. They also vary in the intensity and violence of the enforcement of their tribal rules. But even the most liberal and tolerant tribe is still a tribe and is at its core a totalitarian social structure.

Most tribes of these millennia were totalitarian regimes of the worse type. The control of thought and behavior was absolute and total. The terror used to subjugate and repress the people in these tribes was horrendous, pervasive and unremitting. For 15,000 years mankind was locked down, and locked up, mentally and physically. People think what they are told to think and they do what they are told to do, or they paid with their life. There was no room for deviation or non-conformity. The biggest obstacle to freedom, to escape from these dictatorships was the lack of freedom inside the slave's own mind! These tribal peoples lapped up every edict that emanated from the ruling elite just as if it came directly from God's lips! Because for them it had!

Through such Pagan Gods the tribesmen is enslaved to tribal authority as completely as any dog has ever been enslaved to its human master! The rulers became deified as super humans whereas the slaves were relegated to near worthless status. Slaves could be put to work, or to death, or to war like zombies. They were like zombies because they never slacked in their obedience or in devotion to their duty. They were 'not to reason why, but to do or die!'

For over 15 millenniums theism, Paganism, slavery, or the absolute rule of chiefs and kings could not be questioned. Innovation was therefore very slow. Most technological advances were in military hardware. But it was more innovation than had occurred among the hunter-gatherers. Still the living conditions for the masses were intolerable. The slaves owned nothing, not even their own bodies. The ruling elite owned everything including the slaves lives. The repression was absolute and unremitting.

The most difficult thing for moderns to fully comprehend is the level of absolute evil of these regimes. Tribalism may also be viewed correctly as an elaborate system designed to rationalize, obscure and glorify the practice of stealing. Each tribal member shows his masochistic traits to those superior to him in the tribal hierarchy, and to those beneath him he displays his sadistic traits. Infants and children make up the lowest rung on the power ladder, next is usually women, then the several layers of slaves followed by the ruling elites. The ruling elite steals from everyone in the tribe, each social level within the tribe steals from those below it, and the tribe as a group steals from other tribes. Recognition and respect for private property did not exist anymore than recognition and respect for human life. Everyone and everything was at the mercy of the tribal collective, which is to say at the mercy of the ruling elite. **Tribalism is not just totally evil; it is the engine that is constantly producing evil and exporting it out into the world.**

Tribal peoples were semi-psychotic much of the time; they believed in and spoke to numerous supernatural Gods. They glorified in death, particularly in war, and they considered it a great honor to offer up their children in blood sacrifices to their Gods. The savagery, violence and blood lust of these ancient ancestors is difficult to fully understand.

The Theist Identity

At the start of this Section I described how slaves were necessary if humans were to succeed at establishing permanent settlements at this early time in our technological development. I explained that to create slaves it was necessary to inculcate in children a slave psychology. We know that to create a new psychology would require a new *super*concept and a new **identity**. Indoctrination with special concepts is required to produce these mental alterations. The bundle of special concepts necessary to create the **slave identity** is known as **theism**. Using a fictional character Shama, I illustrated one way these theist concepts may have been discovered.

Then I explained that the Pagan Religion was the institution established to teach these theist concepts to each new generation of children in order to continually produce the slaves needed. I explained how the psychological modifications caused by these theist concepts resulted in the **collectivists' identity** and how this created a psychological need for a collective or tribe. I then referred to such tribal peoples as having a **tribal identity.**

All of these terms: theist, slave, Pagan, religious, tribal, and collectivists, when used to refer to a psychological type, or **identity**, refer to the **same** psychological type! I have used these different terms simply to reflect the differing vantage points from which I have approached the understanding of this type of identity. The theist, slave, Pagan, religious tribesman and tribal collectivist have an identity that is the result of indoctrination with theism. They have a **theist identity**. So to simplify things from here on I will refer to this type of psychology as the **theist identity**. **The theist identity** produces the psychology of the slave. It also produces the tribal psychology. The **theist identity** produces the psychology of the Pagan as well as the psychology of the tribesman. And it is the religious **theist identity** which produces the collectivist's type of psychology.

People with a **theist identity** display a number of unique psychological traits. These psychological traits fall into two groups which correspond to the two major alterations in the human mind caused by theism.

The first group of traits found in those with a **theist identity** result from reification. This cognitive error causes the tribesman to have a collectivist **self**. Tribesmen are therefore unable to separate their **self** from the collective. The tribal mind can never understand 'property rights' or 'civil rights'. 'Reality' for them is the tribe, their supernatural Gods and their deified tribal rulers. The tribesman must have his tribe and his chief to function. Anarchy means 'no king' or without a ruler. This concept of being free of the tribe and its leadership is inconceivable and terrifying for the tribesman. Without a 'leader' there is no one to tell the slave what is 'real' and what he should do! The authority of the tribal rulers is deified, revered, and feared. The tribesman believes in assorted supernatural personages. He believes in the existence of various Gods, angels, saints and cupits as well as Devils, demons, ghost, and monsters. His belief is so profound that he 'hears' and 'sees' these creatures in his imagination. He converses with these supernatural personages, begging for forgiveness, praying for various benefits and seeking interventions for himself and his tribe. He sees miracles and the hands of the Gods and other supernatural beings at work behind natural events and in the outcomes of warfare as well as in the ups and downs of tribal politics. He cannot question or doubt the existence of these supernatural beings as they are sanctioned by the tribal rulers. His tribe, its rulers and through them he himself are the center of and the most important aspect of the entire universe.

The over-powering need of the tribesman is to **conform**. He must conform to the expectations of his tribe and its ruling authority. He must conform in every respect and to every extent. The tribe and its ruling authority is his reality and for the sake of his identity he must conform to its every requirement. In order to do this he must know his station and the tribal requirements of those who occupy that station. And then knowing what is expected of him he must comply and conform to these expectations.

Inside the tribesman's mind he erects a set of requirements adopted from the tribal rules to which his conscience demands adherence. To break a tribal commandment is a personal failure to him eliciting shame, guilt and depression. He sees to it that this seldom happens. He seeks to gain the approval and avoid the disapproval of his betters. He is devoted to his tribe, its rulers and its beliefs. Gaining approval and avoiding disapproval of his beliefs and behavior by himself and by tribal authorities is his goal in life.

His thinking is overly abstract, confusing nouns and verbs and inserting himself into elaborate streams of causality. He has trouble distinguishing between the natural world and his fantasy and superstition. Ritual and revelation are as real to him as his own sensation, and perception. People with a **theist identity** have been schooled in *psychoses!* They are trained in abnormal cognition! They tend to advance up the tribal hierarchy to the degree they can skillfully manipulate the tribal conceptual delusions. The person with a **theist identity** displays such reification prominently in his cognition. Much of what he might consider cognition is really 'magical thinking' and actually not thinking at all!

The second group of traits results from his commitment to the ethic of human sacrifice. The tribesman willingly sacrifices both his **self** and others for the benefit of his tribe. This is observed as sadomasochism. Might makes right in the tribesman's mind. And it is his tribe and its rulers who because they are blessed by the Gods, the mightiest of all. Outwardly, especially toward those of higher rank, he is obedient, obsequious and self-punishing. He displays passive-aggressive behaviors, being overly generous and obsequious toward authority figures above him and then reacting with a sadistic aggression toward those below him in station. When angry he easily goes into a homicidal rage.

The tribesman typically displays his masochistic side to his superiors and 'betters' then he reveals his sadistic motivation toward those subservient or below him. This passive/aggressive paradigm is characteristic and fundamental to tribalism. To understand the **theist identity** requires that you understand this sadomasochistic motivation. Sadomasochism is what it is all about!

The tribesman is concerned with whom he may sadistically dominate and who he must submissively give-way to. Regardless of his caste or which layer of slaves he belongs to, he has those below and those above him. Women and children of each caste generally rank below the men and are treated accordingly. Pain and suffering is valued, one's own and that of others. He easily goes from altruism to narcissism, from self-flagellation to murderous rage, and from guilt and depression to hostile aggression. Only his chief can calm and control him either with the soothing power of rhetoric or by encouraging him to vent his rage in some tribally approved manner. The individual with a **theist identity** is psychotic much of the time, floridly psychotic some of the time and unpredictable emotionally and erratic all the time.

Visit To A Tribe

We began this Section with Shama and the invention of the first religion and tribe. Now let us use our imaginations to visit a more recent tribe. Imagine a time machine like Jules Verne envisioned. In fact, the time machine I wish you to imagine is even better than Jules'. Our time machine not only goes back to any date we choose, but as we descend into the past our appearance and clothing changes to match that of contemporary peoples. And on top of that we automatically become able to speak and understand the language of the people at the time and place of our arrival!

Now imagine that you and I decide to use this time machine to go back to the year 750AD to the city of Uxmal in the Yucatan. Our time machine works flawlessly and in a short time we find ourselves in this Mayan city. This is a time when the Mayan Empire was vibrant and powerful. The Mayan people had a written language as well as distinctive art, architecture, mathematical and astronomical systems. At this time the Mayan civilization was one of the most densely populated and culturally dynamic societies in the world. Advances such as epigraphy, and the calendar did not originate with the Maya; however, their civilization fully developed them. Many outside influences are found in Maya art and architecture, which resulted from trade and cultural exchange.

As we arrive at our destination we direct our time machine to an area of the forest near the city that is thick with underbrush. We have arrived during the New Year crop planting and soil fertility celebration. For Mayan's this is one of the most important times of the year. We exit our machine, make sure that it is securely hidden in the underbrush and walk into the city. We find a great celebration underway. Everyone is jubilant, food and beverage vendors are everywhere, and loud music can be heard down the main street.

Chaac, the Mayan God of fertility is a very important figure in the Mayan religion. Colorful pictures of Chaac are everywhere. He is the God of fertility and especially important at this time of the year. The Mayan's believe that Chaac demands the sacrifice of human life in exchange for a bountiful harvest. The religious ceremonies call for the killing of animals, such as dogs, turkeys, and chickens as well as captives taken in battle. Slaves taken in battle merely acquired new masters. The only captives suitable for sacrifice were high ranking prisoners. Anthropologists tell us that the offering which Chaac greatly preferred was little girls! Virgins of 5 and 6 years old were the preferred sacrifice. These children were killed on the Pagan altars by having their hearts cut out of their chest. The Priest then holds up the beating organ for the entire congregation to see. This brings a deafening roar of excitement from the huge crowd! Each death can be counted from a mile away by marking each roar of the enthusiastic Mayans. Reportedly once over 5,000 young children were sacrificed during a single day! We hear the roars of the excited crowd as we approach the Temple and Sacred Altar of Uxmal.

Then we see it! What an absolutely astounding sight! There are hundreds of steps leading from the base of the Temple complex up to the Altar and sacrificial area. On each step stands a child. Some children are crying. Some are smiling cheerfully. Most are somber and all have their hands folded before them as in prayer. I believe most of the children are praying.

Up at the Altar area a Priest is standing. He is wearing a beautiful purple robe and tall hat gold in color. In his right hand he grasps a beautiful shinny silver handled knife while with his left hand he holds a beating human heart high in the air for all to see. Assisting the Priest are two muscular men gowned in white robes with gold ribbons. They take the limp bloody body of the sacrificed child and throw it into a sink hole called the Sacred Cenote. Then they go to the head of the stairs, take the next child by her hand and lead her to the Sacred Altar. They pick her up and set her on the blood drenched Altar. The Priest goes over to her, he anoints her forehead. The music now reaches its loudest crescendo. The little girl lies down on the Altar. The Priest comes over to the child. He can be seen saying a prayer. The knife is nowhere in sight. As the Priest and his assistants continue with prayer, the music ceases and the crowd is quiet.

Then with lightening speed the Priest has plunged the big shinny knife into the little child's chest. A short cry is heard; the child attempts to sit up, and then falls limp back on the Altar. With a short quick movement of the knife the Priest removes the child's beating heart and holds it up for the crowd to see. Once again the huge crowd roars to life with shouts, screams, whistles, and cries of excitement. The cymbals clash and the horns blast as the music reaches a new high. The people throw all sorts of colored objects into the air and sing out with glee as the Altar men throw another lifeless body into the Sacred Cenote. Then each child takes another step up the long stairway to the cadence of the music. The child on the top step reports to the Altar men and takes her seat on the bloody stone structure.

We can't watch any more of this so we make our way back toward the city center and the market place. But there is no escaping the intermittent roar of the crowd marking the killing as each young life is offered up to Chaac. As we arrive at the market place we are surprised to find everyone there busy with commerce and trade, seemingly unaware or unaffected by the religious activities at the Temple. Is the slaughter of children so common place, and so acceptable that it is ignored? We wish to talk to some of these Mayans and get the answers to these questions.

We approach a shop keeper. He greets us eagerly assuming we are customers seeking to buy some of his straw hats or wool blankets. I ask him how he feels about the Temple murders. He finds this a strange question. He asks us how far away we live implying we suffer a great ignorance. We indicate that we are from the far countryside. He explains that the killing is not murder but Religious Sacrifice of the most Holy order. He says that the children welcome the opportunity to serve God, their family and their country by their noble and glorious death. He says that most families encourage their children to be religious and many elect to serve the Gods and the Empire in this way. He adds that these sacrifices ensure a bountiful harvest and the strength of the Mayan Empire.

We thank the shop keeper for this information. We note the intermittent roar of the crowd now at some distance. After a short walk down the market street we meet a young mother carrying her infant. As we approach her she smiles and appears friendly. We tell her we are from the far country and would like to ask her some questions about her city and its customs. She nods agreeably. I ask her how she feels about all the children that are sacrificed to the God Chaac.

She informs us that the Mayan Empire is so great because the Gods, especially Chaac have chosen the Mayan people. She says that Chaac speaks clearly to the clerics and tells them very specifically what sacrifices he demands. The Mayan Empire has grown strong and great because her people have never failed to do their duty as Chaac requires.

I ask her if she would allow her infant to be sacrificed to Chaac when he is of age. The mother tells me that there would be no greater honor for her, her husband, her family and for her child, if he would be chosen as a sacrifice for Chaac.

We say goodbye to the young mother and proceed down the market street. We then come upon a very old man sitting alone smoking an unusual looking pipe. We approach him and after explaining that we are from a far distant province we ask if we might get some information from him. He cordially agrees.

We mention that we have witnessed the Temple sacrifices and cannot understand how the parents of the children, as well as everyone else can tolerate so much killing. Do Mayan's not see the evil of such a practice?

The old man becomes visibly upset. He puts down his pipe and stands up to address us. Listen he says, "You are poorly educated and even more poorly informed in religious matters."

"How could the sacrifice of these Holy children be evil?" He asks. "This is the very essence of Sacred Holiness! These sacrifices are conducted at the direct command of the God Chaac! All that the Priests do is under the direct orders of the Gods!"

"These ceremonies are tough on the Priests and Altar Men. Everyone who wishes to sacrifice their child has to be accommodated. These sacrificial ceremonies last for days and require dozens of Priests and dozens of altar assistants. Their work is arduous and exhausting and they must take turns so that all may take a needed rest from time to time. Do you think that all this effort is for sport?!"

He continues, "This is very serious work. The life of the Mayan Empire requires it! You have a lot to learn."

I ask the old man if some parents do not object to their child being taken for a sacrificial offering.

"Where are you two from?" The old man asks incredulously.

We explain that we are from way back in the woods and that this is our first time to visit a city. He seems to understand but is obviously suspicious of our purpose.

"No, no parents never object to their child being sacrificed. It is a tremendous honor to the child and his parents, who in their right mind would object?" He asks.

I tell him that I find it hard to believe, considering how many children are reported to have been offered up to the Gods that at least one parent has not objected.

The old man reluctantly responds. "Well there have been rumors that such a thing has happened. But it is extremely rare. When it does occur the entire family is sacrificed."

Still probing I ask him if there have not been some parents that run away with their child rather than have he or she sacrificed?

The old man becomes more upset. He starts looking around to see who may be over hearing our conversation. When he sees no one he answers. "Look, I don't know what you two men are up to, but I try to stay out of trouble. Please let me go, I don't want to answer any more questions!"

I reassure him that this is our very last question.

"Well yes, there is a rumor of a couple that ran away into the forest. They were never found. We assume the Monster Gods of the forest devoured them."

We apologized for upsetting the old gentleman and take our leave. We continue our walk about the huge market place.

After a good while we notice in the distance both the old man and the young woman we interviewed earlier speaking to a very official looking guard of some kind. We decide it would be prudent to return to our time.

We make our way through the forest and back to our well hidden time machine. We enter, set the instruments, turn on the machine and begin our return trip from the land of the Mayans. As we await our arrival we contemplate what we have learned. We have learned that these sacrificial ceremonies are not imposed upon the people. The people demand them. The people see these Temple Sacrifices as essential to their survival and heartedly support them. If you inculcate in a people a **theist identity,** they will see to it themselves that they are enslaved!

Contemporary Perspective On Section II

Evil did not enter the world 4.5 billion years ago with the creation of the earth. Evil did not come into the world with the beginning of life on earth 3.5 billion years ago. Evil did not arrive on earth when Homo sapiens evolved into existence 200,000 years ago. We know that evil was invented by humans and entered man's world approximately 20,000 years ago. At that point the psychology of human beings took a turn into a very deep and dark alley of evil.

The species Homo sapien is not evil. Evil must be learned. Humans created the concept of evil, as well as evil itself. We don't know exactly how this happened; I have given you one narrative, the story of Shama, as to how it may have happened. We know the definition of evil; it is ***the exploitation of human beings by human beings***. Evil results from the belief that I have the right to coerce another to do as I want him to do.

We know the nature of evil; it is tribalism. Tribalism is composed of the psychology of theism, the religion of Paganism, the practice of slavery and human sacrifice, a ruling elite and the societal organization we call the tribe. **Tribalism** is the first example of collectivism that was invented. **Collectivism is a system designed to allow a group of elites to exploit a population of human beings**. Collectivism and tribalism are the essence of evil. They are the very definition of evil! The evil is not only the system. Each component of collectivism is fully and completely evil in and of itself! The evil cannot be subdivided and thereby reduced by removing a component from the total system.

Theism is the technical means or the method by which the human mind is enslaved. Paganism, in its many thousands of forms, is the practical application of theism. Pagan Religions and Pagan Religious Schools are the institutions established to teach and promote Paganism and spread the slave psychology among the masses.

Thousands of Pagan Religions, socioeconomic theories, philosophical belief systems and patriotic nationalist dogmas have but one goal, stealing from and exploiting the great mass of human beings. Theism prepares the human mind for the exploitation.

Then just 5000 years ago, coinciding with the increasing use of writing a few heretics began to question the tribal structure. The disadvantages of the collectivist psychology were slowly being realized. But to challenge tribal authority was extremely difficult. The tribal psychology was, and is, so powerful at enslaving the human mind that the development of a new *super*concept to challenge tribal collectivism was slow to emerge and fraught with even more violence.

Section II
Summary

1. To produce slaves-children's minds must be altered to produce a **theist identity**. This required the invention of evil. This is how evil came into the world. Finally we have an adequate definition of evil.

2. To alter the minds of children-to turn them into slaves-you must first sever their minds from their natural God by substituting artificial man-made animal-gods or man-gods. Once their standard of truth is changed from an objective truth (nature) to a subjective truth (humans) they are locked into obedience to human rather than Godly authority. The technique for accomplishing this is called polytheism. Theism relies upon the creation of prejudice against the **self,** the concepts of Original Sin and revelation. Torn from an objective truth or God, the slave or tribesman is by definition, psychotic much of the time.

3. The indoctrination of children in theism produces the collectivist mind. This is required to produce slaves and obedient tribesmen.

4. The institution designed to produce slaves on an assembly line basis by teaching polytheism is the Pagan religion. The Pagan religion and Pagan schools were the first to teach polytheism. Religious schools are mind-alteration factories.

5. Polytheism teaches the most evil beliefs and behaviors and promotes them as the epitome of virtue! Good and evil are thus turned upside down. But the tribesmen or slaves don't realize that this is the case! They therefore regard a religious morality which is actually the epitome

of evil, to be a virtuous moral code! The violent psychotic tribesman dependent upon his chief is the product.

6. Agriculture, exploited through slave labor, allowed permanent settlements, promoted a higher calorie diet and a more sedentary population. Increased body fat among the women increased fertility and led to a huge increase in the human population. The slave morality of polytheism encouraged reproduction.

7. The evil morality of self-sacrifice damages the Hindbrain and the evil concepts of force and supernaturalism harm the Forebrain. The results are reification and human sacrifice.

8. These mental disabilities cause polytheist to suffer a failure of maturation. They therefore remain dependent, and suggestible. They are dependent upon their tribe and its ruling elite. Their suggestibility is useful to the tribe. Theists are capable of any atrocity that the chief may order.

9. Because children are taught polytheism during their early period of imprint programming, they 'feel' it to be valid and true. The **theist identity** produces a tribal psychology characterized by a self-sacrificing slave morality, a punitive conscience, and a propensity for serious cognitive errors.

10. The societal belief system that is designed around the permanent settlement, slaves, religion and the tribe is called tribalism. Socialism, fascism, communism, state capitalism and welfare statism are all forms of tribalism. They are all forms of collectivism. They are all rationalizations for stealing.

11. The tribe is a collectivist organization, autocratic, and totalitarian in structure that relies upon the use of force and violence throughout society. It lends itself to labor intensive agriculture and warfare. The tribe is the prototypical

totalitarian social system that requires massive delusion and deception to maintain. Tribal peoples blindly worship the tribe and its ruling elite. It is a system designed, maintained and perpetuated by the tribal elites to ensure their rule and financial power.

12. The concentration of power and wealth in the hands of a few tribal leaders allowed them to provide endowments to exceptional individuals. Artists, scientists, musicians, inventors, philosophers and writers could then devote their full time and effort in pursue their interest. This allowed the beginnings of the accumulation of knowledge. With the invention of written language this process was greatly accelerated.

13. The invention of polytheism, the manufacture of slaves, and the development of agriculture which led to the creation of collectivism initiated 20 millenniums of constant warfare. Collectives are constantly at war. A huge percentage of the evil behavior adults are engaged in today around the world is the direct result of the indoctrination and training they have received in collectivism when they were infants and children.

14. This second psychological adaptation is characterized by a collective *super*concept, a morality of human sacrifice and reified conceptual cognition.

15. While this psychological adaptation created slaves of everyone in the tribe, most tribesmen are unaware of their entrapment. In most tribes only the lowest rung or caste of slaves has much awareness of their slave status. These are usually the plantation or farm slaves.

Summary Of The First Half Of This Book

The word atheist means "one who is not a theist." But most dictionaries have been written by theist who incorrectly define atheist as "one who does not believe in God". Many atheists accept this definition themselves. But the word means more than one who rejects supernaturalism and organized religion, as important as that is. It is relatively easy to be a Godless theist. This is not an unpopular position. But to truly be an atheist one must reject *all* the components of theism, including collectivism. But even then atheism is but a negative statement. It says what you are not; it does not say what you are!

I have explained in the first two sections of this book that persons who do not believe in a God are floridly psychotic. To be more technically precise; persons who have not incorporated an adequate *super*concept into their mind will be floridly psychotic.

So the question then is not whether a person believes in a God, but rather what are the properties of the God in which he believes? Some people believe in a supernatural God that controls every aspect of their lives. Other people believe in a God that resides in a collective of people such as a tribe or a State. Some people believe in a God personified as nature, objective, non-interventionist and non-human. And still others believe in a God that is non-human, objective, natural, non-interventionist, totally disinterested, completely unaware of humans and perhaps unconscious as well. Many people label such a *super*concept with the word 'reality'.

Whatever *super*concept one holds will affect ones psychology and ones cognition. People who believe in and worship a collective, whether it be an earthly collective, a heavenly collective, or both will display the attributes of collectivism. Sadomasochism and reified thought processes are the prominent theist attributes. With this in mind we now continue with the last of this book.

SECTION III
Statism

Mankind's Third Psychological Adaptation
2500BC-2100? AD

Chapter 8

The Human Environment

I use the term 'tribalism' to refer to the combination of polytheism, institutionalized as the Pagan religion, its religious schools, slavery and the tribal societal structure. Tribalism was the first example of collectivism invented, and has the fundamental attributes of the more modern forms. Beginning about 20,000 years ago when tribalism made its debut it took several centuries to reach its full development and spread completely around the world. From that point up until about 5000 years ago the tribe ruled as the premier societal structure. Up until this contemporary period of human psychological adaptation, the tribes with their Pagan religions ruled mankind with an iron fist. Religions of this period had none of the merciful, forgiving, benevolent qualities which moderns may associate with religion. Religion with these qualities is a contemporary development. The Pagan religion was the tribe's coercive authority and it was concerned with obedience, punishment, torture and blood sacrifice. Its mission was to prepare the masses mentally for their use by the tribal elite. They were brutal, violent, savage institutions. The view of the religious and tribal rulers toward the masses was similar to that a modern slaughter-house manager might have toward the cattle in his beef production operation.

Until the last 5 millenniums the totalitarian nature of the tribe was absolute. Anyone who doubted any aspect of the tribal culture must have shivered in fear at the thought. Anyone who voiced, or indicated in any way, anything less than enthusiastic support for the tribe, its chief and its religious beliefs was dead so fast that a heretical group could not be sustained. One of the frequent events common to most tribes was the ritual torture and execution of non-believers and other non-conformist. The purpose of ritual torture is to bring out as much pain from the victim as possible before he died. This is what the Gods and the tribal rulers demanded. This is what everyone valued.

Tribes of this period were repressive, violent and blood thirsty. Warfare was incessant, inbreeding among the ruling elite common, palace intrigue routine and the sacrifice of human beings on the Pagan altars a regular fixture of tribal life. Man had no room in his mind or in his life where he could freely think. Everything was ritualized and regimented. Thought and action was conducted as a ritual for the purpose of influencing the Gods. Tribesmen were savage, psychotic and superstitious. Few inventions or discoveries were being made. Nothing was improving. Most things were getting worse. The tribal adaptation was proving to be a disaster. For 15 millenniums there was no appreciable change in this situation.

The Danger of Innovation

In 6000 BC Omar lived in the village of Ur in what today is Iraq. His job was to care for the Chief's cattle. The Chief had many sheep and goats, but cattle were rare and more valuable. It was an honor that Omar had been chosen to care for these animals.

The sheep and goats followed the seasons and the availability of grass. So the shepherds that tended these animals had to follow them, sometimes over great distances. The Chief's cattle however were kept in his stables in the village where they were fed hay and grain harvested near the river. Omar therefore did not have to travel like the other animal caretakers. He stayed in the village and had time to observe his animals closely.

Omar spent a lot of time with one bull in particular. This animal was named Oscar. Oscar was very gentle, especially with Omar. Omar could rub Oscar all over and scratch him behind his ears, and Oscar seemed to enjoy this a great deal. Omar could lean against that huge animal and Oscar did nothing but keep on eating. Oscar was just a big pet and would follow Omar around the compound like a dog, going everywhere Omar went.

One day a skid loaded down with a huge harvest of hay and grain got bogged down in the mud on the bank of the Euphrates River. The hay had been brought in from the fields on a barge and then off-loaded onto the skid. Now the skid seemed hopelessly bogged down in the mud so bad that even with all the workmen pulling on it they couldn't get it to budge. It looked like the skid would have to be unloaded and all the hay carried to the barn by hand.

Omar, with Oscar standing beside him, looked down toward the river at the muddy mess with the skid sunk in the mud and unmovable by all the men's efforts. It was a shame but there was no other choice. Finally they started to off-load the grain hay, carry it up the river bank one person, one arm load at the time and re-stack it on another skid parked on an upper dry area. This took hours and consumed the entire day.

That night Omar had a dream. In his dream he replayed the day's events. He had observed how powerful Oscar was and how he had little difficulty walking through the mud. In the dream while standing on the river bank with Oscar at his side he shouted at the foreman of the work crew, "hand me a pull rope and everyone stand back!"

In his dream the foreman handed Omar the thick pull rope. Then to everyone's dismay, Omar did not proceed to pull on the rope himself, but instead made a loop which he placed over Oscar's neck! Then Omar led Oscar up the river bank with the skid loaded with grain hay following behind. Oscar did not appear to struggle the least. The bogged down skid seemed to leap out of its entrenchment and follow Oscar up the embankment almost by magic. The workmen and their foreman were happily impressed!

When Omar awakened he became very frightened. Had he been taken over by the Devil!? Had Witches usurped his soul?! He shook and broke out in a cold sweat! He knew he was risking his life with such thoughts of using Oscar's power! He had better forget the whole business of using Oscar to help with the farm work. He had better keep his mouth shut!

Weeks went by and Omar was able to keep his dream and his idea of using Oscar's power a secret to himself. Then one day his foreman told him the Chief had ordered that all the old pens and corrals be torn down as a new and larger stable and fencing would be built.

For several days Omar and his crew tore the timbers off the post and stacked them in neat piles for reuse. Then Sunday came, work was set aside so all could attend Temple sacrifices. Many men, women and children were scheduled to be offered to the Gods. There was much excitement in the village as everyone donned their finest attire to attend the event. However Omar didn't want to attend. He stayed out at the stable with the animals, especially with Oscar.

As he leaned against Oscar's stall he looked at all the naked fence post standing where the corrals and pens once stood. He thought how it would take all of next week to dig each post up one at the time by hand. Then the image of Oscar pulling post out of the ground came to his mind! He couldn't keep the idea out of his mind!

In spite of shaking with great fear he led Oscar out of his stall and over to one of the post. He took a rope, made a big loop on one end and placed it over Oscar's head. Then he tied the other end of the rope to a post. He then led Oscar away. The post popped out of the ground with what appeared to be no effort from Oscar at all! Omar took Oscar from post to post eventually pulling down all the post. Then Omar put Oscar back in his stall. Omar then piled up all the post beside the stacked timbers. Finally Omar sat down beside Oscar's stall and fell asleep.

Something caused Omar to awaken. He saw standing over him in awe his foreman and many other workmen. They fell to their knees and holding their arms over their heads began to chant prayers to the Gods! Their loud chanting attracted the attention of one of the Chief's officials. He came over to inquire as to the cause of the chanting.

Omar's foreman talked with the Chief's Officer telling him that Omar had completed over a week's work that day alone, all by himself while everyone was worshiping and sacrificing at Temple. The foreman suggested that Omar must be possessed of some Devil God or maybe Omar was himself a powerful and evil Spirit because no human would have the strength to do so great amount of work so quickly!

The Chief's Officer went over to Omar. He said, "How did you take down those many post, all in one day, alone by yourself?

Omar, visibly frightened, answered, "I didn't do it by myself. Oscar pulled them down for me!"

When he spoke those words all the workmen including the foreman, bowed down and began chanting prayers. They were prayers to ward off evil spirits.

The Chief's Officer spoke. "Omar, you are possessed of evil Spirits and you blaspheme the Gods by this alliance with cattle!"

The Chief's Officer ordered the workmen to take Omar and lock him in the Temple jail. He would take this matter to the Chief.

When the foreman told the Chief of Omar's conspiracy and command of the bull he mentioned that Omar had done all this while missing Temple services. The Chief saw Omar's behavior as a sacrilege and ordered that both Omar and Oscar be tortured and sacrificed to the Gods before the Gods should take offense and punish the entire tribe with a pestilence or some other disaster.

The tribe's people were assembled in the village square where sacrifices were regularly held. Omar and Oscar were each tied to a post in the center of the space. As the villagers cheered them on the executioners went to work. First both Omar and Oscar were castrated. Omar because of his blaspheme then had his tongue pulled out of his mouth. Oscar was thrown to the ground with ropes and his legs were severed. Then as the slaves shouted with glee the animal's intestines were slowly pulled out through his rectum. Omar, to everyone's disappointment died quickly. But the huge bull bellowed in agony for hours as he suffered a slow death. By late afternoon he lay still and silent. Then both bodies were burned in a great fire as the tribesmen celebrated with wine, music, song and dance.

Heretics

"None are more hopelessly enslaved than those who falsely believe they are free".

--Johann Wolfgang von Goethe

Have you ever been frightened by some of your thoughts? Are there thoughts that you are afraid to think? If you have never been afraid to think certain things, then you will have difficulty understanding the level of fear with which most people lived during the tribal millenniums. On the other hand you are a very fortunate and unique individual. The fundamental control tool which tribalism has over the human mind works by the fear of thinking it produces. Societal freedom reflects mental freedom; a free society can only be obtained and maintained by free thinkers.

You must make sure that your fear of certain thoughts is not so great that you repress them even before they reach consciousness. When faced with enormous fear of certain thoughts most people will develop an automatic censoring mechanism that simply edits such thoughts before they ever rise to the level of consciousness. Then they can believe that they are free to think whatever they wish, when in reality, they allow themselves to think only 'within the box' of approved thoughts. Tribal and State peoples become expert at not thinking thoughts that they are not supposed to think. They simply repress all thought that might get them into trouble. Heretics are usually very smart. But they are not smarter than many other people. What heretics have that few of their peers possess is courage!

Just how strong and profound the mental enslavement of tribalism is, can be partially understood by contemplating this fact. Many more slaves committed suicide, either directly or by martyring themselves in combat, because of grief over poor obedience to their polytheist beliefs, than the few in number who rebelled against their enslaving thoughts. Tribal peoples become more upset over losing their loyal slave designation than they do over losing their freedom. Such is the power of polytheist indoctrination at enslaving the mind.

To understand the fear people had of having the wrong thoughts you need to appreciate the terror of tribal life. It is difficult to describe to relatively free contemporaries the constant and unremitting fear and terror of life in these tribes. Tribes were vicious and cruel, unintelligible and constantly changing in their demands. They were totally unpredictable in their behavior. No one could be sure of anything from one day to the next. The clerics could make any claim as a call from one or more of the Gods. Nothing but the whims of the many which-doctors, clerics and chiefs held any import on the lives of the people or the activities of the tribe. Superstition, fantasy, real and imaginary fears as well as real and psychotic ambition motivated the tribal leaders. Whatever political maneuvers and intrigues developed among the ruling elite it was the common people who usually paid the price.

Warfare between tribes was continuous and ubiquitous. Larger tribes conquered smaller ones to grow even larger. Smaller tribes merged together to protect themselves from larger tribes. Alliances and betrayals were incessant. No tribe takes lightly the disloyalty of a tribal member. Spies, heretics and saboteurs have always met with particular torture. No ideas ever conceived by the human mind are more evil than those upon which tribalism is based. But of course polytheist can't see this. The ideas of evil, force, supernaturalism, human sacrifice and self- punishment produced exactly the kinds of societies one would expect. These tribes were as evil as the concepts upon which they were built!

I do not mean to suggest that every human being on earth behaved like a barbarian. There were peoples so isolated on remote islands, mountain tops and scattered across immense deserts that they were not subject to attack by neighboring tribes. They may have been able to avoid polytheism and the tribal structure for centuries and enjoy a freer life. But if and when found by a bellicose tribe their 'softness' would have made them easy to conquer. In the struggle for tribal survival the most ferocious and fierce, that is the most psychotic and homicidal, easily conquered or exterminated more peaceful peoples. This was also the fate of most hunter-gatherers.

And within the typical warring tribes, then as now, good people tried to make the best of a horrible situation. But in this atmosphere it is understandable that few had the courage to openly doubt, verbalize and challenge tribal doctrine. Challenges to the tribal machine and its absolute control over thought and action probably occurred sporadically during these millenniums. Those courageous heroes who dared to think and openly doubt, probably met with an unspeakable end. It is understandable that prior to the invention of written language no record of such a brave individual or group exist. But with the advent of written history we begin to see the recording of discontent. While tribalism might be great at producing a war machine it has always been inimical to human beings.

Writing was invented in the early Bronze Age around 4000 B.C[11]. But the use of written symbols goes back at least to 7000 B.C[11]. There is no direct evidence that the written word contributed to a questioning of tribalism. Yet I think it probably did. Writing allows one to communicate anonymously! You can write heretical ideas and post them where others may see them, without revealing your identity. This meant that you could post your ideas, communicate them to others, and still remain alive!

We began to see questioning of tribalism at least 1000 years before the rise of the Greek civilization. As a point of reference Homer composed his Iliad and Odyssey[10] around 800 B.C. The writings of the Greeks, Plato and Aristotle were around 450 BC[10]. The writings of the Greeks reveal minds dissatisfied with the tribal status quo. It provides us with proof that, at least by this time, a few humans, in a few places, had sufficient mental freedom from the totalitarian tribal mind-set, and the courage, to question polytheism, slavery and the tribe, i.e. tribalism, aka 'collectivism'.

What we see is that the religious context and the tribal mindset were so pervasive that few are able to conceive anything beyond or exempt from it. While the hunter-gatherer period lasted much longer than the tribal period, the tribal period is the more recent. Also there is no written documentation of the hunter-gather period, whereas the tribal period has some manuscripts. The tribal period is the one we are still trying to emerge from. The period of the hunter-gatherer has been largely forgotten. To the extent that any reference at all is made to hunter-gatherers it is made despairingly. The prejudice of tribal peoples toward hunter-gatherers is ubiquitous and intense. So it is understandable that the rebels and reformers we now turn to accepted the tribal context in which they found themselves. These thinkers and reformers approached their tribal situation as if it were the basic level or beginning condition of mankind. This is why reformers and challengers generally attack the particular and the concrete while being oblivious to the larger context. This is how they can be so passionate about freedom in the details and completely miss the greater overarching and encompassing enslavement of tribal life.

The purpose of religion is to produce slaves. For 15 thousand years religion was supremely successful at its work. Over the last 5000 years it has continued to be very successful, but the gradually increasing influence of the heretics has challenged the religious adaptation and its master control over the human race. It was the psychological manipulation by witch-doctors and their creation of polytheism which ushered in the tribal period. And it was the invention of writing which began the process of undermining this Pagan system and its total domination of the human spirit.

The first great heretics did not have the luxury of a hunter-gatherers clear and quietly competent mind, or even the partial clarity of the conflicted modern mind. The first heretics began with a tribal mind, a mind of jumbled feelings, fears, confusion, superstition and "magical thinking."The first heretics began in a world which believed in the absolute validity of revelation.

These first heretics had to fight to obtain any tiny modicum of epistemological clarity within their own mind before they could even begin to think about their situation and the world around them.

When you understand all that these thinkers had working against them you are less critical of reformers who challenged details while neglecting to criticize the essence of revealed religion and tribal life. When even a modest objection to tribalism could mean a hasty death everyone had to be careful what they said or wrote. High on the priority list of any critic of the tribal power structure had to be some strategy for staying alive!

Mind of the Heretic

A heretic is someone who wants to think something, say something, write something or do something that his tribe does not allow. He wishes to think about some subject and perhaps experiment with some objects or ideas. What he wants to do is illegal! Yes, that he wants to think is illegal! It is also immoral! For his own safety he must do these things in secret. Number one on his list of strategies for staying alive is secrecy!

The social situation around the heretic is often very chaotic, disorganized, cluttered and insecure. This may also characterize the environment inside his mind. Imagine living in a boarding house with many rooms; each room is full of active energetic people who are loud and obnoxious. The kitchen and bathrooms must be shared, and there is always an argument going on, often more than one. How could you live in such surroundings and still find some way to think and study?

This is the kind of situation I once encountered when living off campus while attending the university. Say you are determined to tune out all the noise, chaos and intrigue and focus on your studies. One way to do this would be to take the closet in your room and turn it into a secret library or office. Using a good lock in a very secret manner, you could create a hideaway for yourself. You could make a place where you could get away from the world and concentrate on thinking. You could quietly enter your room, shut the door, and then enter your closet. Next you would lock the closet door, turn on your sneakily devised desk lamp, and then read and study undisturbed for as long as you wished. But there is a way to do all this entirely within your mind!

Just like with the closet in the room, you can clear out an area of your mind using your imagination. You just push to one side of your mind all the social niceties, all the demands of work, all the family expectations, and all the religious precautions and tribal taboos. You can clear a little free space there in your mind where you can think! This way you don't have to reject or renounce any of your long standing beliefs! This way you don't have to reject your parent's beliefs, or the cherished ideas of your tribe or community. You do not have to reject God! You will not then have to cast them aside. You then don't have to think badly of yourself since you aren't disbelieving any social or religious doctrines. You're just setting all these socially acceptable beliefs aside and escaping into a little hideaway inside your mind! Now secure in this special hideout in your mind you can think and feel whatever you wish and nobody will ever know!

During the 20,000 years of tribal totalitarianism this was about the only little bit of freedom available to most people! And it was, and is, in this 'space' that heretics begin their work. It was in this mental space that heretics worked out their ideas and plotted their strategy. Keep his ideas secret forever or express them to the world? Speak now or wait for a more auspicious time?

Chapter 9
Disclosure

The **Wikipedia** is a multilingual, Web-based, free-content encyclopedia project based mostly on anonymous contributions. The name "Wikipedia" is a portmanteau of the words *wiki* and *encyclopedia*. Wikipedia was founded as an offshoot of Nupedia, a now-abandoned project to produce a free encyclopedia. Jimmy Wales, founder of Nupedia, and Larry Sanger, decided to develop a Web based encyclopedia with a more open, complementary approach. They developed the idea that a wiki might allow members of the public to contribute material. Wikipedia went online on January 10, 2001.

Wikipedia is written collaboratively by an international (and mostly anonymous) group of volunteers. Anyone with Internet access can write and make changes to Wikipedia articles. There are no requirements to provide one's real name when contributing; rather, each writer's privacy is protected unless they choose to reveal their identity themselves. Since its creation in 2001, Wikipedia has grown rapidly into one of the largest reference web sites, attracting around 65 million visitors monthly as of 2009. There are more than 75,000 active contributors working on more than 13,000,000 articles in more than 260 languages. Every day, hundreds of thousands of visitors from around the world collectively make tens of thousands of edits and create thousands of new articles to augment the knowledge held by the Wikipedia encyclopedia.

This project is in itself heretical! The very idea of making knowledge available to the masses virtually free! All of the people associated with creating Wikipedia and making it work must be added to our list of Heroes. I have made extensive use of this website and have freely plagiarized information from this wonderful source. I recommend that you go to **Wikipedia.org** to continue your own studies of the great heretics.

The Honor Roll

There have been thousands of heretics in every culture and ethnic group. I encourage you to research your own tribal, national, ethnic and family history for ancestors who contributed to man's rebellion against tribalism. I have arbitrarily chosen the heretics to mention here. Most of these thinkers are, because of the biases of my background, from the Western tradition. I have no doubt that a similarly impressive list of heretics could be assembled by the study of many other cultures. Also my list is way too short to do any justice to the selection. Many truly great heretics are omitted. What I tried to do is to present a few individuals from different lines of thought. Please feel free to add to this list your own heroes.

I should remind you that these heretics, especially the earliest ones, are first and foremost tribal savages. Their heretical ideas remove them slightly from their tribal context but they are seldom 'nice' individuals. Indeed heretics as a group are not your 'politically correct' crowd. Heretics are much more likely to be irascible, bombastic and conniving than their average, obedient tribal contemporary. I will not spend much time pointing out their other, often disgusting beliefs, and their sometime evil behaviors. My focus is upon the lone beneficial or revolutionary idea, the liberating concept and the little bit of extra freedom their belief has afforded the masses. I can find no perfect heretics; every heretic has his heel of clay, often much more than just a heel.

Another issue we will encounter will be the heretic who provides the masses a great avenue to freedom, and yet simultaneously in some way supports tribalism! This also will not deter us, inconsistency is but another human foible we will overlook. If however a thinker's principal work has been to support the ruling elite and tribalism, as is the case with most of the intelligentsia of every era, then to me he is not a hero and is not included here.

Also let me say that there are probably many errors in my assignment of important ideas. Scholars of antiquity may point out that the first proponent or true inventor of a concept was someone other than the individual I designate. Moderns have debates and lawsuits over who was the actual first inventor of a contemporary idea, so it is understandable that controversy may surround events that took place thousands of years ago. I will admit to all the mistakes I have made. Let us sidestep such concerns for now and devote ourselves to the ideas and inventions themselves. There is time elsewhere to argue over the details of history.

Also it is not necessary that you read about all of these heretics before completing this book. Read the information about as many as you wish. You will quickly get the idea I am trying to get across. Then you may go to page 365 to continue with my theses.

5000 years ago these heretics began the work of leading us out of the mind-locking evil of man's first collectivist mistake. This long, dark alley of evil would prove a very difficult entanglement to escape!

1. Abraham
1813 - 1638 BC

The very first person I can find who successfully challenged tribalism, and lived long enough for history to take note of it, was Abraham. Over 1500 years before the Greeks, around 1700 B.C. Abraham challenged the entire tribal mind-set. And he did so in ways no one had ever survived doing before. The multiple Pagan Gods had no restrictions on their behavior whatsoever, and therefore the witch-doctors, clerics, chiefs and ruling class could do in their name whatever they pleased. Abraham changed this! Abraham was the first person I can find to demand that the Gods have at least some minimal accountability! This meant that for the first time the ruling elite had some restrictions placed upon them. And they did not like it!

When Abraham was a young man he worked in his Dad's pottery store. In the store were a number of clay figurines of the various Pagan Gods, including one that was quite large. One day Abraham's father had to leave the store for a while, leaving him in charge. While his father was gone Abraham took a hammer and destroyed all the clay figurines of the Pagan Gods except for the big one. He then placed the hammer into this last remaining figures' hand.

When Abraham's father returned and saw the destruction he asked Abraham what had happened. Abraham answered that the Gods had gotten into a fight and the big one had destroyed all the others! Abraham's father responded that those Gods couldn't do that, that they had no power and couldn't do anything. Abraham answered 'Then why do you worship them?'(Don't you immediately love this young man?)

Lucky for Abraham, and for us, his father did not turn him into the tribal authorities whereupon he would have surely been sacrificed. Instead Abraham's father, Terah, warned him that the Chaldean authorities would kill him for such talk. Then Terah took his family, including Abraham and moved away. This story gives you a hint of the audacity and courage of this man. And it shows how with the help of his family he survived.

Abraham laid low for a number of years. Then as an adult he left the country and became a nomad traveling from one desert settlement to another. He preached as he traveled proposing a major change in religious belief. He advocated the belief that there was but one God! Yes he broke with the Pagan belief of polytheism and proposed a single God. This was really looking for trouble!

But monotheism could bring a tremendous relief to mankind. How much better for a slave to have but one master to serve that try to serve dozens, in some tribes hundreds! Plus one God, one conceptual truth, would greatly clarify man's thinking. But Abraham didn't stop with this.

Abraham preached that the one true God was ethical! Imagine God as an ethical being! Who could have conceived of such a thing? Pagans had no concept of an ethical God. To our knowledge this is the first time in all of human history that this idea was conceived!

Yes, Abraham taught that God was singular, and that he was ethical in his dealings with man. Not only was this man brazen he was fearless! Don't you know that Abraham's courage both inspired and frightened the people! Was this ever a tremendous blow to the power of the Pagan church and tribalism! This was a tremendous reduction in the power of the Pagan clerics! This was a declaration of a tiny bit of freedom for mankind! What an enormous achievement!

The followers of Abraham demanded that they be allowed to compartmentalize! Abraham invented the mind-body dichotomy whereby a person was allowed to hold two diametrically opposite belief systems in their mind simultaneously, and pledge allegiance to them both!

The polytheism of Pagan religions demanded total subjugation of all selfish worldly concerns to service and obedience to the witch-doctors and other members of the ruling class. Total rejection of theism would mean total rejection of the claims upon the slaves by the royals. Abraham negotiated a compromise. Monotheism means we'll take one God, but that's it! Compartmentalization allowed Abraham's followers to believe that they continued their loyalty to their chief and ruling elite, and at the same time they could 'pursue commercial enterprise and self gain'.

Compartmentalization is a form of hypocrisy, but it does allow much mental freedom. The *super*concept of monotheism allowed more mental freedom. This psycho-epistemological change allowed the followers of Abraham to enjoy greater mental clarity and greater productivity. It allowed Abraham's followers to be productive on an unprecedented scale. Their escape from the shackles of Paganism was only partial, but it was enough to attract the envy and rage of purer, more fanatic Pagans everywhere.

Abraham's ideas were quickly folded back into the Pagan dogma. But the religions he influenced would never again be quite the same. He set them on a path whereby they slowly morphed into the modern day religions. Contemporary religions such as Judaism, Christianity and Islam claim him. Unfortunately the core of theism was mostly left untouched. Thousands of years of Pagan belief clouded, and continues to cloud, the human mind! Most slaves simply do not want to be liberated!

Abraham was one of the first, perhaps the first, to speak out for the slaves! His teachings allowed the slaves a little personal use of their brains. And by reducing the number of reference standards, Gods, man had to contend with, that brain could work more efficiently. This was a great achievement.

2. Akhenaten
1352-1336

Akhenaten was a Pharaoh of the Eighteenth dynasty of Egypt. He ruled for 17 years and died around 1335BC. He is sometimes called the 'Heretic Pharaoh' for abandoning traditional Egyptian polytheism and introducing worship centered on Aten, which is sometimes described as monotheistic. Henotheism may be a more accurate description, since he ranked Aten above others but did not deny their existence. Indeed, an early inscription likens them to stars as compared with the sun, and later official language avoids calling Aten a God.

Akhenaten tried to bring about a departure from traditional religion, but in the end he was not successful. After his death, traditional religious practice was gradually restored. Some dozen years later rulers without clear rights of succession from the Eighteenth Dynasty founded a new dynasty. They discredited Akhenaten and his immediate successors, referring to Akhenaten as 'the enemy' in archival records. He was all but lost from history until the discovery, in the 19th century, of Amarna, the site of Akhetaten, the city he built for Aten. Early excavations at

Amarna by Flinders Petrie sparked interest in the enigmatic pharaoh, which increased with the discovery in the Valley of the Kings, at Luxor, of the tomb of King Tutankhamen, who may have been his son. Akhenaten remains an interesting figure, as does his Queen, Nefertiti.

Akhenaten may have been trying to do what Abraham did, but he met with less success. It shows that even a member of the ruling elite had trouble trying to bring more freedom to the masses. It is always treacherous to attempt to free the slaves.

3. Moses
1300BC

Moses was born into slavery as a member of the Hebrew tribe during the time of the Egyptian Empire. His life was spared by his mother when she defied the Pharaoh's orders to drown all newborn males. He was secreted out of Egypt.

As an adult Moses presented the world a set of 10 commandments that he represented as being God's rules for mankind! Abraham had said that God was singular and that he was ethical in his dealings with man. The Ten Commandments put the relationship between God and man not just under law, but under written law. The Ten Commandments that Moses produced was akin to a Magna Carter! The rule of written law worked for the people and against tribal leaders and Pagan clerics. Tribal governments now found it harder to arbitrarily do whatever they wished to the slaves in God's name. Moses said that God commanded that all men obey these 10 rules-not 1500 or more that the government might have liked to impose. So Moses continued the work started by Abraham, the placing of limits on tribal power! By limiting the power of the theist God he limited the power of the ruling elite.

The evil theist concepts of supernaturalism and self-sacrifice were again left intact. In was going to take a lot more work to undo thousands of years of brain washing and indoctrination.

Moses' teachings, just like those of Abraham, were soon plowed under and folded back into the Pagan religions of the day. But a single Pagan God with but ten commands was a huge improvement. Moses' teachings just like Abraham's gave the slaves a little more freedom. And when you have none, a little can make a big difference.

4. Thales of Miletus
620-547BC

Thales is considered the Father of Philosophy. Before Thales, the Greeks explained the origin and nature of the world through myths of anthropomorphic Gods and heroes. Phenomena such as lightning or earthquakes were attributed to actions of the Gods. In contrast to these mythological explanations, Thales attempted to find natural explanations for events, without reference to the supernatural. He explained earthquakes by hypothesizing that the Earth floats on water, and he believed that earthquakes occur when the Earth is rocked by waves. He may also have subscribed to a natural non-interventionist God.

Thales began the explicit study of epistemology. And with him man began to think about thinking and how to do so more effectively.

5. Buddha
Siddhārtha Gautama
565BC

Siddhartha Gautama was a very remarkable man. Born into one of the most oppressive tribal systems in the world, one that imposed a rigid caste system upon the slaves, he rose to an unprecedented height of enlightenment. He opposed enslavement of mind and body and tried to bring back a commitment to empiric thinking.

> "Believe nothing, no matter where you read it, or who said it, no matter if I have said it, unless it agrees with your own reason and your own common sense."

> "It is a man's own mind, not his enemy or foe that lures him to evil ways".

Buddha also understood how the morality of self-sacrifice harmed the human mind.

> "You can search throughout the entire universe for someone who is more deserving of your love and affection than you are yourself, and that person is not to be found anywhere. You yourself, as much as anybody in the entire universe deserve your love and affection."

Unfortunately Buddha's teachings failed to spread around the globe. For a long time Buddha's teachings remained limited to the area where he had lived and taught. Buddhism was, and remains, a threat to State power and theism. Buddha was unique among the ancients to see clearly the evil consequences of theism. But Buddha was not confrontational. He did not confront tribalism head-on. This was probably wise as by not doing so he lived to be a very old man. But this allowed Pagan tribal authorities to brush his teachings aside as a mystical, personal philosophy and of no practical consequence. And like the teachings of Abraham and Moses his teachings in time were folded back into the Pagan religious beliefs and customs of the region. This is unfortunate. Buddha's teachings could have built so powerfully upon what Abraham and Moses had accomplished. It shows how painfully slow change can be, particularly when it involves changing imprint programming. When the student does not want to learn, or cannot learn, it takes a very exceptional teacher to make any progress. It is difficult to free slaves that prefer to remain slaves.

Buddha made such a huge jump at such an early time in our history that mankind simply couldn't follow. He blazed a path that thinkers and prophets who followed him were forced to study and his impact on moral history should not be underestimated.

6. Confucius
551-479BC

Confucius emphasized personal and governmental morality, correctness of social relationships, justice and sincerity. These values gained prominence in China over other doctrines, such as Legalism or Taoism during the Han Dynasty (206 BC – 220 AD). Confucius' thoughts have been developed into a system of philosophy known as Confucianism. It was introduced to Europe by the Jesuit Matteo Ricci, who was the first to latinise the name as "Confucius."

His teachings may be found in the ***Analects of Confucius*** a collection of his thoughts, which was compiled many years after his death. Modern historians do not believe that any specific documents can be said to have been written by Confucius. But for nearly 2,000 years he was thought to be the editor or author of the ***Classic of Rites***, and the ***Spring and Autumn Annals***.

In the ***Analects***, Confucius presents himself as a "transmitter who invented nothing". He put the greatest emphasis on the importance of study, and it is the Chinese character for study that opens the text. In this respect, he is seen by Chinese people as the Greatest Master. Far from trying to build a systematic theory of life and society or establish a formalism of rites, he wanted his disciples to think deeply for themselves and relentlessly study the outside world. He taught mostly through the old scriptures and by relating the moral problems of the present to past political events or past expressions of feelings by common people. These ideas are preserved in the poems of the ***Book of Odes***.

In times of division, chaos, and endless wars between feudal states, he wanted to restore the Mandate of Heaven that could unify the world and bestow peace and prosperity on the people. Because his vision of personal and social perfection was framed as a revival of the ordered society of earlier times, Confucius is often considered a great proponent of conservatism. But a closer look at what he proposed shows that he used past institutions and rites to push a new political agenda of his own. He wanted a revival of a unified royal state, whose rulers would succeed to power on the basis of their moral merit, not their parentage. These would be rulers devoted to their people, reaching for personal and social perfection. Such a ruler would spread his own virtues to the people instead of imposing proper behavior with laws and rules.

One of the deepest teachings of Confucius may have been the superiority of personal exemplification over explicit rules of behavior. His moral teachings emphasize self-cultivation, emulation of moral example, and the attainment of skilled judgment rather than knowledge of rules. Confucius's ethics may be considered a type of virtue ethics. His teachings rarely rely on reasoned argument. His ethical ideals and methods are conveyed more indirectly, through allusions, innuendo, and even tautology. This is why his teachings need to be examined and put into proper context in order to be understood. A good example is found in this famous anecdote:

When the stables were burnt down, on returning from court, Confucius said, "Was anyone hurt?" He did not ask about the horses.
Analects X.11, tr. A. Waley

Perhaps his most famous teaching was the Golden Rule stated in the negative form, often called the Silver Rule:

Adept Kung asked: "Is there any one word that could guide a person throughout life?"

The Master replied: "How about 'shu' [reciprocity]: never impose on others what you would not choose for yourself?"

Analects XV.24, tr. David Hinton

Confucius deemphasized the importance of the supernatural and the afterlife. Confucius may have been the first Deist. Deist of the 18th century claimed him. If anyone could have created a 'good' government it was he, and he failed. The myth of 'good government' should have been put to rest but it persists.

7. Xenophanes of Colophon
570-480BC

Xenophanes' poetry criticized and satirized a wide range of ideas, including the belief in the pantheon of anthropomorphic Gods. Xenophanes rejected the idea that the Gods resembled humans in form. One famous, proto-sociological passage ridiculed the idea by claiming that, if oxen were able to imagine Gods, then those Gods would be in the image of oxen. His epistemology, which is still influential today, held that there actually exists a truth of reality, but that humans as mortals are unable to know it. Therefore, it is possible to act only on the basis of working hypotheses - we may act as if we know the truth, as long as we know that this is extremely unlikely. This aspect of Xenophanes was brought out again by the late Karl Popper and is a basis of Critical Rationalism.

Xenophanes continued man's study of epistemology. He is very important to the history of science and he shows the continued improvement in cognition afforded by increasing liberation from Paganism.

8. Anaxagoras
500-428BC

Anaxagoras marked a turning-point in the history of philosophy. By the theory of minute constituents of things, and his emphasis on mechanical processes in the formation of order, he paved the way for the atomic theory. However, his enunciation of the order that comes from an intelligent mind suggested the theory that nature is the work of design. Anaxagoras is considered by some to be the first 'freethinker'. Anaxagoras paved the way to physics.

9. Protagoras
490-420BC

Protagoras was a contemporary of Socrates. He practiced a system of criticism that has been called "pre-Socratic dialectic", an alternative to the Aristotelian demonstrative method which, according to Karl Popper, has the fault of being dogmatic. Protagoras knew that the less appealing argument could hide the best answer, which is why he stated that it was necessary to constantly strengthen the weakest argument. Having been born before Socrates himself, his approach in the development of consensual truth could conceivably have contributed to the improved style of many of the great minds which followed him.

10. Democritus
460-370BC

Democritus believed that everything which *is* must be eternal, but denied that "the void" can be equated with nothing. This makes him the first thinker on record to argue against the existence of an entirely empty "void" of space. In order to explain the change around us from basic, unchangeable substance he created a theory that argued that there are various basic elements which always existed but can be rearranged into many different forms. Democritus' theory argued that atoms had several properties, particularly size, shape, and (perhaps) weight; all other properties that we attribute to matter, such as color and taste, are but the result of complex interactions between the atoms in our bodies and the atoms of the matter that we are examining. Furthermore, he believed that the real properties of atoms determine the perceived properties of matter--for example, something that is solid is made of small, pointy atoms, while something that has water like properties is made of large, round atoms. Some types of matter are particularly solid because their atoms have hooks to attach to each other; some are oily because they are made of very fine, small atoms which can easily slip past each other.

Modern scholars credit Democritus with being "the earliest thinker reported as having explicitly posited a supreme good or goal, which he called 'cheerfulness' or 'wellbeing', and which he appears to have identified with the untroubled enjoyment of life. Joy and sorrow are the distinguishing mark of things beneficial and harmful. According to Democritus' philosophy, this supreme good was to be achieved through moderation in the pursuit of pleasure, distinguishing useful pleasures from harmful ones, and conforming to conventional morality. This seems to constitute "a recommendation to a life of moderate, enlightened hedonism" similar to that presented by Socrates in Plato's ***Protagoras*** and later made famous by Epicurus. This is very different from the tribal standard of sadomasochism.

11. Thucydides
460-395BC

Thucydides was a Greek historian and author of the History of the Peloponnesian War, which recounts the 5th century B.C. war between Sparta and Athens to the year 411 B.C. Thucydides has been dubbed the father of "scientific history" due to his strict standards of evidence-gathering and analysis in terms of cause and effect without reference to intervention by the Gods.

12. Socrates
470-399BC

"I know you won't believe me, but the highest form of human excellence is to question oneself and others."

Socrates was a Greek philosopher and Plato's teacher. He is considered the founder of Western philosophy. During the Peloponnesian War he served with some distinction as a soldier. After the war he returned home to work as a stonemason and to raise his children with his wife, Xanthippe. He then inherited a modest fortune from his father, the sculptor Sophroniscus. He used his financial independence as an opportunity to dabble in the political turmoil that consumed Athens following the War.

For the rest of his life, Socrates devoted himself to freewheeling discussions with the young aristocratic citizens of Athens. He continually questioned his students' unwarranted confidence in the truth of popular opinions, including their religious beliefs. His students were fanatically loyal to him. Their parents, however, were often displeased with his influence on their youngsters. He constantly clashed with the current course of Athenian politics and society. He praised Sparta, arch rival to Athens. Rather than upholding a status quo and accepting the development of immorality, Socrates worked to undermine the collective notion of "might makes right" so common to Greece during this period. Plato says that Socrates irritated the establishment with considerations of justice and the pursuit of goodness. He tried to improve the Athenian's sense of justice.

Socrates' earlier association with opponents of the democratic regime had already made him a controversial political figure. When he flaunted his lack of belief in the Olympian Gods many Athenians felt he had gone too far.

Although an amnesty forestalled his direct prosecution for his political activities, an Athenian jury found other charges—

corrupting the youth and interfering with the religion of the city—upon which to convict him. They sentenced him to death in 399 B.C.E.

Accepting this outcome with remarkable grace, Socrates drank hemlock and died in the company of his friends and disciples.

Socrates is remembered for his pursuit of virtue and his strict adherence to truth. Socrates serves as an excellent example of the freethinkers clash with tribalism. The tribe, its leaders and followers, do not take kindly to having truth highlighted. Truth always exposes tribalism as the villain it is; there is no way to white wash it. Truth is the state's greatest enemy. So whether the freethinker exposes the lack of freedom of speech, lack of the freedom of thought, lack of justice, false and delusional ideas, or simply the presence of ignorance, he is not someone the tribe or the state appreciates.

13. Plato
428-348BC

Plato was also a Greek philosopher. Together with his teacher, Socrates, and his student, Aristotle, Plato helped to lay the philosophical foundations of Western culture. Plato was also a mathematician, writer and founder of the Academy in Athens. His academy was the first institution of higher learning in the Western world. Plato was a student of Socrates, and was much influenced by his thinking.

Plato's brilliance as a writer and thinker can be witnessed by reading his Socratic dialogues. The dialogues have been used to teach a range of subjects, including philosophy, logic, rhetoric, mathematics, and other subjects.

The reason I include Plato here is to point out the heretical act of intellectual pursuits in and of themselves! The fact that men like Plato could live, think, write, and establish a university is testament to the amount of freedom enjoyed in Athens at this time. The tribe and state saw right away that it would have to take over and control academia. There was just too much potential for heresy at such institutions!

14. Aristotle
384-322BC

Aristotle was also a Greek philosopher, a student of Plato. He wrote on many different subjects, including physics, metaphysics, poetry, theater, logic, rhetoric, politics, government, ethics, biology and zoology.

Aristotle together with Socrates and Plato are three of the most important figures in Western thought. He was one of the first to systematize philosophy and science. His thinking on physics and science had a profound impact on medieval thought, which lasted until the Renaissance, and the accuracy of some of his biological observations was only confirmed in the last century. His logical works contain the earliest formal study of logic that we have and were not superseded until the late nineteenth century. All aspects of Aristotle's philosophy continue to be the object of active academic study today.

Aristotle not only studied almost every subject possible at the time, but made significant contributions to most of them. In physical science, Aristotle studied anatomy, astronomy, economics, embryology, geography, geology, meteorology, physics and zoology. In philosophy, he wrote on aesthetics, ethics, government, metaphysics, politics, psychology, rhetoric and theology. He also studied education, foreign customs, literature and poetry. His combined works constitute a virtual encyclopedia of Greek knowledge

For Aristotle philosophic method implies the ascent from the study of particular phenomena to the knowledge of essences, while for Plato philosophic method used the descent from knowledge of universal ideas to a contemplation of particular imitations of those ideas. In a certain sense, Aristotle's method is both inductive and deductive, while Plato's is essentially deductive.

Aristotle's "natural philosophy" was a branch of philosophy examining the phenomena of the natural world, and included fields that would be regarded today as physics, biology and other natural sciences. In modern times, the scope of philosophy has become limited to more generic or abstract inquiries, such as ethics and metaphysics, in which logic plays a major role. In contrast, Aristotle's philosophical endeavors encompassed virtually all facets of intellectual inquiry.

In the larger sense of the word, Aristotle makes philosophy coextensive with reasoning, which he also would describe as "science". So we see Aristotle making a great effort to study and try to understand perceptual or empiric thought processes. Indeed Aristotle tried to return man to a percept-derived reasoning and away from ideologically derived conclusions.

We can contrast Aristotle's cognitive approach with Plato's. Platonic reasoning was from concept to conclusion whereas Aristotle's approach was from perception to conclusion. The essential point here for me, is that a person be free to think anyway he chooses. It is results that matter. But being forced to accept any conceptual model through intimidation is the antithesis of reason. 'Up in your head' thinking may be fine, but only if you are allowed to 'come down' any time you choose.

It is remarkable that these two men, and many of their contemporaries, felt free enough to think, study, and learn the way they did. In addition to their brilliance it is obvious that they were very courageous.

15. Epicurus
341– 270 BC

"That which creates unsurpassable joy is the removal of a great evil."

"Justice's greatest reward is peace of mind."

"It is impossible for the one who instills fear to remain free from fear."

"Happiness and blessedness do not belong to abundance of riches or exalted position or offices or power, but to freedom from pain and gentleness of feeling and a state of mind that sets limits that are in accordance with nature."

Epicurus was a Greek philosopher and the founder of Epicureanism, a school of thought that was popular for over 600 years. Of his over 300 written works only a few fragments and letters survive; much of what we know about Epicureanism comes from later followers or commentators.

For Epicurus, the purpose of philosophy was to attain the happy, tranquil life, characterized by the absence of pain and fear, by living a self-sufficient life surrounded by friends. He taught that pleasure and pain are the measures of what is good and bad.

Epicurus founded his school in 306 BC. It was based in Epicurus' home and garden. An inscription on the gate to the garden said:

"Stranger, here you will do well to tarry; here our highest good is pleasure".

Epicurus played an important part in what is known as the "Greek miracle": when men first tried to explain the nature of the world, not with the aid of myths or religion, but with material principles. He is a key figure in the development of science and the scientific method because of his insistence that nothing should be believed except that which was tested through direct observation and logical deduction.

Many of his ideas about nature and physics presaged important scientific concepts of our time. He was a key figure in the Axial Age, the period from 800 BC to 200 BC, during which similarly revolutionary thinking appeared in China, India, Iran, the Near East, and Ancient Greece. His statement of the Ethic of Reciprocity as the foundation of ethics is the earliest in Ancient Greece, and differs from the usual formulation by emphasizing the minimization of harm to oneself or others as the source of happiness.

Epicurus accepted both women and slaves into his school introducing the new concept of fundamental human egalitarianism into Greek thought. He was one of the first Greeks to break from the god-fearing and god-worshiping tradition common at the time. "The gods do not punish the bad and reward the good as the common man believes."

Epicurus' philosophy is based on the theory that all good and bad derive from the sensations of pleasure and pain. What is good is what is pleasurable, and what is bad is what is painful. Pleasure and pain were ultimately, for Epicurus, the basis for the moral distinction between good and bad.

In his epistemology he emphasized the senses, and his Principle of Multiple Explanations is an early contribution to the philosophy of science: if several theories are consistent with the observed data, retain them all.

His emphasis minimizing harm and maximizing happiness in his formulation of the Ethic of Reciprocity was later picked up by the democratic thinkers of the French Revolution, and others, like John Locke, who wrote that people had a right to "life, liberty, and property." To Locke, one's own body was part of their property, and thus one's right to property would theoretically guarantee safety for their persons, as well as their possessions.

This triad, as well as the egalitarianism of Epicurus, was carried forward into the American freedom movement and Declaration of Independence, by the American founding father, Thomas Jefferson, as "all men are created equal" and endowed with certain "inalienable rights such as life, liberty, and the pursuit of happiness." Epicurus was therefore a key influence on the foundation of the American legal system.

I need comment little on Epicurus. What did he have wrong? He is a towering figure in man's slow emergence from Pagan theist tyranny.

16. Jesus Christ
7-36AD

Jesus, also known as Jesus of Nazareth, is the central figure of Christianity, and is also an important figure in several other religions. He is also called Jesus Christ, where "Christ" is a title derived from the Greek Christós, meaning the "Anointed One," which corresponds to the Hebrew-derived "Messiah". The name "Jesus" is an Anglicization of the Greek Iēsous, itself a Hellenization of the Hebrew Yehoshua, meaning "YHWH rescues".

The main sources of information regarding Jesus' life and teachings are the four canonical **_Gospels of the New Testament_**: Matthew, Mark, Luke, and John. Most scholars in the fields of history and biblical studies agree that Jesus was a Galilean Jew, a teacher and healer. He taught in parables and aphorisms, challenged pious traditions, legalism and social hierarchy, He was crucified in Jerusalem on orders of the Roman Governor Pontius Pilate under the accusation of sedition against the Roman Empire.

Jesus went with his followers to Jerusalem during a Passover festival where a large crowd came to meet him, shouting, "Hosanna! Blessed is he who comes in the name of the Lord! Blessed is the King of Israel! His triumphal entry into the city brought him to the attention of the Jewish clerics. He then created a disturbance at Herod's Temple by overturning the tables of the moneychangers who set up shop there, claiming that they had made the Temple a "den of robbers." (Mark 11:17). Later that week, Jesus celebrated the Passover meal with his disciples — an event subsequently known as the Last Supper — in which he prophesied that he would be betrayed by one of his disciples, and would then be executed. In this ritual he took bread and wine in hand, saying: "this is my body which is given for you" and "this cup which is poured out for you is the New Covenant in my blood," and instructed them to "do this in remembrance of me" (Luke 22:7–20). Following the supper, Jesus and his disciples went to pray in the Garden of Gethsemane.

While in the Garden, Jesus was arrested by Temple guards on the orders of the Jewish clerics. The arrest took place clandestinely at night to avoid a riot, as Jesus was popular with the people at large (Mark 14:2). Judas Iscariot, one of his apostles, betrayed Jesus by identifying him to the guards with a kiss. Simon Peter, another one of Jesus' apostles, used a sword to attack one of Jesus' captors, cutting off his ear, which, according to Luke, Jesus immediately healed miraculously. Jesus rebuked the apostle, stating "all they that take the sword shall perish by the sword" (Matthew 26:52). After his arrest, Jesus' apostles went into hiding.

During the Trial of Jesus, the high priests and elders asked Jesus, "Are you the Son of God?," and after he replied, "You are right in saying I am," they condemned Jesus for blasphemy (Luke 22:70–71). The high priests then turned him over to the Roman procurator Pontius Pilate, based on an accusation of sedition for claiming to be King of the Jews. When Jesus came before Pilate, Pilate asked him, "Are you the king of the Jews?" to which he replied, "It is as you say." According to the Gospels, Pilate personally felt that Jesus was not guilty of any crime against the Romans, and since there was a custom at Passover for the Roman governor to free a prisoner, Pilate offered the crowd a choice between Jesus of Nazareth and an insurrectionist named Barabbas. The crowd chose to have Barabbas freed and Jesus crucified. Pilate washed his hands to indicate that he was innocent of the injustice of the decision. According to all four Gospels, Jesus died before late afternoon at Calvary, which was also called Golgotha.

The theme of Jesus' teachings was that of repentance, unconditional love, forgiveness of sin, grace, and the coming of the Kingdom of God. Other legacies include a view of God as more lovingly parental, merciful, and more forgiving, and the growth of a belief in a blissful afterlife and in the resurrection of the dead. His teaching promoted the value of those who had commonly been regarded as inferior: women, the poor, ethnic outsiders, children, prostitutes, the sick, prisoners, etc.

What I find most remarkable is Jesus' stand against force. Force, as I mentioned earlier is one of the basic pillars of theism. When you survey world history very few religious or political leaders have promoted non-violent protest. I count only Jesus, Gandhi, Martin Luther King, Jr. and Nelson Mandela. Jesus would not even allow force to be used to save his own life!

Another very important message, not original to Jesus but powerfully reinforced by him was belief that kings or Caesars were not divine! The Roman Empire was based on the 'Divine Right of Kings'. Within 300 years of Jesus' death the Roman Empire was on its way out.

Jesus promoted mercy, love, forgiveness and the goodness of people at a time when those beliefs were considered weaknesses. Jesus has been a powerful force against Paganism.

The teachings of Jesus, like those of Abraham, Moses and Buddha were immediately folded back into the Pagan religions of the times. Most of the Pagan rituals and holidays were retained and the tradition of self-sacrifice and supernaturalism persisted unchecked. But his powerful enduring legacy against divine human authority and violence remain.

17. Sextus Empiricus
100-200AD

Sextus Empiricus was a physician and philosopher. He belonged to the "empiric" school of medicine as reflected by his name. His philosophical work is the most complete surviving account of ancient Greek and Roman skepticism.

He advises that we should suspend judgment about virtually all beliefs, that is, we should neither affirm any belief as true nor deny any belief as false. Sextus did not deny the possibility of knowledge. He criticizes the Academic skeptic's claim that nothing is knowable as being an affirmative belief. Instead, Sextus advocates simply giving up belief: that is, suspending judgment about whether or not anything is knowable. Remember that this advice is in the context of the tribal theist psychology.

Sextus also allowed that we might affirm claims about our experience. However, he pointed out that this does not imply any objective knowledge of external reality. For while I might know that the honey I eat tastes sweet to me, this is merely a subjective judgment, and as such may not tell me anything true about the honey itself.

My interest in Sextus Empiricus here is to note that his work is the first on record of a conscious attempt to nullify the destructive effects of force and supernaturalism on cognition. His admonition to avoid beliefs and judgment is his way of warning against ideology. He de-legitimizes concept down or top down Platonic thinking directing us toward perceptual reasoning. Sextus' advice went largely unheeded until the Scientific Age.

18. Pelagius
354-420AD

Pelagius was an ascetic Roman Catholic Monk who denied the doctrine of Original Sin. He was declared a heretic by the Council of Carthage. He was well educated, fluent in both Greek and Latin, and learned in theology. He spent time as an ascetic, focusing on practical asceticism, which his teachings clearly reflect. He was not, however, a cleric. He was certainly well known in Rome, both for the harsh asceticism of his public life as well as the power and persuasiveness of his speech. His reputation in Rome earned him praise early in his career even from such pillars of the Church as Augustine, who referred to him as a "saintly man." However, he was later accused of lying about his own teachings in order to avoid public condemnation. Most of his later life was spent defending himself against other theologians and the Catholic Church.

19. Martin Luther
1483-1546 AD

"I am more afraid of my own heart than of the Pope and all his Cardinals. I have within me the great pope, Self."

Martin Luther was a German monk, theologian, and church reformer. He is considered to be the founder of Protestantism. Luther's theology challenged the authority of the Holy Roman Catholic Church by holding that the **_Bible_** is the sole source of religious authority and that all baptized Christians are a general priesthood. According to Luther, salvation was attainable only by faith in Jesus as the Messiah, a faith unmediated by the Church. These ideas helped to inspire the Protestant Reformation and changed the course of Western civilization.

Luther became convinced that the Church had lost sight of what he saw as several of the central truths of Christianity, the most important of which, for Luther, was the doctrine of justification — God's act of declaring a sinner righteous — by faith alone. He began to teach that salvation or redemption is a gift of God's grace, attainable only through faith in Jesus as the messiah.

In 1516-17, Johann Tetzel, a Dominican friar and Papal commissioner for indulgences, was sent to Germany by the Roman Catholic Church to sell indulgences to raise money to rebuild St Peter's Basilica in Rome. In Roman Catholic theology, an "indulgence" is the remission of punishment because a sin already committed has been forgiven; the indulgence is granted by the church when the sinner confesses and receives absolution. When an indulgence is given, the church is extending merit to a sinner from its Treasure House of Merit, an accumulation of merits it has collected based on the good deeds of the saints. These merits could be bought and sold.

On October 31, 1517, Luther wrote to Albert, Archbishop of Mainz and Magdeburg, protesting the sale of indulgences. He enclosed in his letter a copy of his ***Disputation of Martin Luther on the Power and Efficacy of Indulgences,*** which came to be known as ***The 95 Theses.*** Luther had no intention of confronting the church, but saw his disputation as a scholarly objection to church practices, and the tone of the writing is accordingly "searching, rather than doctrinaire." There is nevertheless an undercurrent of challenge in several of the theses, particularly in Thesis 86, which asks: "Why does not the Pope, whose wealth today is greater than the wealth of the richest Crassus, build the basilica of St. Peter with his own money rather than with the money of poor believers?"

Luther objected to a saying attributed to Johann Tetzel that "as soon as the coin in the coffer rings, the soul from purgatory springs," insisting that, since forgiveness was God's alone to grant, those who claimed that indulgences absolved buyers from all punishments and granted them salvation were in error. Christians, he said, must not slacken in following Christ on account of such false assurances.

According to Philip Melanchthon, writing in 1546, Luther nailed a copy of ***The 95 Theses*** to the door of the Castle Church in Wittenberg that same day — church doors acting as the bulletin boards of his time — an event now seen as sparking the Protestant Reformation, and celebrated every October 31 as Reformation Day.

On June 15, 1520, the Pope warned Luther with the edict Exsurge Domine that he risked excommunication unless he recanted 41 sentences drawn from his writings, including ***The 95 Theses***, within 60 days.

Luther, who had sent the Pope a copy of ***On the Freedom of a Christian*** in October, publicly set fire to the edict at Wittenberg on December 10, 1520. As a consequence, Luther was excommunicated by Leo X on January 3, 1521.

"We want him to be apprehended and punished as a notorious heretic". The excommunication made it a crime for anyone in Germany to give Luther food or shelter. It permitted anyone to kill Luther without legal consequence.

The apprehension of Luther was the last thing Frederick III, Elector of Saxony wanted, so he had him discreetly intercepted on his way home by masked horsemen and escorted to the security of the Wartburg Castle at Eisenach, where Luther grew a beard and lived incognito for nearly eleven months.

In a private Mass, in the summer of 1521, Luther widened his target from individual pieties like indulgences and pilgrimages to doctrines at the heart of Church practices. His essay concerning Confession rejected the Roman Catholic Church's requirement of confession, although he affirmed the value of private confession and absolution. In the introduction to his **_New Testament_** — published in September 1522 and selling 5,000 copies in two months — he explained that good works spring from faith; they do not produce it.

Martin Luther is unquestionably one of the greatest hero's of mans' efforts to free himself of the mental chains of theism. He sides with his own mind against the Holy Roman Empire! He advocates that everyone determine for himself God's teachings by reading the Bible for themselves! He comes close to explicitly advocating compartmentalization, with religion kept mentally separated from reason. In so doing Luther not only founded the Lutheran Church and launched the Protestant Reformation, but more than anyone else ushered in the Age of Science and the Industrial Revolution! Lassie-faire, free enterprise and capitalism owe a great debt to Martin Luther. And the principle of separation of Church and State which forms an integral aspect of the Constitution of the United States is a direct result of Luther's enormous contribution to the Western World and to freedom everywhere.

20. Nicolaus Copernicus
1473-1543

Copernicus formulated a scientifically based heliocentric cosmology that displaced the Earth from the center of the universe.

His epochal book **_On the Revolutions of the Celestial Spheres,_** is often regarded as the starting point of modern astronomy and the defining epiphany that began the Scientific Revolution.

Greek, Indian and Muslim thinkers had published heliocentric hypotheses centuries before Copernicus. But his publication was an observation-based, mathematically-supported scientific theory of heliocentrism. He demonstrated that the motions of celestial objects can be explained without putting the Earth at rest in the center of the universe. This was a landmark in the history of modern science that became known as the Copernican Revolution.

In 1616, in connection with the Galileo affair, the Roman Catholic Church's Congregation of the Index suspended Copernicus' book until it could be "corrected," on the grounds that it was "false and altogether opposed to Holy Scripture." The corrections, omitting or altering nine sentences, were issued four years later, in 1620. The same edict also prohibited any work that defended the mobility of the Earth or the immobility of the Sun, or that attempted to reconcile these assertions with Scripture.

In 1633 Galileo Galilei was convicted of grave suspicion of heresy for "following the position of Copernicus, which is contrary to the true sense and authority of Holy Scripture," and was placed under house arrest for the rest of his life. The Catholic Church's **_Index of Prohibited Books_** issued in 1758 omitted the general prohibition of works defending heliocentrism, but retained the specific prohibitions of the original uncensored versions of Copernicus' book and Galileo's work. Those prohibitions were not dropped from the Index until 1835.

Copernicus's theory is of extraordinary importance in the history of human knowledge. Many authors suggest that few other persons have exerted a comparable influence on human culture in general and on science in particular. There are parallels with the life of Charles Darwin, in that both men produced a short early description of their theories, but held back on a definitive publication until late in life, against a backdrop of controversy, particularly with regard to religion.

Many meanings have been ascribed to Copernicus's theory, apart from its strictly scientific import. His work affected religion as well as science, religious belief as well as freedom of scientific inquiry. Copernicus's rank as a scientist is often compared with that of Galileo.

Copernicus' life and work speak for themselves. Just speaking ones opinion on astronomy threatened the State! Doing anything which threatened tribalism could get you killed!

21. Galileo Galilee
1564 -1642AD

"I do not feel obliged to believe that the same God who has endowed us with sense, reason and intellect has intended us to forgo their use."

In 1610 Galileo published an account of his telescopic observations of the moons of Jupiter, using this observation to argue in favor of the sun-centered, Copernican theory of the universe against the dominant earth-centered Ptolemaic and Aristotelian theories. The next year Galileo visited Rome in order to demonstrate his telescope to the influential philosophers and mathematicians of the Jesuit College, and to let them see with their own eyes the reality of the four moons of Jupiter.

In 1612, opposition arose to the Sun-centered solar system which Galileo supported. In 1614, from the pulpit of Santa Maria Novella, Father Tommaso Caccini denounced Galileo's opinions on the motion of the Earth, judging them dangerous and close to heresy. Galileo went to Rome to defend himself against these accusations, but, in 1616, Cardinal Bellarmino personally handed Galileo an admonition enjoining him neither to advocate nor teach Copernican astronomy. During 1621 and 1622 Galileo wrote his first book, ***The Assayer****,* which was approved and published in 1623. In 1630, he returned to Rome to apply for a license to print the ***Dialogue Concerning the Two Chief World***

Systems, published in Florence in 1632. In October of that year, however, he was ordered to appear before the Holy Office in Rome. Although Galileo tried to remain loyal to the Catholic Church, his adherence to experimental results, and their most honest interpretation, led to a rejection of blind allegiance to philosophical and religious authority. In broader terms, this aided to separate science from both philosophy and religion; making compartmentalization official!

Because of his ***Dialogue Concerning the Two Chief World Systems,*** Galileo was ordered to stand trial on suspicion of heresy in 1633. He was found guilty by the Inquisition arrested and forced to recant his heliocentrism.

After a period with the friendly Archbishop of Siena, Galileo was allowed to return to his villa at Florence. Here he spent the remainder of his life under house arrest, and where he later became blind. It was while Galileo was under house arrest that he dedicated his time to one of his finest works, ***Two New Sciences.*** Here he summarized work he had done some forty years earlier, on the two sciences now called kinematics and strength of materials. This book has received high praise from both Sir Isaac Newton and Albert Einstein. As a result of this work, Galileo is often called, the "father of modern physics."

The Inquisition's ban on reprinting Galileo's works was lifted in 1718 when permission was granted to publish an edition of his works excluding the condemned ***Dialogue***. In 1741 Pope Benedict XIV authorized the publication of an edition of Galileo's complete scientific works which included a mildly censored version of the ***Dialogue***. In 1758 the general prohibition against works advocating heliocentrism was removed from the ***Index of prohibited books***, although the specific ban on uncensored versions of the ***Dialogue*** and Copernicus's book remained. All traces of official opposition to heliocentrism by the Church disappeared in 1835 when these works were finally dropped from the Index.

On 31 October 1992, Pope John Paul II expressed regret for how the Galileo affair was handled, as the result of a study conducted by the Pontifical Council for Culture.

I can find no better example of the necessity of mental compartmentalization for the tribal scientist that this story of Galileo. He tried to believe, and remain loyal to two diametrically opposite belief and cognitive systems. It tore him apart. Yet he held to his own perceptions even when forced to recant the results of those perceptions; his discoveries. The religious authorities here again demonstrate their lack of any desire to know God. Their concern is with upholding their man-made ideology. Galileo is truly a hero and a model for all who fight the tyranny of theism.

22. John Locke
1632–1704

Locke was an English philosopher. Locke is considered the first of the British Empiricists. His ideas had enormous influence on the development of epistemology and political philosophy, and he is widely regarded as one of the most influential Enlightenment thinkers and contributors to liberal theory. His writings influenced Voltaire and Rousseau, many Scottish Enlightenment thinkers, as well as the American revolutionaries. His influence is reflected in the American Declaration of Independence.

Locke's theory of mind is often cited as the origin for modern conceptions of identity and "the self", figuring prominently in the later works of philosophers such as David Hume, Jean-Jacques Rousseau and Immanuel Kant. Locke was the first philosopher to define the *self* through a continuity of "consciousness." He also postulated that the mind was a "blank slate" or "tabula rasa"; that is, contrary to Cartesian or Christian philosophy, Locke maintained that people are born without innate sin.

Locke composed the bulk of the ***Two Treatises of Government*** to defend the Glorious Revolution of 1688, and also to counter the absolutist political philosophy of Sir Robert Filmer and Thomas Hobbes.

In 1683, Locke fled to the Netherlands, Holland, under strong suspicion of involvement in the Rye House Plot, which was a plot against King James II. In the Netherlands Locke had time to return to his writing, spending a great deal of time composing the ***Letter on Toleration.*** Locke did not return home until after the Glorious Revolution. The bulk of Locke's publishing took place after his arrival back in England — his ***Essay Concerning Human Understanding***, the ***Two Treatises of Civil Government*** and ***A Letter Concerning Toleration*** all appearing in quick succession upon his return from exile.

Locke exercised a profound influence on philosophy and politics, in particular on liberalism. Most modern libertarians claim him as an influence. He was a strong influence on Voltaire, while his arguments concerning liberty and the social contract later influenced the written works of Alexander Hamilton, James Madison, Thomas Jefferson, and other Founding Fathers of the United States. In addition, Locke's views influenced the American and French Revolutions. But Locke's influence may have been even more profound in the realm of epistemology. Locke redefined subjectivity, or *self* and intellectual historians such as Charles Taylor and Jerrold Seigel argue that Locke's ***Essay Concerning Human Understanding*** (1690) marks the beginning of the modern conception of the *self*.

Unlike Thomas Hobbes, Locke believed that human nature is characterized by reason and tolerance. Like Hobbes, Locke believed that human nature allowed men to be selfish. This is apparent with the introduction of currency. In a natural state all people were equal and independent, and none had a right to harm another's "life, health, liberty, or possessions." Locke never refers to Hobbes by name, however, and may instead have been responding to other writers of the day. Locke also advocated governmental checks and balances and believed that revolution is not only a right but an obligation in some circumstances. These ideas would come to have profound influence on the Constitution of the United States and its Declaration of Independence.

Locke defines the *self* as "that conscious thinking thing, (whatever substance, made up of whether spiritual, or material, simple, or compounded, it matters not) which is sensible, or conscious of pleasure and pain, capable of happiness or misery, and so is concerned for itself, as far as that consciousness extends," but Locke does not ignore the "substance." He writes "the body too goes to the making the man." The Lockean *self* is therefore a self-aware, self-reflective consciousness that is fixed in a body. In his Essay, Locke explains the gradual unfolding of this conscious mind. Arguing against both the Augustinian view of man as originally sinful and the Cartesian position which holds that man innately knows basic logical propositions, Locke posits an "empty" mind—a tabula rasa—that is shaped by experience; sensations and reflections being the two sources of all our ideas.

Locke's **Some Thoughts Concerning Education** is an outline for how to educate this mind; he expresses his belief that education makes the man, or more fundamentally, that the mind is an "empty cabinet" with the statement, "I think I may say that of all the men we meet with, nine parts of ten are what they are, good or evil, useful or not, by their education."

Locke also suggested that "the little and almost insensible impressions on our tender infancies have very important and lasting consequences." He argued that the "associations of ideas" that one makes when young are more important than those made later because they are the foundation of the self—they are what first mark the tabula rasa. In the Essay, in which he introduces both of these concepts, Locke warns against, for example, letting "a foolish maid" convince a child that "goblins and sprites" are associated with the night for "darkness shall ever afterwards bring with it those frightful ideas, and they shall be so joined, that he can no more bear the one than the other." "Associationism," as this theory would come to be called, exerted a very powerful influence over eighteenth-century thought, particularly educational theory. Nearly every educational writer would warn parents not to allow their children to develop negative associations. It also led to the development of psychology and other new disciplines with David Hartley's attempt to discover a biological mechanism for associationism in his ***Observations on Man*** (1749).

Locke's concepts mark the beginning of psychology and thus concern for man as man! We can thank Locke as much as any one person for the humanism of our modern world. We owe him a huge debt of gratitude.

23. Isaac Newton
1643–1727

Newton was an English physicist, mathematician, astronomer, natural philosopher, alchemist, and theologian who is perceived and considered by a substantial number of scholars and the general public as one of the most influential men in history. His ***Philosophiæ Naturalis Principia Mathematica,*** published in 1687, is by itself considered to be among the most influential books in the history of science, laying the groundwork for most of classical mechanics. In this work, Newton described universal gravitation and the three laws of motion which dominated the scientific view of the physical universe for the next three centuries. Newton showed that the motions of objects on Earth and of celestial bodies are governed by the same set of natural laws by demonstrating the consistency between Kepler's laws of planetary motion and his theory of gravitation, thus removing the last doubts about heliocentrism and advancing the scientific revolution.

In mechanics, Newton enunciated the principles of conservation of both momentum and angular momentum. In optics, he built the first practical reflecting telescope and developed a theory of color based on the observation that a prism decomposes white light into the many colors which form the visible spectrum. He also formulated an empirical law of cooling and studied the speed of sound.

In mathematics, Newton shares the credit with Gottfried Leibniz for the development of the differential and integral calculus. He also demonstrated the generalized binomial theorem, developed the so-called "Newton's method" for approximating the zeroes of a function, and contributed to the study of power series.

Newton remains influential to scientists, as demonstrated by a 2005 survey of scientists and the general public in Britain's Royal Society asking who had the greater effect on the history of science, Newton or Albert Einstein. Newton was deemed to have made the greater overall contribution to science, although the two men were closer when it came to contributions to humanity.

Newton was also highly religious, though an unorthodox Christian, writing more on Biblical hermeneutics than the natural science he is remembered for today.

Most modern historians believe that Newton and Leibniz developed infinitesimal calculus independently, using their own unique notations. Newton is generally credited with the generalized binomial theorem, valid for any exponent. He discovered Newton's identities, Newton's method, classified cubic plane curves (polynomials of degree three in two variables), made substantial contributions to the theory of finite differences, and was the first to use fractional indices and to employ coordinate geometry to derive solutions to Diophantine equations. He approximated partial sums of the harmonic series by logarithms (a precursor to Euler's summation formula), and was the first to use power series with confidence and to revert power series.

From 1670 to 1672, Newton lectured on optics. During this period he investigated the refraction of light, demonstrating that a prism could decompose white light into a spectrum of colors, and that a lens and a second prism could recompose the multicolored spectrum into white light.

He also showed that the colored light does not change its properties by separating out a colored beam and shining it on various objects. Newton noted that regardless of whether it was reflected or scattered or transmitted, it stayed the same color. Thus, he observed that color is the result of objects interacting with already-colored light rather than objects generating the color themselves. This is known as Newton's theory of color.

From this work he concluded that the lens of any refracting telescope would suffer from the dispersion of light into colors (chromatic aberration), and as a proof of the concept he constructed a telescope using a mirror as the objective to bypass that problem. Actually building the design, the first known functional reflecting telescope, today known as a Newtonian telescope, involved solving the problem of a suitable mirror material and shaping technique. Newton ground his own mirrors out of a custom composition of highly reflective speculum metal, using Newton's rings to judge the quality of the optics for his telescopes. In late 1668 he was able to produce this first reflecting telescope. In 1671 the Royal Society asked for a demonstration of his reflecting telescope. Their interest encouraged him to publish his notes **_On Colour_**, which he later expanded into his **_Opticks._** When Robert Hooke criticized some of Newton's ideas, Newton was so offended that he withdrew from public debate. Newton and Hooke had brief exchanges in 1679-80, when Hooke, appointed to manage the Royal Society's correspondence, opened up a correspondence intended to elicit contributions from Newton to Royal Society transactions. This had the effect of stimulating Newton to work out a proof that the elliptical form of planetary orbits would result from a centripetal force inversely proportional to the square of the radius vector. But the two men remained generally on poor terms until Hooke's death.

Newton argued that light is composed of particles or corpuscles, which were refracted by accelerating into a denser medium. He verged on sound-like waves to explain the repeated pattern of reflection and transmission by thin films, but still retained his theory of 'fits' that disposed corpuscles to be reflected or transmitted. Later physicists instead favored a purely wavelike explanation of light to account for the interference patterns, and the general phenomenon of diffraction. Today's quantum mechanics, photons and the idea of wave–particle duality bear only a minor resemblance to Newton's understanding of light.

In his ***Hypothesis of Light*** of 1675, Newton posited the existence of the ether to transmit forces between particles. The contact with the theosophist Henry More, revived his interest in alchemy. He replaced the ether with occult forces based on Hermetic ideas of attraction and repulsion between particles. John Maynard Keynes, who acquired many of Newton's writings on alchemy, stated that "Newton was not the first of the age of reason: he was the last of the magicians."-Newton's interest in alchemy cannot be isolated from his contributions to science; however, he did apparently abandon his alchemical researches. (This was at a time when there was no clear distinction between alchemy and science.) Had he not relied on the occult idea of action at a distance, across a vacuum, he might not have developed his theory of gravity.

In 1704 Newton published ***Opticks***, in which he expounded his corpuscular theory of light. He considered light to be made up of extremely subtle corpuscles, that ordinary matter was made of grosser corpuscles and speculated that through a kind of alchemical transmutation "Are not gross Bodies and Light convertible into one another, …and may not Bodies receive much of their Activity from the Particles of Light which enter their Composition?" Newton also constructed a primitive form of a frictional electrostatic generator, using a glass globe.

In 1679, Newton returned to his work on mechanics, i.e., gravitation and its effect on the orbits of planets, with reference to Kepler's laws of planetary motion, after stimulation by a brief exchange of letters in 1679-80 with Hooke, who had been appointed to manage the Royal Society's correspondence, and who opened up a correspondence intended to elicit contributions from Newton to Royal Society transactions. Newton's reawakening interest in astronomical matters received further stimulus by the appearance of a comet in the winter of 1680/1681, on which he corresponded with John Flamsteed. After the exchanges with Hooke, Newton worked out a proof that the elliptical form of planetary orbits would result from a centripetal force inversely proportional to the square of the radius vector. Newton communicated his results to Edmond Halley and to the Royal Society in ***De motu corporum in gyrum***, a tract written on about 9 sheets which was copied into the Royal Society's Register Book in December 1684.-This tract contained the nucleus that Newton developed and expanded to form the ***Principia***.

The ***Philosophiae Naturalis Principia Mathematica*** (now known as the ***Principia***) was published on 5 July 1687 with encouragement and financial help from Edmond Halley. In this work Newton stated the three universal laws of motion that were not to be improved upon for more than two hundred years. He used the Latin word *gravitas* (weight) for the effect that would become known as gravity, and defined the law of universal gravitation. In the same work Newton presented a calculus-like method of geometrical analysis by 'first and last ratios', gave the first analytical determination, based on Boyle's law, of the speed of sound in air, inferred the oblateness of the spheroidal figure of the Earth, accounted for the precession of the equinoxes as a result of the Moon's gravitational attraction on the Earth's oblateness, initiated the gravitational study of the irregularities in the motion of the moon, provided a theory for the determination of the orbits of comets, and much else.

Newton's postulate of an invisible force able to act over vast distances led to him being criticized for introducing "occult agencies" into science. Later, in the second edition of the ***Principia*** (1713), Newton firmly rejected such criticisms in a concluding General Scholium, writing that it was enough that the phenomena implied a gravitational attraction, as they did; but they did not so far indicate its cause, and it was both unnecessary and improper to frame hypotheses of things that were not implied by the phenomena. (Here Newton used what became his famous expression 'Hypotheses non fingo').

With the ***Principia***, Newton became internationally recognized. He acquired a circle of admirers, including the Swiss-born mathematician Nicolas Fatio de Duillier, with whom he formed an intense relationship that lasted until 1693, when it abruptly ended, at the same time that Newton suffered a nervous breakdown.

In the 1690s, Newton wrote a number of religious tracts dealing with the literal interpretation of the Bible. Henry More's belief in the Universe and rejection of Cartesian dualism may have influenced Newton's religious ideas. A manuscript he sent to John Locke in which he disputed the existence of the Trinity was never published. Later works – ***The Chronology of Ancient Kingdoms Amended*** (1728) and ***Observations upon the Prophecies of Daniel and the Apocalypse of St. John*** (1733) – were published after his death. He also devoted a great deal of time to alchemy.

Historian Stephen D. Snobelen says of Newton, "Isaac Newton was a heretic. But ... he never made a public declaration of his private faith — which the orthodox would have deemed extremely radical. He hid his faith so well that scholars are still unraveling his personal beliefs." Snobelen concludes that Newton was at least a Socinian sympathizer (he owned and had thoroughly read at least eight Socinian books), possibly an Arian and almost certainly an antitrinitarian . In an age notable for its religious intolerance there are few public expressions of Newton's radical views, most notably his refusal to take holy orders and his refusal, on his death bed, to take the sacrament when it was offered to him.

Although the laws of motion and universal gravitation became Newton's best-known discoveries, he warned against using them to view the Universe as a mere machine, as if akin to a great clock. He said, "Gravity explains the motions of the planets, but it cannot explain who set the planets in motion. God governs all things and knows all that is or can be done."

His scientific fame notwithstanding, Newton's studies of the Bible and of the early Church Fathers were also noteworthy. Newton wrote works on textual criticism, most notably ***An Historical Account of Two Notable Corruptions of Scripture***. He also placed the crucifixion of Jesus Christ at 3 April, AD 33, which agrees with one traditionally accepted date. He also attempted, unsuccessfully, to find hidden messages within the Bible.

In his own lifetime, Newton wrote more on religion than he did on natural science. He believed in a rationally immanent world, but he rejected the hylozoism implicit in Leibniz and Baruch Spinoza. Thus, the ordered and dynamically informed Universe could be understood, and must be understood, by an active reason. In his correspondence, Newton claimed that in writing the ***Principia*** "I had an eye upon such Principles as might work with considering men for the belief of a Deity".-He saw evidence of design in the system of the world: "Such a wonderful uniformity in the planetary system must be allowed the effect of choice". But Newton insisted that divine intervention would eventually be required to reform the system, due to the slow growth of instabilities.-For this Leibniz lampooned him: "God Almighty wants to wind up his watch from time to time: otherwise it would cease to move. He had not, it seems, sufficient foresight to make it a perpetual motion."-Newton's position was vigorously defended by his follower Samuel Clarke in a famous correspondence.

Newton and Robert Boyle's mechanical philosophy was promoted by rationalist pamphleteers as a viable alternative to the pantheists and was accepted hesitantly by orthodox preachers as well as dissident preachers like the latitudinarians. Thus, the clarity and simplicity of science was seen as a way to combat the emotional and metaphysical superlatives of both superstitious enthusiasm and the threat of atheism, and, at the same time, the second wave of English deists used Newton's discoveries to demonstrate the possibility of a "Natural Religion".

The attacks made against pre-Enlightenment "magical thinking", and the mystical elements of Christianity, were given their foundation with Boyle's mechanical conception of the Universe. Newton gave Boyle's ideas their completion through mathematical proofs and, perhaps more importantly, was very successful in popularizing them. Newton refashioned the world governed by an interventionist God into a world crafted by a God that designs along rational and universal principles. These principles were available for all people to discover, allowed people to pursue their own aims fruitfully in this life, not the next, and to perfect themselves with their own rational powers.

Newton saw God as the master creator whose existence could not be denied in the face of the grandeur of all creation. His spokesman, Clarke, rejected Leibniz' theodicy which cleared God from the responsibility for *l'origine du mal* by making God removed from participation in his creation, since as Clarke pointed out, such a deity would be a king in name only, and but one step away from atheism. But the unforeseen theological consequence of the success of Newton's system over the next century was to reinforce the deist position advocated by Leibniz. The understanding of the world was now brought down to the level of simple human reason, and humans, as Odo Marquard argued, became responsible for the correction and elimination of evil.

On the other hand, latitudinarian and Newtonian ideas taken too far resulted in the millenarians, a religious faction dedicated to the concept of a mechanical Universe, but finding in it the same enthusiasm and mysticism that the Enlightenment had fought so hard to extinguish. Enlightenment philosophers chose a short history of scientific predecessors — Galileo, Boyle, and Newton principally — as the guides and guarantors of their applications of the singular concept of Nature and Natural Law to every physical and social field of the day.

It was Newton's conception of the Universe based upon Natural and rationally understandable laws that became one of the seeds for Enlightenment ideology. Locke and Voltaire applied concepts of Natural Law to political systems advocating intrinsic rights; the physiocrats and Adam Smith applied Natural conceptions of psychology and self-interest to economic systems and the sociologists criticized the current social order for trying to fit history into Natural models of progress. Monboddo and Samuel Clarke resisted elements of Newton's work, but eventually rationalized it to conform with their strong religious views of nature.

The famous three laws of motion (stated in modernized form):

Newton's First Law (also known as the Law of Inertia) states that an object at rest tends to stay at rest and that an object in uniform motion tends to stay in uniform motion unless acted upon by a net external force.

Newton's Second Law states that an applied force, $\vec{F}$, on an object equals the rate of change of its momentum, $\vec{p}$, with time. Mathematically, this is expressed as

$$\vec{F} = \frac{d\vec{p}}{dt} = \frac{d}{dt}(m\vec{v}) = \vec{v}\frac{dm}{dt} + m\frac{d\vec{v}}{dt}.$$

Since the second law applies to an object with constant mass ($dm/dt = 0$), the first term vanishes, and by substitution using the definition of acceleration, the equation can be written in the iconic form

$$\vec{F} = m\vec{a}.$$

The first and second laws represent a break with the physics of Aristotle, in which it was believed that a force was necessary in order to maintain motion. They state that a force is only needed in order to *change* an object's state of motion. The SI unit of force is the *newton*, named in Newton's honor.

Newton's Third Law states that for every action there is an equal and opposite reaction. This means that any force exerted onto an object has a counterpart force that is exerted in the opposite direction back onto the first object. A common example is of two ice skaters pushing against each other and sliding apart in opposite directions. Another example is the recoil of a firearm, in which the force propelling the bullet is exerted equally back onto the gun and is felt by the shooter. Since the objects in question do not necessarily have the same mass, the resulting acceleration of the two objects can be different (as in the case of firearm recoil).

Unlike Aristotle's, Newton's physics is meant to be universal. For example, the second law applies both to a planet and to a falling stone.

The vector nature of the second law addresses the geometrical relationship between the direction of the force and the manner in which the object's momentum changes. Before Newton, it had typically been assumed that a planet orbiting the sun would need a forward force to keep it moving. Newton showed instead that all that was needed was an inward attraction from the sun. Even many decades after the publication of the ***Principia***, this counter-intuitive idea was not universally accepted, and many scientists preferred Descartes' theory of vortices

The question was not whether gravity existed, but whether it extended so far from Earth that it could also be the force holding the moon to its orbit. Newton showed that if the force decreased as the inverse square of the distance, one could indeed calculate the Moon's orbital period, and get good agreement. He guessed the same force was responsible for other orbital motions, and hence named it "universal gravitation".

Writings by Newton

Method of Fluxions (1671)

Of Natures Obvious Laws & Processes in Vegetation (unpublished, c. 1671–75)

De Motu Corporum in Gyrum (1684)

Philosophiae Naturalis Principia Mathematica (1687)

Opticks (1704)

Reports as Master of the Mint (1701–25)

Arithmetica Universalis (1707)

The System of the World, Optical Lectures, The Chronology of Ancient Kingdoms, (Amended) and De mundi systemate (published posthumously in 1728)

Observations on Daniel and The Apocalypse of St. John (1733)

An Historical Account of Two Notable Corruptions of Scripture (1754)

24. Thomas Paine
1737–1809

"It is the responsibility of the patriot to protect his country from its government."

"I do not believe in the creed professed by the Jewish church, by the Roman church, by the Greek Church, by the Turkish church, by the Protestant church, nor by any church that I know of. My own mind is my own church."

"All national institutions of churches, whether Jewish, Christian or Turkish, appear to me no other than human inventions, set up to terrify and enslave mankind, and monopolize power and profit."

"The most formidable weapon against errors of every kind is reason. I have never used any other and I trust I never shall."

"The Creation is the Bible of the Deist. He there reads, in the handwriting of the Creator himself, the certainty of His existence and the immutability of His power, and all other Bibles and Testaments are to him forgeries."

"Is it because you are sunk in the cruelty of superstition, or feel no interest in the honor of

Creator, that you listen to the horrid tales of the Bible, or hear them with callous indifference?"

Thomas Paine was a pamphleteer, revolutionary, radical, classical liberal and intellectual. Born in Great Britain, he lived and worked there until his late thirties. He migrated to the American colonies just in time to take part in the American Revolution. His main contribution was as the author of the powerful, widely read pamphlet, **<u>Common Sense</u>** (1776), advocating independence for the American Colonies from the Kingdom of Great Britain, and of The American Crisis, supporting the Revolution.

<u>Common Sense</u> was published anonymously on 10 January 1776, and spread quickly among literate colonists. Within three months, 120,000 copies are alleged to have been distributed throughout the colonies, which themselves totaled only four million free inhabitants, making it the best-selling work in 18th-century America. Its total sales in both America and Europe reached 500,000 copies. It convinced many colonists, including George Washington and John Adams, to seek redress in political independence from the Kingdom of Great Britain, and argued strongly against any compromise short of independence. The work was greatly influenced by the equally controversial pro-independence writer Benjamin Rush and was instrumental in bringing about the Declaration of Independence.

Later, Paine was a great influence on the French Revolution. He wrote the **<u>Rights of Man</u>** (1791) as a guide to the ideas of the Enlightenment. Despite an inability to speak French, he was elected to the French National Assembly in 1792. Regarded as an ally of the Girondists, he was seen with increasing disfavor by the Montagnards and in particular by Robespierre.

Before his arrest and imprisonment, knowing that he would likely be arrested and executed, Paine wrote the first part of **<u>The Age of Reason</u>**, an assault on organized "revealed" religion combining a compilation of inconsistencies he found in the Bible with his own advocacy of Deism. He was arrested in Paris and imprisoned in December 1793. Paine escaped execution apparent-

chance. A guard walked through the prison placing a chalk mark on the doors of the prisoners who were due to be condemned that day. He placed one on the door of the cell that Paine shared with three other prisoners, which, because Paine was ill at the time, he had asked to be left open. The prisoners in the cell then closed the door so that the chalk mark faced into the cell when they were due to be rounded up. They were overlooked, and survived the few vital days needed to be spared by the fall of Robespierre. Paine was released in November 1794.

Paine remained in France during the early Napoleonic era, but condemned Napoleon's moves towards dictatorship, calling him "the completest charlatan that ever existed." Paine remained in France until 1802, when he returned to America on an invitation from Thomas Jefferson, who had been elected president.

Paine's strength lay in his ability to present complex ideas in clear and concise form, as opposed to the more philosophical approaches of his Enlightenment contemporaries in Europe, and it was Paine who proposed the name United States of America for the new nation. When the war arrived, Paine published a series of important pamphlets, ***The Crisis,*** credited with inspiring the early colonists during the ordeals faced in their long struggle with the British. To inspire the troops, General George Washington ordered Paine's "***The American Crisis***" to be read out loud to his men. The first Crisis paper began with the famous words:

"These are the times that try men's souls: The summer soldier and the sunshine patriot will, in this crisis, shrink from the service of their country; but he that stands it now, deserves the love and thanks of man and woman. Tyranny, like hell, is not easily conquered; yet we have this consolation with us, that the harder the conflict, the more glorious the triumph. What we obtain too cheap, we esteem too lightly: it is dearness only that gives every thing its value."

– Published on 23 December 1776

He described himself as a "Deist" and commented:

"How different is Christianity to the pure and simple profession of Deism! The true Deist has but one Deity, and his religion consists in contemplating the power, wisdom, and benignity of the Deity in his works, and in endeavoring to imitate him in everything moral, scientificalui, and mechanical. The opinions I have advanced... are the effect of the most clear and long-established conviction that the Bible and the Testament are impositions upon the world, that the fall of man, the account of Jesus Christ being the Son of God, and of his dying to appease the wrath of God, and of salvation by that strange means, are all fabulous inventions, dishonorable to the wisdom and power of the Almighty; that the only true religion is Deism, by which I then meant, and mean now, the belief of one God, and an imitation of his moral character, or the practice of what are called moral virtues—and that it was upon this only (so far as religion is concerned) that I rested all my hopes of happiness hereafter. So say I now—and so help me God."

The first article published in America advocating the emancipation of slaves and the abolition of slavery was written by Paine. Titled "***African Slavery in America***," it appeared on March 8, 1775 in the Postscript to the Pennsylvania Journal and Weekly Advisor.

Thomas Paine's writings had great influence on his contemporaries, especially the American revolutionaries. His books inspired both philosophical and working-class radicals in the United Kingdom; and he is often claimed as an intellectual ancestor by United States liberals, libertarians, anarchists, progressives and radicals. Both Abraham Lincoln and Thomas Edison read his works with respect.

Edison said of Paine:

"I have always regarded Paine as one of the greatest of all Americans. Never have we had a sounder intelligence in this republic... It was my good fortune to encounter Thomas Paine's works in my boyhood... it was, indeed, a revelation to me to read that great thinker's views on political and theological subjects. Paine educated me then about many matters of which I had never before thought. I remember very vividly the flash of enlightenment that shone from Paine's writings and I recall thinking at that time, 'What a pity these works are not today the schoolbooks for all children!' My interest in Paine was not satisfied by my first reading of his works. I went back to them time and again, just as I have done since my boyhood days."

Thomas Paine is the best known and perhaps the best spokesman for modern Deism. He is a hero by every measure.

25. Voltaire
François-Marie Arouet
1694–1778

"Those who can make you believe absurdities can make you commit atrocities."

"Paper money eventually reaches its intrinsic value-zero"

Voltaire was a French Enlightenment writer, essayist, deist and philosopher known for his wit, philosophical sport, and defense of civil liberties, including freedom of religion and the right to a fair trial. He was an outspoken supporter of social reform despite strict censorship laws and harsh penalties for those who broke them. A satirical polemicist, he frequently made use of his works to criticize Christian Church dogma and the French institutions of his day.

Many of Voltaire's works and ideas would influence important thinkers of both the American and French Revolutions, an honor that he would share with other political theorists such as John Locke.

Voltaire was educated by Jesuits at the Collège Louis-le-Grand where he learned Latin and Greek; later in life he became fluent in Italian, Spanish, and English. From 1711 to 1713 he studied law. In 1726, Voltaire insulted the powerful young nobleman, "Chevalier De Rohan," and was given two options: imprisonment or exile. He chose exile and from 1726 to 1729 lived in England.

While in England Voltaire was attracted to the philosophy of John Locke and ideas of mathematician and scientist, Sir Isaac Newton. He studied England's Constitutional Monarchy and its religious tolerance. Voltaire was particularly interested in the philosophical rationalism of the time, and in the study of the natural sciences. After returning to Paris he wrote a book praising English customs and institutions. It was interpreted as criticism of the French government and in 1734; Voltaire was forced to leave Paris again.

Most of Voltaire's early life revolved around Paris until his exile. From the beginning Voltaire had trouble with the authorities for his energetic attacks on the government and the Catholic Church. These activities were to result in numerous imprisonments and exiles. One of his writings, about Louis XV's regent, Philippe II, Duke of Orléans, led to his being imprisoned in the Bastille. While there, he wrote his debut play, ***Edipe,*** and adopted the name Voltaire which came from his hometown in southern France. ***Edipe***'s success began Voltaire's influence and brought him into the French Enlightenment. Voltaire was a prolific writer, and produced works in almost every literary form, authoring plays, poetry, novels, essays, historical and scientific works, over 20,000 letters and over two thousand books and pamphlets.

Many of Voltaire's prose works and romances, usually composed as pamphlets, were written as polemics. ***Candide*** attacks religious and philosophical optimism, certain social and political ways of the time, the received forms of moral and metaphysical orthodoxy, and some were written to deride the Bible. In these works, Voltaire's ironic style is apparent, particularly the restraint and simplicity of the verbal treatment.

Voltaire's works, especially his private letters, frequently contain the word "l'infâme" and the expression "écrasez l'infâme, or "crush the infamy". The phrase refers to abuses of the people by royalty and the clergy that Voltaire saw around him. He had felt these effects in his own exiles, and in the confiscations of his books.

Voltaire's largest philosophical work is the ***Dictionnaire Philosophique,*** comprising articles contributed by him to the ***Encyclopédie*** and several minor pieces. It directed criticism at French political institutions, Voltaire's personal enemies, the Bible, and the Roman Catholic Church.

Like many other key figures during the European Enlightenment, Voltaire considered himself a Deist. He did not believe that absolute faith, based upon any particular or singular religious text or tradition of revelation, was needed to believe in God. In fact, Voltaire's focus instead on the idea of a universe based on reason and a respect for nature reflected the contemporary Pantheism, increasingly popular throughout the seventeenth and eighteenth centuries and which continues in a form of deism today known as "Voltairean Pantheism."

In terms of religious texts, Voltaire was largely of the opinion that the Bible was 1) an outdated legal and/or moral reference, 2) by and large a metaphor, but one that still taught some good lessons, and 3) a work of Man, not a divine gift. These beliefs did not hinder his religious practice, however, though it did gain him somewhat of a bad reputation in the Catholic Church. It may be noted that Voltaire was indeed seen as somewhat of a nuisance to many believers.

From translated works on Confucianism and Legalism, Voltaire drew on Chinese concepts of politics and philosophy - which were based on rational principles, to look critically at European organized religion and hereditary aristocracy.

Voltaire also displayed, as part of his ***Dictionnaire Philosophique,*** an inclination towards the ideas of Hinduism and the works of Brahmin priests, asking, "Is it not probable that the Brahmins were the first legislators of the earth, the first philosophers, the first theologians?" His attitudes towards religious institutions are further shown in the criticisms he made of Christian missionaries in India.

Voltaire perceived the French bourgeoisie to be too small and ineffective, the aristocracy to be parasitic and corrupt, the commoners as ignorant and superstitious, and the church as a static force useful only as a counterbalance since its "religious tax" or the tithe helped to create a strong backing for revolutionaries.

He is best known today for his novel, ***Candide*** which satirized the philosophy of Leibniz. ***Candide*** was also subject to censorship and Voltaire jokingly claimed that the actual author was a certain "Dr DeMad" in a letter, where he reaffirmed the main polemical stances of the text.

Voltaire is also known for many memorable aphorisms, such as:"If God did not exist, it would be necessary to invent him", contained in a verse epistle from 1768, addressed to the anonymous author of a controversial work, ***The Three Impostors***.

Voltaire is remembered and honored in France as a courageous polemicist who indefatigably fought for civil rights — the right to a fair trial and freedom of religion — and who denounced the hypocrisies and injustices of the ancien régime. The ancien régime involved an unfair balance of power and taxes between the First Estate (the clergy), the Second Estate (the nobles), and the Third Estate (the commoners and middle class, who were burdened with most of the taxes).

We can never have too many Voltaire's!

26. David Hume
1711 –1776

Hume was an 18th-century Scottish philosopher, economist, and historian, considered among the most important figures in the history of Western philosophy and the Scottish Enlightenment.

He first gained recognition and respect as a historian. His ***History of England*** was the standard work on English history for sixty or seventy years until Macaulay's. Interest in Hume's work has in recent years centered on his philosophical writing.

Hume was the first great philosopher of the modern era to carve out a thoroughly naturalistic philosophy. This philosophy partly consisted in the rejection of the historically prevalent concept of human minds as being miniature versions of the Divine mind; a notion Edward Craig called the 'Image of God' doctrine. This doctrine was associated with a trust in the powers of human reason and insight into reality. Hume's skepticism came in his rejection of this 'insight ideal', and the confidence derived from it that the world is as we represent it. Instead, the best we can do Hume explained in '***Science of Man***' is to apply the best explanatory and empirical principles available to the investigation of human mental phenomena.

Hume completed ***A Treatise of Human Nature*** at the age of twenty-six. Many scholars today consider the ***Treatise*** to be Hume's most important work and one of the most important books in the history of philosophy,

After the publication of ***Essays Moral and Political*** in 1744, he applied for the Chair of Pneumatics and Moral Philosophy at the University of Edinburgh. However, the position was given to William Cleghorn, after the majority of Edinburgh ministers petitioned the town council not to appoint Hume because of his atheism. He then wrote ***Philosophical Essays concerning Human Understanding.***

Hume was then charged with heresy, but he was defended by his young clerical friends who argued that as an atheist he lay outside the jurisdiction of the Church. Despite his acquittal—and, possibly, due to the opposition of Thomas Reid of Aberdeen, who that year launched a Christian critique of his metaphysics—Hume failed to gain the Chair of Philosophy at the University of Glasgow. It was after returning to Edinburgh in 1752, as he wrote in ***My Own Life***, that "the Faculty of Advocates chose me their Librarian, an office from which I received little or no emolument, but which gave me the command of a large library." It was this resource that enabled him to continue his historical research for his book ***The History of Great Britain***.

Hume achieved great literary fame as a historian. His enormous **_History of Great Britain_** from the Saxon kingdoms to the Glorious Revolution was a best-seller in its day

Hume's early essay **_Of Superstition and Religion_** laid the foundations for nearly all subsequent secular thinking about the history of religion. Critics of religion during Hume's time were required to express themselves cautiously. Less than 15 years before Hume was born, 18-year-old college student Thomas Aikenhead was put on trial for saying openly that he thought Christianity was nonsense; he was later convicted and hanged for blasphemy. Hume followed the common practice of expressing his views obliquely, through characters in dialogues. Hume did not acknowledge authorship of **_Treatise_** until the year of his death, in 1776. His essays **_On Suicide,_** and **_On the Immortality of the Soul_** and his **_Dialogues concerning Natural Religion_** were held from publication until after his death, and they still bore neither author's nor publisher's name. So masterly was Hume in disguising his own views that debate continues to this day over whether Hume was actually a deist or an atheist. Regardless, in his own time Hume's alleged atheism caused him to be passed over for many positions.

Hume's solution to the problem of explaining our inductions is Nature. Nature has determined us to expect more of the same, for: "this operation of the mind, by which we infer like effects from like causes, and vice versa, is so essential to the subsistence of all human creatures, it is not probable, that it could be trusted to the fallacious deductions of our reason, which is slow in its operations; appears not, in any degree, during the first years of infancy; and at best is, in every age and period of human life, extremely liable to error and mistake." This is the closest thing possible during his (pre-Darwinian) time to an evolutionary account of our inductive tendencies, and Hume here has lit on a central feature in any properly naturalistic Science of Man.

Hume advocated a moral theory based on human actions and human responsibility rather than metaphysics and religion. As part of his theory, Hume created historically influential arguments for compatibilism, the idea that causal determinism is compatible with free will.

In opposition to Christian thinkers who argued that in order for a person to be morally responsible, his actions must not be determined by any physical cause, Hume wrote that moral responsibility requires determinism: Hume argued that it makes sense to hold a person responsible for an action only if the action was caused by his character, rather than by random events or external compulsion.

Hume said that moral responsibility requires an action to be (1) caused by the person's character (e.g. stealing a bag of cash because you don't care about the victim), and (2) not caused by external compulsion or force (e.g. stealing a bag of cash because a man with a gun forces you to). In line with this, Hume defines free will to be when one gets to act according to one's character. Hume said that thinkers who believe moral responsibility requires actions to be uncaused are mistakenly thinking of the first kind of cause (character) as being the same as the second kind (compulsion), and that Hume's theory of causation makes the situation clear.

For Hume, the only way to support theistic religion beyond strict fideism is by an appeal to miracles saying, in ***On Miracles*** "...we may conclude, that the Christian religion not only was first attended with miracles, but even at this day cannot be believed by any reasonable person without one. Mere reason is insufficient to convince us of its veracity: and whoever is moved by faith to assent to it, is conscious of a continued miracle in his own person, which subverts all the principles of his understanding, and gives him a determination to believe what is most contrary to custom and experience."

Hume argued that, at minimum, miracles could never give religion much support. There are several arguments suggested by Hume's essay, all of which turn on his conception of a miracle: namely, a violation of the laws of nature. His very definition of miracles from his ***An Enquiry concerning Human Understanding*** states that miracles are violations of the laws of nature and consequently have a very low probability of occurring. In a slogan, extraordinary claims require extraordinary evidence. But far from that, Hume observes, "The gazing populace receive greedily, without examination, whatever soothes superstition and promotes wonder."

Another claim is his argument that human testimony could never be reliable enough to countermand the evidence we have for the laws of nature. This point on miracles has mostly been applied to the question of the resurrection of Jesus, where Hume would no doubt ask, "Which is more likely – that a man rose from the dead or that this testimony is mistaken in some way?"

One of the oldest and most popular arguments for the existence of God is the design argument – that all the order and 'purpose' in the world bespeaks a divine origin. A modern manifestation of this belief is creationism. Hume gave the classic criticism of the design argument in ***Dialogues concerning Natural Religion*** and here are some of his points.

"For the design argument to be feasible, it must be true that order and purpose are observed only when they result from design. But order is observed regularly, resulting from presumably mindless processes like snowflake or crystal generation. Design accounts for only a tiny part of our experience with order and "purpose".

We must ask therefore if it is right to compare the world to a machine — as in Paley's watchmaker argument — when perhaps it would be better described as a giant inert animal.

Even if the design argument is completely successful, it could not (in and of itself) establish a robust theism; one could easily reach the conclusion that the universe's configuration is the result of some morally ambiguous, possibly unintelligent agent or agents whose method bears only a remote similarity to human design. In this way it could be asked if the designer was God, or further still, who designed the designer?

If a well-ordered natural world requires a special designer, then God's mind (being so well-ordered) also requires a special designer. And then this designer would likewise need a designer, and so on ad infinitum. We could respond by resting content with an inexplicably self-ordered divine mind but then why not rest content with an inexplicably self-ordered natural world?

Often, what appears to be purpose, where it looks like object X has feature F in order to secure some outcome O, is better explained by a filtering process: that is, object X wouldn't be around did it not possess feature F, and outcome O is only interesting to us as a human projection of goals onto nature. This mechanical explanation of teleology anticipated natural selection. The design argument does not explain pain, suffering, and natural disasters.

Hume is one of our hero's because he had the courage to ask important, difficult questions. He also offered some very good answers!

27. Adam Smith
1723–1790

Smith was a Scottish moral philosopher and a pioneering political economist. One of the key figures of the intellectual movement known as the Scottish Enlightenment, he is known primarily as the author of two treatises: ***The Theory of Moral Sentiments*** (1759), and ***An Inquiry into the Nature and Causes of the Wealth of Nations*** (1776). The latter was one of the earliest attempts to systematically study the historical development of industry and commerce in Europe, as well as a sustained attack on the doctrines of mercantilism. Smith's work helped to create the modern academic discipline of economics and provided one of the best-known intellectual rationales for free trade, capitalism, and libertarianism.

In about 1750 he met the philosopher David Hume, who was his senior by over a decade. The alignments of opinion that can be found within the details of their respective writings covering history, politics, philosophy, economics, and religion indicate that they both shared a close intellectual alliance and friendship.

There has been considerable scholarly debate about the nature of Adam Smith's religious views. At Oxford, Smith rejected Christianity and it is generally believed that he returned to Scotland as a Deist.

The Wealth of Nations was Smith's most influential work, and is considered to be very important in the creation of the field of economics and its development into an autonomous systematic discipline. In the Western world, it is arguably the most influential book on the subject ever published. When the book, which has become a classic manifesto against mercantilism (the theory that large reserves of bullion are essential for economic success), appeared in 1776, there was a strong sentiment for free trade in both Britain and America. This new feeling had been born out of the economic hardships and poverty caused by the American War of Independence. However, at the time of publication, not everybody was immediately convinced of the advantages of free trade: the British public and Parliament still clung to mercantilism.

The Wealth of Nations also rejects the Physiocratic school's emphasis on the importance of land; instead, Smith believed labor was paramount, and that a division of labor would affect a great increase in production. One example he used was the making of pins. One worker could probably make only twenty pins per day. But if ten people divided up the eighteen steps required to make a pin, they could make a combined amount of 48,000 pins in one day. However, it is less well known that Smith also concluded that excessive division of labor would lead man to his most ignorant state possible. Nations was so successful, in fact, that it led to the abandonment of earlier economic schools, and later economists, such as Thomas Malthus and David Ricardo, focused on refining Smith's theory into what is now known as classical economics. Both Modern economics and, separately, Marxian economics owe significantly to classical economics. Malthus expanded Smith's ruminations on overpopulation, while Ricardo believed in the "iron law of wages" — that overpopulation would prevent wages from topping the subsistence level.

Smith postulated an increase of wages with an increase in production, a view considered more accurate today.

One of the main points of ***The Wealth of Nations*** is that the free market, while appearing chaotic and unrestrained, is actually guided to produce the right amount and variety of goods by a so-called "invisible hand" (an image that Smith had previously employed in ***Theory of Moral Sentiments***, but which has its original use in his essay, "***The History of Astronomy***"). If a product shortage occurs, for instance, its price rises, creating a profit margin that creates an incentive for others to enter production, eventually curing the shortage. If too many producers enter the market, the increased competition among manufacturers and increased supply would lower the price of the product to its production cost, the "natural price". Even as profits are zeroed out at the "natural price," there would be incentives to produce goods and services, as all costs of production, including compensation for the owner's labor, are also built into the price of the goods. If prices dip below a zero profit, producers would drop out of the market; if they were above a zero profit, producers would enter the market. Smith believed that while human motives are often selfishness and greedy, the competition in the free market would tend to benefit society as a whole by keeping prices low, while still building in an incentive for a wide variety of goods and services. Nevertheless, he was wary of businessmen and argued against the formation of monopolies.

Smith vigorously attacked the antiquated government restrictions which he thought were hindering industrial expansion. In fact, he attacked most forms of government interference in the economic process, including tariffs, arguing that this creates inefficiency and high prices in the long run. This theory, now referred to as "laissez-faire", which means "let them do" or more relevant to the study of economics, "let the market set supply and demand with no interference". It is believed that this theory influenced government legislation in later years, especially during the 19th century.

Two of the most famous and often-quoted passages in ***The Wealth of Nations*** are:

"It is not from the benevolence of the butcher, the brewer, or the baker that we expect our dinner, but from their regard to their own interest. We address ourselves, not to their humanity but to their self-love, and never talk to them of our own necessities but of their advantages.

As every individual, therefore, endeavors as much as he can both to employ his capital in the support of domestic industry, and so to direct that industry that its produce may be of the greatest value; every individual necessarily labors to render the annual value of society as great as he can. He generally, indeed, neither intends to promote the public interest, nor knows how much he is promoting it. By preferring the support of domestic to that of foreign industry, he intends only his own security; and by directing that industry in such a manner as its produce may be of the greatest value, he intends only his own gain, and he is in this, as in many other cases, led by an invisible hand to promote an end which was no part of his intention. Nor is it always the worse for the society that it was no part of it. By pursuing his own interest he frequently promotes that of society more effectually than when he really intends to promote it. I have never known much good done by those who affected to trade for the public good. It is an affectation, indeed, not very common among merchants, and very few words need be employed in dissuading them from it."

Another favorite quote, usually recited by economists, also from ***The Wealth of Nations*** is:

"People of the same trade seldom meet together, even for merriment and diversion, but the conversation ends in a conspiracy against the public, or in some contrivance to raise prices. It is impossible indeed to prevent such meetings, by any law which either could be executed, or would be consistent with liberty and justice. But though the law cannot hinder people of the same trade from sometimes assembling together, it ought to do nothing to facilitate such assemblies; much less to render them necessary."

In the ***Wealth of Nations*** Smith claims that self-interest alone can lead to socially beneficial results. But in his ***Theory of Moral Sentiments*** Smith argues that sympathy is required to achieve socially beneficial results. On the surface it appears that a contradiction exists. Adam Smith himself cannot have seen any contradiction, since he produced a revised edition of ***Moral Sentiments*** after the publication of ***Wealth of Nations***. Both sets of ideas are to be found in his ***Lectures on Jurisprudence***. In recent years most students of Adam Smith's work have argued that no contradiction exists. In the ***Theory of Moral Sentiments***, Smith develops a theory of psychology in which individuals in society find it in their self-interest to develop sympathy as they seek approval of what he calls the "impartial spectator." The self-interest he speaks of is not a narrow selfishness but something that involves sympathy.

In any case, Adam Smith apparently believed that moral sentiments and self-interest would always add up to the same thing. One possible line of reasoning he might have employed in reaching this conclusion is as follows: the invisible hand cannot operate if there is no society, for precluding a societal construct precludes division of labor, and thus, the efficiency which comes with its manifestation. Now for society to exist, justice is a necessary condition (as pointed out in Smith's ***Theory of Moral Sentiments***). For justice to exist in any social setting, individuals must harbor the passions of gratitude and resentment governed by a sense of 'merit' and 'demerit' (again from Smith's ***Theory of Moral Sentiments***). And finally, as Smith himself would have so vehemently argued, the sense of 'merit' and 'demerit' is almost exclusively engendered by human sympathy. In conclusion, the invisible hand of the market is, at some level, contingent upon the ability of humans to sympathize: Smith's self-interest is indeed in consonance with the notion of sympathy.

On June 25, 2006, when Warren Buffet announced that he would donate his wealth to The Bill and Melinda Gates Foundation, he was presented with a copy of Adam Smith's ***Wealth of Nations*** by Bill Gates.

28. Jacques Rousseau
1712 -1778

Rousseau was a philosopher, literary figure, and composer of the Enlightenment whose political philosophy influenced the French Revolution and the development of liberal, conservative and socialist theory. With his ***Confessions, Reveries of a Solitary Walker*** and other writings, he invented modern autobiography and encouraged a new focus on the building of subjectivity that bore fruit in the work of thinkers as diverse as Hegel and Freud. His novel ***Julie, ou la nouvelle Héloïse*** was one of the best-selling fictional works of the eighteenth century and was important to the development of romanticism. Rousseau made important contributions to music as a theorist and a composer.

In 1749, as Rousseau was walking to visit Diderot in a Vincennes prison, he read an essay competition entry sponsored by the Académie de Dijon, named the Mercure de France. The work asked whether the development of the arts and sciences had been morally beneficial. Rousseau said this question caused him to immediately perceive the principle of the natural goodness of humanity on which all his later philosophical works were based. He answered the competition question in the negative, in his 1750 "***Discourse on the Arts and Sciences***", which won him first prize and gained him significant fame.

Rousseau returned to Geneva where he reconverted to Calvinism and regained his official Genevan citizenship in 1754. In 1755, Rousseau completed his second major work, the ***Discourse on the Origin and Basis of Inequality Among Men*** (the ***Discourse on Inequality***).

In 1762, he published two major books ***Of the Social Contract, Principles of Political Right*** and then ***Émile, or On Education*** in May. The books criticized religion and were banned in France and Geneva. Rousseau was forced to flee arrest and made stops in Bern and Môtiers in Switzerland, where he enjoyed the protection of Frederick the Great of Prussia .While in Môtiers, Rousseau wrote the ***Constitutional Project for Corsica.***

His house in Motiers was stoned on the night of September 6, 1765 – he took refuge with the philosopher David Hume in Great Britain. Isolated at Wootton on the borders of Derbyshire and Staffordshire, Rousseau suffered a serious decline in his mental health and began to experience paranoid fantasies about plots against him involving Hume and others. Rousseau's letter to Hume, in which he articulates the perceived misconduct, sparked an exchange which was published in and received with great interest in contemporary Paris.

While he was not allowed to return to France before 1770, Rousseau returned under the name "Renou," in 1767. As a condition of his return, he was not allowed to publish any books, but after completing his ***Confessions,*** Rousseau began private readings in 1771. At the request of Madame d'Epinay the police ordered him to stop, and the ***Confessions*** was only partially published in 1782, four years after his death. All his subsequent works were only to appear posthumously.

In 1772, he was invited to present recommendations for a new constitution for the Polish-Lithuanian Commonwealth, resulting in the ***Considerations on the Government of Poland,*** which was to be his last major political work. In 1776 he completed ***Dialogues: Rousseau Judge of Jean-Jacques*** and began work on the ***Reveries of the Solitary Walker***

Theory of Natural Man

Rousseau saw a fundamental divide between society and human nature. Rousseau believed that man was good when in the state of nature (the state of all other animals, and the condition humankind was in before the creation of civilization and society), but is corrupted by society. This idea has often led to attributing the idea of the noble savage to Rousseau, an expression first used by John Dryden in ***The Conquest of Granada***(1672). Rousseau, however, never used the expression himself and it does not adequately render his idea of the natural goodness of humanity. Rousseau's idea of natural goodness is complex and easy to misunderstand. Contrary to what might be suggested by a casual reading, the idea does not imply that humans in the state of nature act morally; in fact, terms such as 'justice' or 'wickedness' are simply inapplicable to pre-political society as Rousseau understands it. Humans there may act with all of the ferocity of an animal. They are good because they are self-sufficient and thus not subject to the vices of political society. He viewed society as artificial and held that the development of society, especially the growth of social interdependence, has been inimical to the well-being of human beings.

In Rousseau's philosophy, society's negative influence on men centers on its transformation of amour de soi, a positive self-love, into amour-propre, or pride. Amour de soi represents the instinctive human desire for self-preservation, combined with the human power of reason. In contrast, amour-propre is artificial and forces man to compare himself to others, thus creating unwarranted fear and allowing men to take pleasure in the pain or weakness of others. Rousseau was not the first to make this distinction; it had been invoked by, among others, Vauvenargues.

In "**_Discourse on the Arts and Sciences_**" Rousseau argued that the arts and sciences had not been beneficial to humankind because they were not human needs, but rather a result of pride and vanity. Moreover, the opportunities they created for idleness and luxury contributed to the corruption of man. He proposed that the progress of knowledge had made governments more powerful and had crushed individual liberty. He concluded that material progress had actually undermined the possibility of true friendship by replacing it with jealousy, fear and suspicion.

His subsequent **_Discourse on Inequality_** tracked the progress and degeneration of mankind from a primitive state of nature to modern society. He suggested that the earliest human beings were solitary and differentiated from animals by their capacity for free will and their perfectibility. He also argued that these primitive humans were possessed of a basic drive to care for themselves and a natural disposition to compassion or pity. As humans were forced to associate together more closely by the pressure of population growth, they underwent a psychological transformation and came to value the good opinion of others as an essential component of their own well-being. Rousseau associated this new self-awareness with a golden age of human flourishing. However, the development of agriculture, metallurgy, private property, and the division of labor led to humans becoming increasingly dependent on one another, and led to inequality. The resulting state of conflict led Rousseau to suggest that the first state was invented as a kind of social contract made at the suggestion of the rich and powerful. This original contract was deeply flawed as the wealthiest and most powerful members of society tricked the general population, and thus instituted inequality as a fundamental feature of human society. Rousseau's own conception of the social contract can be understood as an alternative to this fraudulent form of association. At the end of the ***Discourse on Inequality***, Rousseau explains how the desire to have value in the eyes of others, which originated in the golden age, comes to undermine personal integrity and authenticity in a society marked by interdependence, hierarchy, and inequality.

The Social Contract

Perhaps Rousseau's most important work is ***The Social Contract***, which outlines the basis for a legitimate political order within a framework of classical republicanism. Published in 1762, it became one of the most influential works of political philosophy in the Western tradition. It developed some of the ideas mentioned in an earlier work, the article "***Economie Politique***", featured in Diderot's ***Encyclopédie***. The treatise begins with the dramatic opening lines, "Man is born free, and everywhere he is in chains. One man thinks himself the master of others, but remains more of a slave than they." Rousseau claimed that the state of nature was a primitive condition without law or morality, which human beings left for the benefits and necessity of cooperation. As society developed, division of labor and private property required the human race to adopt institutions of law. In the degenerate phase of society, man is prone to be in frequent competition with his fellow men while at the same time becoming increasingly dependent on them. This double pressure threatens both his survival and his freedom. According to Rousseau, by joining together into civil society through the social contract and abandoning their claims of natural right, individuals can both preserve themselves and remain free. This is because submission to the authority of the general will of the people as a whole guarantees individuals against being subordinated to the wills of others and also ensures that they obey themselves because they are, collectively, the authors of the law.

While Rousseau argues that sovereignty should be in the hands of the people, he also makes a sharp distinction between sovereignty and government. The government is charged with implementing and enforcing the general will and is composed of a smaller group of citizens, known as magistrates. Rousseau was bitterly opposed to the idea that the people should exercise sovereignty via a representative assembly. Rather, they should make the laws directly. It was argued that this would prevent Rousseau's ideal State from being realized in a large society, such as France was at the time. Much of the subsequent controversy about Rousseau's work has hinged on disagreements concerning his claims that citizens constrained to obey the general will are thereby rendered free.

Education

Rousseau set out his views on education in *Émile*, a semi-fictitious work detailing the growth of a young boy of that name, presided over by Rousseau himself. He brings him up in the countryside, where, he believes, humans are most naturally suited, rather than in a city, where we only learn bad habits, both physical and intellectual. The aim of education, Rousseau says, is to learn how to live righteously. This is accomplished by following a guardian who can guide his pupil through various contrived learning experiences.

Religion

Rousseau was most controversial in his own time for his views on religion. His view that man is good by nature conflicts with the doctrine of Original Sin and his theology of nature expounded by the Savoyard Vicar in *Émile* led to the condemnation of the book in both Calvinist Geneva and Catholic Paris. In the *Social Contract* he claims that true followers of Jesus would not make good citizens. This was one of the reasons for the book's condemnation in Geneva.

In his main writings, Rousseau identifies nature with the primitive state of savage man. Later he took nature to mean the spontaneity of the process by which man builds his egocentric, instinct based character and his little world. Nature thus signifies interiority and integrity, as opposed to that imprisonment and enslavement which society imposes in the name of progressive emancipation from cold-hearted brutality.

Hence, to go back to nature means to restore to man the forces of this natural process, to place him outside every oppressing bond of society and the prejudices of civilization. It is this idea that made his thought particularly important in Romanticism, though Rousseau himself is sometimes regarded as a figure of The Enlightenment.

You will immediately recognize the debt I owe to Rousseau.

29. Benjamin Franklin
1706 –1790

"How exact and regular is everything in the natural world! How wisely in every part contriv'd! We cannot here find the least defect! Those who have studied the mere animal and vegetable creation, demonstrate that nothing can be more harmonious and beautiful! All the heavenly bodies, the stars and planets, are regulated with the utmost Wis-

dom! And can we suppose less care to be taken in the order of the moral than in the natural system?"

Franklin was one of the most important and influential Founding Fathers of the United States. He was a leading author, political theorist, politician, printer, scientist, inventor, civic activist, and diplomat. As a scientist he was a major figure in the history of physics for his discoveries and theories regarding electricity. As a political writer and activist he, more than anyone, invented the idea of an American nation, and as a diplomat during the American Revolution, he secured the French alliance that helped to make independence possible.

Franklin was famous for his curiosity, his writings (popular, political and scientific), his inventions, and his diversity of interests. As a leader of the Enlightenment, he gained the recognition of scientists and intellectuals across Europe. An agent in London before the Revolution, and Minister to France during the war, he, more than anyone else, defined the new nation in the minds of Europe. His success in securing French military and financial aid was a great contributor to the American victory over Britain. He invented the lightning rod, bifocals, and the iron furnace Stove. He was an early proponent of colonial unity. Many historians hail him as the "First American."

Franklin became a national hero in America when he spearheaded the effort to have Parliament repeal the unpopular Stamp Act. An accomplished diplomat, he was widely admired among the French as American minister to Paris and was a major figure in the development of positive Franco-American relations. From 1775 to 1776, Franklin was Postmaster General under the Continental Congress and from 1785 to 1788 was President of the Supreme Executive Council of Pennsylvania. Toward the end of his life, he became one of the most prominent abolitionists.

In 1773, Franklin published two of his most celebrated pro-American satirical essays: "***Rules by Which a Great Empire May***

Be Reduced to a Small One", and "***An Edict by the King of Prussia***".

While in England Franklin obtained private letters of Massachusetts governor Thomas Hutchinson and lieutenant governor Andrew Oliver which proved they were encouraging London to crack down on the rights of the Bostonians. Franklin sent them to America where they escalated the tensions. Franklin now appeared to the British as the fomenter of serious trouble. Hopes for a peaceful solution ended as he was systematically ridiculed and humiliated by the Privy Council. He left London in March 1775.

By the time Franklin arrived in Philadelphia on May 5, the American Revolution had begun with fighting at Lexington and Concord. The New England militia had trapped the main British army in Boston. The Pennsylvania Assembly unanimously chose Franklin as their delegate to the Second Continental Congress. In 1776, he was a member of the Committee of Five that drafted the Declaration of Independence and made several small changes to Thomas Jefferson's draft.

In December 1776, Franklin was dispatched to France as commissioner for the United States. He conducted the affairs of his country towards the French nation with great success, which included securing a critical military alliance in 1778 and negotiating the Treaty of Paris (1783).

When he finally returned home in 1785, Franklin occupied a position only second to that of George Washington as the champion of American independence. Franklin became an abolitionist, freeing both of his slaves. He eventually became president of Pennsylvania Abolition Society.

Franklin was the only Founding Father who is a signatory of all four of the major documents of the founding of the United States: the Declaration of Independence, the Treaty of Paris, the Treaty of Alliance with France, and the United States Constitution.

Like the other advocates of republicanism, Franklin emphasized that the new republic could survive only if the people were virtuous in the sense of attention to civic duty and rejection of corruption. All his life he had been exploring the role of civic and personal virtue, as expressed in Poor Richard's aphorisms. Although Franklin's parents had intended for him to have a career in the church, Franklin became disillusioned with organized religion after discovering Deism.

"I soon became a thorough Deist." He went on to attack Christian principles of free will and morality in a 1725 pamphlet, "A Dissertation on Liberty and Necessity, Pleasure and Pain". He consistently attacked religious dogma, arguing that morality was more dependent upon virtue and benevolent actions than on strict obedience to religious orthodoxy: "I think opinions should be judged by their influences and effects; and if a man holds none that tend to make him less virtuous or more vicious, it may be concluded that he holds none that are dangerous, which I hope is the case with me."

A few years later, Franklin repudiated his 1725 pamphlet as an embarrassing "erratum." In 1790, just about a month before he died, Franklin wrote the following in a letter to Ezra Stiles, president of Yale, who had asked him his views on religion...:

> "As to Jesus of Nazareth, my Opinion of whom you particularly desire, I think the System of Morals and his Religion, as he left them to us, the best the world ever saw or is likely to see; but I apprehend it has received various corrupt changes, and I have, with most of the present Dissenters in England, some Doubts as to his divinity; tho' it is a question I do not dogmatize upon, having never studied it, and I think it needless to busy myself with it now, when I expect soon an Opportunity of knowing the Truth with less Trouble"

Like most Enlightenment intellectuals, Franklin separated virtue, morality, and faith from organized religion, although he felt that if religion in general grew weaker, morality, virtue, and society in general would also decline. Thus he wrote Thomas Paine,

"If men are so wicked with religion, what would they be if without it." According to David Morgan, Franklin was a proponent of all religions. He prayed to "Powerful Goodness" and referred to God as the "INFINITE." John Adams noted that Franklin was a mirror in which people saw their own religion: "The Catholics thought him almost a Catholic. The Church of England claimed him as one of them. The Presbyterians thought him half a Presbyterian, and the Friends believed him a wet Quaker.

"Whatever else Benjamin Franklin was, concludes Morgan, "he was a true champion of generic religion." Ben Franklin was noted to be "the spirit of the Enlightenment."

30. Thomas Jefferson
1743-1826

"For here we are not afraid to follow truth wherever it may lead, nor to tolerate any error so long as reason is left free to combat it."

"May(the American Revolution) be to the world, what I believe it will be, the signal of arousing men to burst the chains under which monkish ignorance and superstition had persuaded them to bind themselves."

"The future inhabitants of [both] the Atlantic and Mississippi states will be our sons. We think we see their happiness in their union, and we wish it. Events may prove otherwise; and if they see their interest in separating why should we take sides? God bless them both, and keep them in union if it be for their good, but separate them if it be better."

> *"Your sect (Judaism) by its sufferings has furnished a remarkable proof of the universal spirit of religious intolerance inherent in every sect, disclaimed by all while feeble, and practiced by all when in power. Our laws have applied the only antidote to this vice, protecting all on an equal footing. But more remains to be done, for although we are free by law, we are not so in practice; public opinion erects itself into an Inquisition, and exercises its offices with as much fanaticism as fans the flames of an Auto-da-Fe."*

Thomas Jefferson was the third President of the United States (1801–1809), the principal author of the ***Declaration of Independence*** (1776), and one of the most influential Founding Fathers for his promotion of the ideals of Republicanism in the United States. As a political philosopher, Jefferson was a man of the Enlightenment and knew many intellectual leaders in Britain and France. He idealized the independent yeoman farmer as exemplar of republican virtues, distrusted cities and financiers, and favored states' rights and a strictly limited federal government. Jefferson supported the separation of church and state and was the author of the ***Virginia Statute for Religious Freedom*** (1779, 1786). He was the eponym of Jeffersonian democracy and the co-founder and leader of the Democratic-Republican Party, which dominated American politics for a quarter-century and was the precursor of the modern-day Democratic Party. Jefferson served as the wartime Governor of Virginia (1779–1781), first United States Secretary of State (1789–1793) and second Vice President (1797–1801).

Jefferson achieved distinction as, among other things, a horticulturist, statesman, architect, archaeologist, paleontologist, author, inventor and founder of the University of Virginia.

In 1774, Jefferson wrote, "***A Summary View of the Rights of British America***" which was intended as instructions for the Virginia delegates to a national congress. The pamphlet was a powerful argument of American terms for a settlement with Britain. It helped speed the way to independence, and marked Jefferson as one of the most thoughtful patriot spokesmen. He was appointed by the Continental Congress of the United Colonies to the Committee of Five to prepare a draft of the proposed Declaration of Independence. The committee assigned Thomas Jefferson the task of producing a draft Declaration for its consideration.

Jefferson's republican political principles were heavily influenced by the Country Party of 18th century British opposition writers. He was influenced by John Locke (particularly relating to the principle of inalienable rights). Historians find few traces of any influence by his French contemporary, Jean-Jacques Rousseau.

His opposition to the Bank of the United States was fierce: "I sincerely believe, with you, that banking establishments are more dangerous than standing armies; and that the principle of spending money to be paid by posterity, under the name of funding, is but swindling futurity on a large scale." Nevertheless Madison and Congress, seeing the financial chaos caused by the lack of a national bank in the War of 1812, disregarded his advice and created the Second Bank of the United States in 1816.

Jefferson believed that each individual has "certain inalienable rights." That is, these rights exist with or without government; man cannot create, take, or give them away. It is the right of "liberty" on which Jefferson is most notable for expounding. He defines it by saying "rightful liberty is unobstructed action according to our will within limits drawn around us by the equal rights of others. I do not add 'within the limits of the law,' because law is often but the tyrant's will, and always so when it violates the rights of the individual." Hence, for Jefferson, though government cannot create a right to liberty, it can indeed violate it. And the limit of an individual's rightful liberty is not what law says it is but is simply a matter of stopping short of prohibiting other individuals from having the same liberty. A proper government, for Jefferson, is one that not only prohibits individuals in society from infringing on the liberty of other individuals, but also restrains itself from diminishing individual liberty.

Jefferson's commitment to equality was expressed in his successful efforts to abolish primogeniture in Virginia, the rule by which the first born son inherited all the land. Jefferson believed that individuals have an innate sense of morality that prescribes right from wrong when dealing with other individuals—that whether they choose to restrain themselves or not, they have an innate sense of the natural rights of others. He even believed that moral sense to be reliable enough that an anarchist society could function well, provided that it was reasonably small. On several occasions, he expressed admiration for tribal, communal way of living of Native Americans: In fact, Jefferson is sometimes seen as a philosophical anarchist.

He said in a letter to Colonel Carrington: "I am convinced that those societies (as the Indians) which live without government enjoy in their general mass an infinitely greater degree of happiness than those who live under the European governments." However, Jefferson believed anarchism to be "inconsistent with any great degree of population." Hence, he did advocate government for the American expanse provided that it exists by "consent of the governed."

In the Preamble to his original draft of the **_Declaration of Independence_**, Jefferson wrote:

> "We hold these truths to be sacred & undeniable; that all men are created equal & independent, that from that equal creation they derive rights inherent & inalienable, among which are the preservation of life, & liberty, & the pursuit of happiness; that to secure these ends, governments are instituted among men, deriving their just powers from the consent of the governed; that whenever any form of government shall become destructive of these ends, it is the right of the people to alter or to abolish it, & to institute new government, laying its

foundation on such principles & organizing its powers in such form, as to them shall seem most likely to effect their safety & happiness."

Jefferson's dedication to "consent of the governed" was so thorough that he believed that individuals could not be morally bound by the actions of preceding generations. This included debts as well as law. He said that "no society can make a perpetual constitution or even a perpetual law. The earth belongs always to the living generation." He even calculated what he believed to be the proper cycle of legal revolution: "Every constitution then, and every law, naturally expires at the end of nineteen years. If it is to be enforced longer, it is an act of force, and not of right." He arrived at nineteen years through calculations with expectancy of life tables, taking into account what he believed to be the age of "maturity"—when an individual is able to reason for himself. He also advocated that the national debt should be eliminated. He did not believe that living individuals had a moral obligation to repay the debts of previous generations. He said that repaying such debts was "a question of generosity and not of right."

Jefferson's very strong defense of States' rights, especially in the Kentucky and Virginia Resolutions of 1798, set the tone for hostility to expansion of Federal powers. However, some of his foreign policies did in fact strengthen the government. Most important was the Louisiana Purchase in 1803, when he used the implied powers to annex a huge foreign territory and all its French and Indian inhabitants. His enforcement of the Embargo Act of 1807, while it failed in terms of foreign policy, demonstrated that the federal government could intervene with great force at the local level in controlling trade that might lead to war.

During the presidential campaign of 1800, the Federalists attacked Jefferson as an infidel, claiming that Jefferson's intoxication with the religious and political extremism of the French Revolution disqualified him from public office. But Jefferson wrote at length on religion and many scholars agree with the claim that Jefferson was a deist, a common position held by intellectuals in the late 18th century, at least for much of his life. As Avery Dulles, a leading Catholic theologian reports, "In his college years at William and Mary [Jefferson] came to admire Francis Bacon, Isaac Newton, and John Locke as three great paragons of wisdom. Under the influence of several professors he converted to the deist philosophy." Dulles concludes:

> "In summary, then, Jefferson was a deist because he believed in one God, in divine providence, in the divine moral law, and in rewards and punishments after death; but did not believe in supernatural revelation. He was a Christian deist because he saw Christianity as the highest expression of natural religion and Jesus as an incomparably great moral teacher. He was not an orthodox Christian because he rejected, among other things, the doctrines that Jesus was the promised Messiah and the incarnate Son of God. Jefferson's religion is fairly typical of the American form of deism in his day."

Biographer Merrill Peterson summarizes Jefferson's theology:

> "First, that the Christianity of the churches was unreasonable, therefore unbelievable, but that stripped of priestly mystery, ritual, and dogma, reinterpreted in the light of historical evidence and human experience, and substituting the Newtonian cosmology for the discredited Biblical one, Christianity could be conformed to reason. Second, morality required no divine sanction or inspiration, no appeal beyond reason and nature, perhaps not even the hope of heaven or the fear of hell; and so the whole edifice of Christian revelation came tumbling to the ground."

The Declaration of Independence incorporates concepts from Deism. Jefferson used deist terminology in repeatedly stating his belief in a creator, and in the United States Declaration of Independence used the terms "Creator" and "Nature's God." Jefferson believed, furthermore, it was this Creator that endowed humanity with a number of inalienable rights, such as "life, liberty, and the pursuit of happiness." His experience in France just before the French Revolution made him deeply suspicious of Catholic priests and bishops as a force for reaction and ignorance. Similarly, his experience in America with inter-denominational intolerance served to reinforce this skeptical view of religion. In a letter to Willam Short, Jefferson wrote: "the serious enemies are the priests of the different religious sects, to whose spells on the human mind its improvement is ominous."

Jefferson was raised in the Church of England, at a time when it was the established church in Virginia and only denomination funded by Virginia tax money. Before the Revolution, Jefferson was a vestryman in his local church, a lay position that was part of political office at the time. He also had friends who were clergy, and he supported some churches financially. During his Presidency, Jefferson attended the weekly church services held in the House of Representatives. Jefferson later expressed general agreement with his friend Joseph Priestley's Unitarianism that is the rejection of the doctrine of Trinity. In a letter to a pioneer in Ohio he wrote,

"I rejoice that in this blessed country of free inquiry and belief, which has surrendered its conscience to neither kings nor priests, the genuine doctrine of only one God is reviving, and I trust that there is not a young man now living in the United States who will not die a Unitarian."

Jefferson refused to issue proclamations calling for days of prayer and thanksgiving during his Presidency, yet he did do so as Governor in Virginia. His private letters indicate he was skeptical of too much interference by clergy in matters of civil government. His letters contain the following observations:

"History, I believe, furnishes no example of a priest-ridden people maintaining a free civil government," and, "In every country and in every age, the priest has been hostile to liberty. He is always in alliance with the despot, abetting his abuses in return for protection to his own."

While opposed to the institutions of organized religion, Jefferson invoked the notion of divine justice in his opposition to slavery:

"Can the liberties of a nation be thought secure when we have removed their only firm basis, a conviction in the minds of the people that these liberties are of the gift of God? That they are not to be violated but with his wrath? Indeed I tremble for my country when I reflect that God is just: that his justice cannot sleep forever: that considering numbers, nature and natural means only, a revolution of the wheel of fortune, an exchange of situation is among possible events: that it may become probable by supernatural interference!"

"Believing with you that religion is a matter which lies solely between man and his God, that he owes account to none other for his faith or his worship, that the legislative powers of government reach actions only, and not opinions, I contemplate with sovereign reverence that act of the whole American people which declared that their legislature should "make no law respecting an establishment of religion, or prohibiting the free exercise thereof," thus building a wall of separation between church and State."

In 1962 when President John F. Kennedy welcomed forty-nine Nobel Prize winners to the White House he said, "I think this is the most extraordinary collection of talent and of human knowledge that has ever been gathered together at the White House – with the possible exception of when Thomas Jefferson dined alone."

31. Charles Lyell
1797-1875

Lyell was a Scottish lawyer and geologist. In 1832, he married Mary Horner of Bonn, daughter of Leonard Horner (1785-1864), also associated with the Geological Society of London. The new couple spent their honeymoon in Switzerland and Italy on a geological tour of the area. During the 1840s, he traveled to the United States and Canada, which resulted in his writing two popular travel-and-geology books: ***1845's Travels in North America*** and ***A Second Visit to the United States*** (from 1849).

Principles of Geology, Lyell's first book, was also his most famous, most influential, and most important. First published in three volumes in 1830-33, it established Lyell's credentials as an important geological theorist and introduced the doctrine of uniformitarianism. The central argument in ***Principles*** was that "the present is the key to the past:" That geological remains from the distant past can, and should, be explained by reference to geological processes now in operation and thus directly observable. Lyell's interpretation of geologic change as the steady accumulation of minute changes over enormously long spans of time was also a central theme in the Principles, and a powerful influence on the young Charles Darwin, who was given Volume 1 of the first edition by Robert FitzRoy, captain of HMS Beagle, just before they set out on the voyage of the Beagle. On their first stop ashore at St Jago Darwin found rock formations which seen "through Lyell's eyes" gave him a revolutionary insight into the geological history of the island, an insight he applied throughout his travels. While in South America Darwin received Volume 2 which firmly rejected the idea of organic evolution, proposing "Centres of Creation" to explain diversity and territory of species.

Darwin's ideas gradually moved beyond this, but in geology he was very much Lyell's disciple and sent home extensive evidence and theorizing supporting Lyell's uniformitarianism, including Darwin's ideas about the formation of atolls. On his return they became close friends. Lyell continued to firmly reject the idea of organic evolution in each of the first nine editions of the Principles. Confronted with Darwin's ***On the Origin of Species***, he finally offered a tepid endorsement of evolution in the tenth edition.

Elements of Geology began as the fourth volume of the third edition of Principles: Lyell intended Elements to act as a suitable field guide for students of geology. The systematic, factual description of geological formations of different ages contained in Principles grew so unwieldy, however, that Lyell split it off into a single volume under the Elements title in 1838. The book went through six editions, eventually growing to two volumes and ceasing to be the inexpensive, portable handbook that Lyell had originally envisioned. Late in his career, therefore, Lyell produced a condensed version ***titled Student's Elements of Geology*** that fulfilled the original purpose.

Geological Evidences of the Antiquity of Man brought together Lyell's views on three key themes from the geology of the Quaternary Period of Earth history: glaciers, evolution, and the age of the human race. First published in 1863, it went through three editions that year, with a fourth and final edition appearing in 1873.

Lyell's geological interests ranged from volcanoes and geological dynamics through stratigraphy, paleontology and glaciology to topics that would now be classified as prehistoric archaeology and paleoanthropology. He is best known, however, for his role in popularizing the doctrine of uniformitarianism.

From 1830 to 1833 his multi-volume ***Principles of Geology*** was published. The work's subtitle was "An Attempt to explain the former changes of the Earth's surface by reference to causes now in operation", and this explains Lyell's impact on science. He drew his explanations from field studies conducted directly before he went to work on the founding geology text. He was, along with the earlier John Playfair, the major advocate of the then-controversial idea of uniformitarianism that the earth was shaped entirely by slow-moving forces acting over a very long period of time. This was in contrast to catastrophism, a geologic idea that went hand-in-hand with age of the earth as implied by biblical chronology. In various revised editions (twelve in all, through 1872), ***Principles of Geology*** was the most influential geological work in the middle of the 19th century, and did much to put geology on a modern footing.

Before the work of Lyell, phenomena such as earthquakes were understood by the destruction that they wrought. One of the contributions that Lyell made in Principles was to explain the cause of earthquakes. Lyell, in contrast focused on recent earthquakes (150 yrs), evidenced by surface irregularities such as faults, fissures, stratigraphic displacements and depressions.

Lyell's work on volcanoes focused largely on Vesuvius and Etna, both of which he had earlier studied. His conclusions supported gradual building of volcanoes, so-called "backed up-building," as opposed to the upheaval argument supported by other geologists.

Lyell's most important specific work was in the field of stratigraphy. From May 1828, until February 1829, he traveled with Roderick Impey Murchison (1792-1871) to the south of France (Auvergne volcanic district) and to Italy. In these areas he concluded that the recent strata (rock layers) could be categorized according to the number and proportion of marine shells encased within. Based on this he proposed dividing the Tertiary period into three parts, which he named the Pliocene, Miocene, and Eocene.

In ***Principles of Geology*** (first edition, vol. 3, Ch. 2, 1833) Lyell proposed that icebergs could be the means of transport for erratics. During periods of global warming, ice breaks off the poles and floats across submerged continents, carrying debris with it, he conjectured. When the iceberg melts, it rains down sediments upon the land. Because this theory could account for the presence of diluvium, the word "drift" became the preferred term for the loose, unsorted material, today called "till." Furthermore, Lyell believed that the accumulation of fine angular particles covering much of the world (today called loess) was a deposit settled from mountain flood water. Today some of Lyell's mechanisms for geologic processes have been disproven, though many have stood the test of time. His observational methods and general analytical framework remain in use today as foundational principles in geology.

Charles Darwin was a close personal friend, and Lyell was one of the first prominent scientists to support ***On the Origin of Species***; he also fully accepted natural selection as the driving engine behind evolution in his tenth edition of Principles. In fact, Lyell was instrumental in arranging the peaceful co-publication of the theory of natural selection by Darwin and Alfred Wallace in 1858, reflecting the fact that each had arrived at the theory independently (Darwin long before Wallace, however). Lyell's own ***The Geological Evidence of the Antiquity of Man*** followed a few years later in 1863. Lyell's data was important because Darwin thought that populations of an organism changed very slowly, requiring what is now known as "geologic time".

Mankind believed for millennia that the earth was static, fixed and unchanging. Lyell and the early geologists taught us that this belief was untrue. The earth is dynamic, and ever-changing. The earth and the universe are much older than the ancients appreciated.

32. Charles Darwin
1809 –1882

> *"It appears to me (whether rightly or wrongly) that direct arguments against Christianity and theism produce hardly any effect on the public; and freedom of thought is best promoted by the gradual illumination of men's minds which follows from the advance of science."*

> *"If the misery of the poor be caused not by the laws of nature, but by our institutions, great is our sin."*

Darwin was an English naturalist. After becoming eminent among scientists for his field work and inquiries into geology, he proposed and provided scientific evidence that all species of life have evolved over time from one or a few common ancestors through the process of natural selection. The fact that evolution occurs became accepted by the scientific community and the general public in his lifetime, while his theory of natural selection came to be widely seen as the primary explanation of the process of evolution in the 1930s, and now forms the basis of modern evolutionary theory. In modified form, Darwin's scientific discovery remains the foundation of biology, as it provides a unifying logical explanation for the diversity of life.

Darwin developed his interest in natural history while studying first medicine at Edinburgh University, then theology at Cambridge. His five-year voyage on the Beagle established him as a geologist whose observations and theories supported Charles Lyell's uniformitarian ideas, and publication of his journal of the voyage made him famous as a popular author. Puzzled by the geographical distribution of wildlife and fossils he collected on the voyage; Darwin investigated the transmutation of species and conceived his theory of natural selection in 1838. Having seen others attacked as heretics for such ideas, he confided only in his closest friends and continued extensive research to meet anticipated objections. His research was still in progress in 1858 when Alfred Russel Wallace sent him an essay which described a similar theory, prompting immediate joint publication of both of their theories.

His 1859 book ***On the Origin of Species*** established evolution by common descent as the dominant scientific explanation of diversification in nature. He examined human evolution and sexual selection in ***The Descent of Man***, and ***Selection in Relation to Sex***, followed by ***The Expression of the Emotions in Man and Animals.*** His research on plants was published in a series of books, and in his final book, he examined earthworms and their effect on soil.

The Beagle survey took five years, two-thirds of which Darwin spent on land. He carefully noted a rich variety of geological features, fossils and living organisms, and methodically collected an enormous number of specimens, many of them new to science.[1] At intervals during the voyage he sent specimens to Cambridge together with letters about his findings, and these established his reputation as a naturalist. His extensive detailed notes showed his gift for theorizing and formed the basis for his later work. The journal he originally wrote for his family, published as The Voyage of the Beagle, summarizes his findings and provides social, political and anthropological insights into the wide range of people he met, both native and colonial.

Before they set out, FitzRoy gave Darwin the first volume of Charles Lyell's **_Principles of Geology_**, which explained landforms as the outcome of gradual processes over huge periods of time. On their first stop ashore at St Jago, Darwin found that a white band high in the volcanic rock cliffs consisted of baked coral fragments and shells. This matched Lyell's concept of land slowly rising or falling, giving Darwin a new insight into the geological history of the island which inspired him to think of writing a book on geology. He went on to make many more discoveries, some of them particularly dramatic. He saw stepped plains of shingle and seashells in Patagonia as raised beaches, and after experiencing an earthquake in Chile saw mussel-beds stranded above high tide showing that the land had just been raised. High in the Andes he saw several fossil trees that had grown on a sand beach, with seashells nearby. He theorized that coral atolls form on sinking volcanic mountains, and confirmed this when the Beagle surveyed the Cocos (Keeling) Islands.

In South America, Darwin found and excavated rare fossils of gigantic extinct mammals in strata with modern seashells, indicating recent extinction and no change in climate or signs of catastrophe. Though he correctly identified one as a Megatherium and fragments of armour reminded him of the local armadillo, he assumed his finds were related to African or European species and it was a revelation to him after the voyage when Richard

Owen showed that they were closely related to living creatures exclusively found in the Americas.

Lyell's second volume, which argued against evolutionism and explained species distribution by "centres of creation", was sent out to Darwin. He puzzled over all he saw, and his ideas went beyond Lyell. In Argentina, he found that two types of rhea had separate but overlapping territories. On the Galápagos Islands, he collected mockingbirds and noted that they were different depending on which island they came from. He also heard that local Spaniards could tell from their appearance on which island tortoises originated, but thought the creatures had been imported by buccaneers. In Australia, the marsupial rat-kangaroo and the platypus seemed so unusual that Darwin thought it was almost as though two distinct Creators had been at work.

In Cape Town he and FitzRoy met John Herschel, who had recently written to Lyell about that "mystery of mysteries", the origin of species. When organizing his notes on the return journey, Darwin wrote that if his growing suspicions about the mockingbirds and tortoises were correct, "such facts undermine the stability of Species", then cautiously added "would" before "undermine". He later wrote that such facts "seemed to me to throw some light on the origin of species".

Three natives who had been taken from Tierra del Fuego on the Beagle's previous voyage were taken back there to become missionaries. They had become "civilised" in England over the previous two years, yet their relatives appeared to Darwin to be "miserable, degraded savages". A year on, the mission had been abandoned and only Jemmy Button spoke with them to say he preferred his harsh previous way of life and did not want to return to England. Because of this experience, Darwin came to think that humans were not as far removed from animals as his friends then believed, and saw differences as relating to cultural advances towards civilization rather than being racial. He detested the slavery he saw elsewhere in South America, and was saddened by the effects of European settlement on Aborigines in Australia and Maori in New Zealand.

Captain FitzRoy was committed to writing the official Narrative of the Beagle voyages, and near the end of the voyage, he read Darwin's diary and asked him to rewrite this Journal to provide the third volume, on natural history.

Continuing his research in London, Darwin's wide reading now included "for amusement" the 6th edition of Malthus's ***An Essay on the Principle of Population*** which calculates from the birth rate that human population could double every 25 years, but in practice growth is kept in check by death, disease, wars and famine. Darwin was well prepared to see at once that this also applied to de Candolle's "warring of the species" of plants and the struggle for existence among wildlife, explaining how numbers of a species kept roughly stable. As species always breed beyond available resources, favorable variations would make organisms better at surviving and passing the variations on to their offspring, while unfavorable variations would be lost. This would result in the formation of new species. On 28 September 1838 he noted this insight, describing it as a kind of wedging, forcing adapted structures into gaps in the economy of nature as weaker structures were thrust out. He now had a theory by which to work, and over the following months compared farmers picking the best breeding stock to a Malthusian Nature selecting from variants thrown up by "chance" so that "every part of every newly acquired structure is fully practiced and perfected", and thought this analogy "the most beautiful part of my theory".

Early in 1842, Darwin sent a letter about his ideas to Lyell, who was dismayed that his ally now denied "seeing a beginning to each crop of species". In May, Darwin's book on coral reefs was published after more than three years of work, and he then wrote a "pencil sketch" of his theory. To escape the pressures of London, the family moved to rural Down House in November. On 11 January 1844 Darwin wrote to his botanist friend Joseph Dalton Hooker about his theory, saying it was like confessing "a murder", but to his relief Hooker thought that "there might have been a gradual change of species" and expressed interest in Darwin's explanation. By July, Darwin had expanded his "sketch" into a 230-page "Essay". His fears that his ideas would be dismissed as Lamarckian Radicalism were reawakened by controversy over the anonymous publication in October of ***Vestiges of the Natural History of Creation***, which was severely attacked by establishment scientists. However, the book was a best-seller and widened middle-class interest in transmutation, paving the way for Darwin as well as reminding him of the need to answer all difficulties before making his theory public. Darwin completed his third geological book in 1846, and embarked on a huge study of barnacles with the assistance of Hooker. In 1847, Hooker read the "Essay" and sent notes that provided Darwin with the calm critical feedback that he needed, but would not commit himself and questioned Darwin's opposition to continuing acts of Creation.

By the start of 1856, Darwin was investigating whether eggs and seeds could survive travel across seawater to spread species across oceans. Hooker increasingly doubted the traditional view that species were fixed, but their young friend Thomas Huxley was firmly against evolution. Lyell was intrigued by Darwin's speculations without realizing their extent. When he read a paper by Alfred Wallace on the Introduction of species, he saw similarities with Darwin's thoughts and urged him to publish to establish precedence. Though Darwin saw no threat, he began work on a short paper. Finding answers to difficult questions held him up repeatedly, and he expanded his plans to a "big book on species" titled Natural Selection. He continued his researches, obtaining information and specimens from naturalists worldwide including Wallace who was working in Borneo. In December 1857, Darwin received a letter from Wallace asking if the book would examine human origins. He responded that he would avoid that subject, "so surrounded with prejudices", while encouraging Wallace's theorizing and adding that "I go much further than you."

Darwin's book was half way when, on 18 June 1858, he received a paper from Wallace describing natural selection. Shocked that he had been "forestalled", Darwin sent it on to Lyell, as requested, and, though Wallace had not asked for publication, offered to send it to any journal that Wallace chose. His family was in crisis with children in the village dying of scarlet fever, and he put matters in the hands of Lyell and Hooker. They agreed on a joint presentation at the Linnean Society on 1 July of **On the Tendency of Species to form Varieties**; and on the **Perpetuation of Varieties and Species by Natural Means of Selection**; however, Darwin's baby son died of the scarlet fever and he was too distraught to attend.

There was little immediate attention to this announcement of the theory; the president of the Linnean remarked in May 1859 that the year had not been marked by any revolutionary discoveries. Later, Darwin could only recall one review; Professor Haughton of Dublin claimed that "all that was new in them was false, and what was true was old." Darwin struggled for thirteen months to produce an abstract of his "big book", suffering from ill health but getting constant encouragement from his scientific friends. Lyell arranged to have it published by John Murray.

On the Origin of Species by Means of Natural Selection, or **The Preservation of Favoured Races in the Struggle for Life** (usually abbreviated to **The Origin of Species**) proved unexpectedly popular, with the entire stock of 1,250 copies oversubscribed when it went on sale to booksellers on 22 November 1859. In the book, Darwin set out "one long argument" of detailed observations, inferences and consideration of anticipated objections. His only allusion to human evolution was the understatement that "light will be thrown on the origin of man and his history". He avoided the then controversial term "evolution", but at the end of the book concluded that "endless forms most beautiful and most wonderful have been, and are being, evolved." His theory is simply stated in the introduction:

As many more individuals of each species are born than can possibly survive; and as, consequently, there is a frequently recurring struggle for existence, it follows that any being, if it vary however slightly in any manner profitable to itself, under the complex and sometimes varying conditions of life, will have a better chance of surviving, and thus be naturally selected. From the strong principle of inheritance, any selected variety will tend to propagate its new and modified form.

There was wide public interest in Charles Darwin's book and a controversy which he monitored closely, keeping press cuttings of reviews, articles, satires, parodies and caricatures. Critical reviewers were quick to pick out the unstated implications of "men from monkeys", while amongst favorable responses Huxley's reviews included swipes at Richard Owen, leader of the scientific establishment Huxley was trying to overthrow. Owen's verdict was unknown until his April review condemned the book.

The Church of England scientific establishment, including Darwin's old Cambridge tutors Sedgwick and Henslow, reacted against the book, though it was well received by a younger generation of professional naturalists. In 1860, the publication of Essays and Reviews by seven liberal Anglican theologians diverted clerical attention away from Darwin. An explanation of higher criticism and other heresies, it included the argument that miracles broke God's laws, so belief in them was atheistic—and praise for "Mr Darwin's masterly volume [supporting] the grand principle of the self-evolving powers of nature".

The most famous confrontation took place at a meeting of the British Association for the Advancement of Science in Oxford. Professor John William Draper delivered a long lecture about Darwin and social progress, then Samuel Wilberforce, the Bishop of Oxford, argued against Darwin. In the ensuing debate Joseph Hooker argued strongly for Darwin and Thomas Huxley established himself as "Darwin's bulldog" – the fiercest defender of evolutionary theory on the Victorian stage. Both sides came away feeling victorious, but Huxley went on to make much of his claim that on being asked by Wilberforce whether he was descended from monkeys on his grandfather's side or his grandmother's side, Huxley muttered:

"The Lord has delivered him into my hands" and replied that he "would rather be descended from an ape than from a cultivated man who used his gifts of culture and eloquence in the service of prejudice and falsehood".

Darwin's illness kept him away from the public debates, though he read eagerly about them and mustered support through correspondence. Asa Gray persuaded a publisher in the United States to pay royalties, and Darwin imported and distributed Gray's pamphlet Natural Selection is not inconsistent with Natural Theology. In Britain, friends including Hooker and Lyell took part in the scientific debates which Huxley pugnaciously led to overturn the dominance of clergymen and aristocratic amateurs under Owen in favor of a new generation of professional scientists. Owen made the mistake of (wrongly) claiming certain anatomical differences between ape and human brains, and accusing Huxley of advocating "Ape Origin of Man". Huxley gladly did just that, and his campaign over two years was devastatingly successful in ousting Owen and the "old guard". Darwin's friends formed The X Club and helped to gain him the honor of the Royal Society's Copley Medal in 1864.

Broader public interest had already been stimulated by **_Vestiges_**, and the **_Origin of Species_** was translated into many languages and went through numerous reprints, becoming a staple scientific text accessible both to a newly curious middle class and to "working men" who flocked to Huxley's lectures. Darwin's theory also resonated with various movements at the time and became a key fixture of popular culture.

Despite repeated bouts of illness during the last twenty-two years of his life, Darwin pressed on with his work. He had published an abstract of his theory, but more controversial aspects of his "big book" were still incomplete, including explicit evidence of humankind's descent from earlier animals, and exploration of possible causes underlying the development of society and of human mental abilities. He had yet to explain features with no obvious utility other than decorative beauty. His experiments, research and writing continued.

When Darwin's daughter fell ill, he set aside his experiments with seedlings and domestic animals to accompany her to a seaside resort where he became interested in wild orchids. This developed into an innovative study of how their beautiful flowers served to control insect pollination and ensure cross fertilization. As with the barnacles, homologous parts served different functions in different species. Back at home, he lay on his sickbed in a room filled with experiments on climbing plants. A reverent Ernst Haeckel who had spread the gospel of Darwinism in Germany visited him. Wallace remained supportive, though he increasingly turned to spiritualism.

Variation of Plants and Animals Under Domestication, the first part of Darwin's planned "big book" (expanding on his "abstract" published as **_The Origin of Species_**), grew to two huge volumes, forcing him to leave out human evolution and sexual selection, and sold briskly despite its size. A further book of evidence, dealing with natural selection in the same style, was largely written, but remained unpublished until transcribed in 1975.

Punch's almanac for 1882, published shortly before Darwin's death, depicts him amidst evolution from chaos to Victorian gentleman with the title **_Man Is But A Worm_**.

The question of human evolution had been taken up by his supporters (and detractors) shortly after the publication of The **_Origin of Species_**, but Darwin's own contribution to the subject came more than ten years later with the two-volume **_The Descent of Man_**, and **_Selection in Relation to Sex_** published in 1871. In the second volume, Darwin introduced in full his concept of sexual selection to explain the evolution of human culture, the differences between the human sexes, and the differentiation of human races, as well as the beautiful (and seemingly non-adaptive) plumage of birds. A year later Darwin published his last major work, **_The Expression of the Emotions in Man and Animals_**, which focused on the evolution of human psychology and its continuity with the behavior of animals. He developed his ideas that the human mind and cultures were developed by natural and sexual selection, an approach which has been revived in the last three decades with the emergence of evolutionary psychology. As he concluded in **_Descent of Man_**, Darwin felt that, despite all of humankind's "noble qualities" and "exalted powers":

"Man still bears in his bodily frame the indelible stamp of his lowly origin."

His evolution-related experiments and investigations culminated in books on the movement of climbing plants, insectivorous plants, the effects of cross and self fertilization of plants, different forms of flowers on plants of the same species, and **_The Power of Movement in Plants_**. In his last book, he returned to the effect earthworms have on soil formation.

The 1851 death of Darwin's daughter, Annie, was the final step in pushing an already doubting Darwin away from the idea of a beneficent God.

When investigating transmutation of species he knew that his naturalist friends thought this a bestial heresy undermining miraculous justifications for the social order, the kind of radical argument then being used by Dissenters and atheists to attack the Church of England's privileged position as the established church. Though Darwin wrote of religion as a tribal survival strategy, he still believed that God was the ultimate lawgiver. His belief dwindled, and with the death of his daughter Annie in 1851, Darwin finally lost all faith in Christianity. He continued to help the local church with parish work, but on Sundays would go for a walk while his family attended church. He now thought it better to look at pain and suffering as the result of general laws rather than direct intervention by God. When asked about his religious views, he wrote that he had never been an atheist in the sense of denying the existence of a God, and that generally "an Agnostic would be the more correct description of my state of mind."

The "Lady Hope Story", published in 1915, claimed that Darwin had reverted back to Christianity on his sickbed. The claims were refuted by Darwin's children and have been dismissed as false by historians. His daughter, Henrietta, who was at his deathbed, said that he did not convert to Christianity. His last words were, in fact, directed at Emma: "Remember what a good wife you have been."

Darwin's theories and writings, combined with Gregor Mendel's genetics (the "modern synthesis"), form the basis of all modern biology.

33. Lysander Spooner
1808-1887

Lysander Spooner was a libertarian, individualist, anarchist, entrepreneur, political philosopher, abolitionist, supporter of the labor movement, and legal theorist of the 19th century. He is also known for competing with the U.S. Post Office with his American Letter Mail Company, which was forced out of business by the United States Government. He has been identified by some contemporary writers as an anarcho-capitalist.

Later known as an early individualist anarchist, Spooner advocated what he called Natural Law — or the "Science of Justice" — wherein acts of initiatory coercion against individuals and their property were considered "illegal" but the so-called criminal acts that violated only man-made legislation were not.

He believed that the price of borrowing capital could be brought down by competition of lenders if the government deregulated banking and money. This he believed would stimulate entrepreneurship. In his ***Letter to Cleveland***, Spooner argued, "All the great establishments, of every kind, now in the hands of a few proprietors, but employing a great number of wage laborers, would be broken up; for few or no persons, who could hire capital and do business for themselves would consent to labor for wages for another." Spooner took his own advice and started his own business called American Letter Mail Company which competed with the U.S. Post Office.

His activism began with his career as a lawyer, which itself violated Massachusetts law. Spooner had studied law under the prominent lawyers and politicians John Davis and Charles Allen, but he had never attended college. According to the laws of the state, college graduates were required to study with an attorney for three years, while non-graduates were required to do so for five years.

With the encouragement of his legal mentors, Spooner set up his practice in Worcester after only three years, openly defying the courts. He saw the three-year privilege for college graduates as a state-sponsored discrimination against the poor and also providing a monopoly income to those who met the requirements. He argued that such discrimination was "so monstrous a principle as that the rich ought to be protected by law from the competition of the poor." In 1836, the legislature abolished the restriction. He opposed all licensing requirements for lawyers, doctors or anyone else that was prevented from being employed by such requirements. To prevent a person from doing business with a person without a professional license he saw as a violation of the natural right to contract.

Postal rates were notoriously high in the 1840s, and in 1844, Spooner founded the American Letter Mail Company, which had offices in various cities, including Baltimore, Philadelphia, and New York. Stamps could be purchased and then attached to letters which could be sent to any of its offices. From here agents were dispatched who traveled on railroads and steamboats, and carried the letters in hand bags. Letters were transferred to messengers in the cities along the routes who then delivered the letters to the addressees. This was a challenge to the United States Post Office's monopoly. As he had done when challenging the rules of the Massachusetts bar, he published a pamphlet titled "***The Unconstitutionality of the Laws of Congress Prohibiting Private Mails.***" Although Spooner had finally found commercial success with his mail company, legal challenges by the government eventually exhausted his financial resources. He closed up shop without ever having had the opportunity to fully litigate his constitutional claims. The lasting legacy of Spooner's challenge to the postal service was the 3-cent stamp, adopted in response to the competition his company provided.

Spooner attained his greatest fame as a figure in the abolitionist movement. His most famous work, a book titled ***The Uncon-***

stitutionality of Slavery, was published in 1846 to great acclaim among many abolitionists but criticism from others. Spooner's book contributed to a controversy within the abolitionist movement over whether the United States Constitution supported the institution of slavery. The "disunionist" faction, led by William Lloyd Garrison and Wendell Phillips, argued the Constitution legally recognized and enforced the oppression of slaves (as, for example, in the provisions for the capture of fugitive slaves in Article IV, Section 2). They also cited the frequent appeals to Constitutional compromise by Southern politicians, who insisted that protection of the "peculiar institution" was part of the sectional compromise on which the Constitution was based. The disunionists thus argued that keeping the free states in a political union with the slave states made the citizens of the free states complicit in the slave system, and denounced the Constitution as "a covenant with death and an agreement with hell."

Spooner challenged the claim that the text of the Constitution supported slavery. Although he recognized that the Founders had probably not intended to outlaw slavery when writing the Constitution, he argued that only the meaning of the text, not the private intentions of its writers, was enforceable. Spooner used a complex system of legal and natural law arguments in order to show that the clauses usually interpreted as supporting slavery did not, in fact, support it, and that several clauses of the Constitution prohibited the states from establishing slavery under the law. Spooner's arguments were cited by other pro-Constitution abolitionists, such as Gerrit Smith and the Liberty Party, which adopted it as an official text in its 1848 platform. Frederick Douglass, originally a Garrisonian disunionist, later came to accept the pro-Constitution position, and cited Spooner's arguments to explain his change of mind.

From the publication of this book until 1861, Spooner actively campaigned against slavery. He published subsequent pamphlets on ***Jury Nullification*** and other legal defenses for escaped slaves and offered his legal services, often free of charge, to fugitives. In the late 1850s, copies of his book were distributed to members of Congress sparking some debate over their contents. Even Senator Albert Gallatin Brown of Mississippi, a slavery proponent, praised the argument's intellectual rigor and conceded it was the most formidable legal challenge he had seen from the abolitionists to date. In 1858, Spooner circulated a ***Plan for the Abolition of Slavery,*** calling for the use of guerrilla warfare against slaveholders by black slaves and non-slaveholding free Southerners, with aid from Northern abolitionists. Spooner also participated in an aborted plot to free John Brown after his capture following the failed raid on Harper's Ferry, Virginia.

In 1860, Spooner was actively courted by William Seward to support the fledgling Republican Party. An admitted sympathizer with the Jeffersonian political philosophy, Spooner adamantly refused the request and soon became an outspoken abolitionist critic of the party. To Spooner, the Republicans were hypocrites for purporting to oppose slavery's expansion but refusing to take a strong, consistent moral stance against slavery itself. Although Spooner had advocated the use of violence to abolish slavery, he denounced the Republicans' use of violence to prevent the Southern states from seceding during the American Civil War. He published several letters and pamphlets about the war, arguing that the Republican war aim was not the overthrow of slavery, but rather to maintain the Union by force. He blamed the bloodshed on Republican political leaders such as Secretary of State Seward and Senator Charles Sumner, who often spoke out against slavery but would not attack it on a constitutional basis, and who pursued military policies seen as vengeful and abusive.

Though denouncing its embrace of slavery, Spooner sided with the Confederate States of America's right to secede on the basis that they were choosing to exercise government by consent - a fundamental constitutional and legal principle to Spooner's philosophy. The North, by contrast, was trying to deny the Southerners their inherent right to be governed by their consent. He believed they were attempting to coerce the obedience of the southern states to a union they did not wish to enter. He believed that Compensated Emancipation was a preferable way to end slavery, something many nations had done. He argued that the right for states to secede derives from the same right of the slaves to be free. This argument was not popular in the North or South once the war started, as it was contrary to the government positions held on both sides.

Spooner harshly condemned the Civil War and the Reconstruction period that followed. Though he approved of the fact that black slavery was abolished, he criticized the North for failing to make this the purpose of their cause. Instead of fighting to abolish slavery, they fought to "preserve the union" and, according to Spooner, to bolster business interests behind that union. Spooner believed a war of this type was hypocritical and dishonest, especially on the part of Radical Republicans like Sumner who were by then claiming to be abolitionist heroes for ending slavery. Spooner also argued that the war came at a great cost to liberty and proved that the rights expressed in the ***Declaration of Independence*** no longer held true - the people could not "dissolve the political bands" that tie them to a government that "becomes destructive" of the consent of the governed because if they did so, as Spooner believed the south had attempted to do, they would be met by the bayonet to enforce their obedience to the former government.

The Union government's actions during the war caused Spooner to radicalize his views to an anarchistic view. In response, Spooner published one of his most famous political tracts, ***No Treason***. In this lengthy essay, Spooner argued that the Constitution was a contract of government which had been irreparably violated during the war and was thus void. Furthermore, since the government now existing under the Constitution pursued coercive policies that were contrary to the Natural Law and to the consent of the governed, it had been demonstrated that document was unable to adequately stop many abuses against liberty or to prevent tyranny from taking hold. Spooner bolstered his argument by noting that the Federal government, as established by a legal contract, could not legally bind all persons living in the nation since none had ever signed their names or given their consent to it - that consent had always been assumed, which fails the most basic burdens of proof for a valid contract in the courtroom.

Spooner widely circulated the ***No Treason*** pamphlets, which also contained a legal defense against the crime of treason itself intended for former Confederate soldiers (hence the name of the pamphlet, arguing that "no treason" had been committed in the war by the south). These excerpts were published in DeBow's Review and some other well known southern periodicals of the time.

Spooner continued to write and publish extensively in the decades following Reconstruction, producing works such as ***Natural Law or The Science of Justice*** and ***Trial By Jury***. In ***Trial By Jury*** he defended the doctrine of Jury Nullification, which holds that in a free society a trial jury not only has the authority to rule on the facts of the case, but also on the legitimacy of the law under which the case is tried, and which would allow juries to refuse to convict if they regard the law they are asked to convict under as illegitimate. He became closely associated with Benjamin Tucker's anarchist journal Liberty, which published all of his later works in serial format, and for which he wrote several editorial columns on current events. He argued that ". . . almost all fortunes are made out of the capital and labor of other men than those who realize them. Indeed, except by his sponging capital and labor from others."

Spooner's influence extends to the wide range of topics he addressed during his lifetime. He is remembered today primarily for his abolitionist activities and for his challenge to the post office monopoly, which had a lasting influence of significantly reducing postal rates. Spooner's writings contributed to the development of libertarian political theory in the United States, and were often reprinted in early libertarian journals such as the Rampart Journal. His writings were also a major influence on Austrian School economist Murray Rothbard and libertarian law professor and legal theorist Randy Barnett.

Spooner's ***The Unconstitutionality of Slavery*** was cited in the 2008 District of Columbia v. Heller case which struck down the federal district's ban on handguns. Justice Antony Scalia, writing for the court, quotes Spooner as saying the right to bear arms was necessary for those who wanted to take a stand against slavery.

34. Louis Pasteur
1822–1895

Pasteur was a French chemist and microbiologist born in Dole. He is best known for his remarkable breakthroughs in the causes and preventions of disease. His discoveries reduced mortality from puerperal fever, and he created the first vaccine for rabies. His experiments supported the germ theory of disease. He was best known to the general public for inventing a method to stop milk and wine from causing sickness - this process came to be called pasteurization. He is regarded as one of the three main founders of microbiology, together with Ferdinand Cohn and Robert Koch. Pasteur also made many discoveries in the field of chemistry, most notably the molecular basis for the asymmetry of certain crystals.

Pasteur demonstrated that fermentation is caused by the growth of microorganisms, and that the emergent growth of bacterium in nutrient broths is not due to spontaneous generation but rather to biogenesis

He exposed boiled broths to air in vessels that contained a filter to prevent all particles from passing through to the growth medium, and even in vessels with no filter at all, with air being admitted via a long tortuous tube that would not allow dust particles to pass. Nothing grew in the broths unless the flasks were broken open; therefore, the living organisms that grew in such broths came from outside, as spores on dust, rather than spontaneously generated within the broth. This was one of the last and most important experiments disproving the theory of spontaneous generation. The experiment also supported germ theory.

While Pasteur was not the first to propose germ theory (Girolamo Fracastoro, Agostino Bassi, Friedrich Henle and others had suggested it earlier), he developed it and conducted experiments that clearly indicated its correctness and managed to convince most of Europe it was true. Today he is often regarded as the father of germ theory and bacteriology, together with Robert Koch.

Pasteur's research also showed that the growth of microorganisms was responsible for spoiling beverages, such as beer, wine and milk. With this established, he invented a process in which liquids such as milk were heated to kill most bacteria and molds already present within them. He and Claude Bernard completed the first test on April 20, 1862. This process was soon afterwards known as pasteurization.

Beverage contamination led Pasteur to the idea that microorganisms infecting animals and humans cause disease. He proposed preventing the entry of microorganisms into the human body, leading Joseph Lister to develop antiseptic methods in surgery.

In 1865, two parasitic diseases called pébrine and flacherie were killing great numbers of silkworms at Alais (now Alès). Pasteur worked several years proving it was a microbe attacking silkworm eggs which caused the disease, and that eliminating this microbe within silkworm nurseries would eradicate the disease.

Pasteur also discovered anaerobiosis, whereby some microorganisms can develop and live without air or oxygen, called the Pasteur effect.

Pasteur's later work on diseases included work on chicken cholera. During this work, a culture of the responsible bacteria had spoiled and failed to induce the disease in some chickens he was infecting with the disease. Upon reusing these healthy chickens, Pasteur discovered that he could not infect them, even with fresh bacteria; the weakened bacteria had caused the chickens to become immune to the disease, even though they had only caused mild symptoms.

His assistant Charles Chamberland (of French origin) had been instructed to inoculate the chickens after Pasteur went on holiday. Chamberland failed to do this, but instead went on holiday himself. On his return, the month old cultures made the chickens unwell, but instead of the infection being fatal, as it usually was, the chickens recovered completely. Chamberland assumed an error had been made, and wanted to discard the apparently faulty culture when Pasteur stopped him. Pasteur guessed the recovered animals now might be immune to the disease, as were the animals at Eure-et-Loir that had recovered from anthrax.

In the 1870s, he applied this immunization method to anthrax, which affected cattle, and aroused interest in combating other diseases.

Pasteur publicly claimed he had made the anthrax vaccine by exposing the bacillus to oxygen. His laboratory notebooks, now in the Bibliotheque Nationale in Paris, in fact show Pasteur used the method of rival Jean-Joseph-Henri Toussaint, a Toulouse veterinary surgeon, to create the anthrax vaccine. This method used the oxidizing agent potassium dichromate. Pasteur's oxygen method did eventually produce a vaccine but only after he had been awarded a patent on the production of an anthrax vaccine.

The notion of a weak form of a disease causing immunity to the virulent version was not new; this had been known for a long time for smallpox. Inoculation with smallpox was known to result in far less scarring, and greatly reduced mortality, in comparison to the naturally acquired disease. Edward Jenner had also discovered vaccination, using cowpox to give cross-immunity to smallpox (in 1796), and by Pasteur's time this had generally replaced the use of actual smallpox material in inoculation. The difference between smallpox vaccination and cholera and anthrax vaccination was that the weakened form of the latter two disease organisms had been generated artificially, and so a naturally weak form of the disease organism did not need to be found.

This discovery revolutionized work in infectious diseases, and Pasteur gave these artificially weakened diseases the generic name of vaccines, to honor Jenner's discovery. Pasteur produced the first vaccine for rabies by growing the virus in rabbits, and then weakening it by drying the affected nerve tissue.

The rabies vaccine was initially created by Emile Roux, a French doctor and a colleague of Pasteur who had been working with a killed vaccine produced by desiccating the spinal cords of infected rabbits. The vaccine had only been tested on eleven dogs before its first human trial.

This vaccine was first used on 9-year old Joseph Meister, on July 6, 1885, after the boy was badly mauled by a rabid dog. This was done at some personal risk for Pasteur, since he was not a licensed physician and could have faced prosecution for treating the boy. However, left without treatment, the boy faced almost certain death from rabies. After consulting with colleagues, Pasteur decided to go ahead with the treatment. The treatment proved to be a spectacular success, with Meister avoiding the disease; thus, Pasteur was hailed as a hero and the legal matter was not pursued. The treatment's success laid the foundations for the manufacture of many other vaccines. The first of the Pasteur Institutes was also built on the basis of this achievement.

Because of his study in germs, Pasteur encouraged doctors to sanitize their hands and equipment before surgery. Prior to this, few doctors or their assistants practiced the procedure of washing their hands and equipment.

In his triumphal lecture at the Sorbonne in 1864, Pasteur said "Never will the doctrine of spontaneous generation recover from the mortal blow struck by this simple experiment" (referring to his swan-neck flask experiment wherein he proved that fermenting microorganisms would not form in a flask containing fermentable juice until an entry path was created for them.)

Although the belief that disease and pestilence is punishment from God persist, no one has done more than Pasture to destroy the superstition.

35. Sigmund Freud
1856-1939

"Quite frequently I encounter people who equate lack of certitude with giant inferential leaps. Science deals with probabilities, often quite high probabilities, but not certitudes. It is one of the strengths of the scientific method as it acknowledges a chance of error (while maintaining rigorous standards to establish provisional acceptance of propositions).

It is a mistake to believe that a science consists in nothing but conclusively proved propositions, and it is unjust to demand that it should. It is a demand only made by those who feel a craving for authority in some form and a need to replace the religious catechism by something else, even if it be a scientific one. Science in its catechism has but few apodictic precepts; it consists mainly of statements which it has developed to varying degrees of probability. The capacity to be content with these approximations to certainty and the ability to carry on constructive work despite the lack of final confirmation are actually a mark of the scientific habit of mind." -- Sigmund Freud

"The weakness of my argument does not imply the strength of yours" - Sigmund Freud

Freud was an Austrian neurologist and psychiatrist who founded the psychoanalytic school of psychology. Freud is best known for his theories of the unconscious mind and the defense mechanism of repression. He is also renowned for his redefinition of sexual desire as the primary motivational energy of human life which is directed toward a wide variety of objects; as well as his therapeutic techniques, including his theory of transference in the therapeutic relationship and the presumed value of dreams as sources of insight into unconscious desires.

Freud is commonly referred to as "the father of psychoanalysis" and his work has been highly influential — popularizing such notions as the unconscious, the Oedipus complex, defense mechanisms, Freudian slips and dream symbolism — while also making a long-lasting impact on fields as diverse as literature, film, Marxist and feminist theories, and psychology.

After the publication of Freud's books in 1900 and 1901, interest in his theories began to grow, and a circle of supporters developed in the following period. Freud often chose to disregard the criticisms of those who were skeptical of his theories, however, which earned him the animosity of a number of individuals, the most famous being Carl Jung, who originally supported Freud's ideas. Part of the reason for their fallout was due to Jung's growing commitment to religion and mysticism, which conflicted with Freud's atheism.

In 1930, Freud received the Goethe Prize in appreciation of his contribution to psychology and to German literary culture. Three years later the Nazis took control of Germany and Freud's books featured prominently among those burned by the Nazis. In March 1938, Nazi Germany annexed Austria in the Anschluss. This led to violent outbursts of anti-Semitism in Vienna, and Freud and his family received visits from the Gestapo. Freud decided to go into exile "to die in freedom". He and his family left Vienna in June 1938 and traveled to London.

Freud has been influential in two related but distinct ways. He simultaneously developed a theory of how the human mind is organized and operates internally, and how human behavior both conditions and results from this particular theoretical understanding. This led him to favor certain clinical techniques for attempting to help cure psychopathology.

Freudian therapy, or psychoanalysis, was to bring to consciousness repressed thoughts and feelings. According to some of his successors, including his daughter Anna Freud, the goal of therapy is to allow the patient to develop a stronger ego; according to others, notably Jacques Lacan, the goal of therapy is to lead the analysand to a full acknowledgment of his or her inability to satisfy the most basic desires.

Classically, the bringing of unconscious thoughts and feelings to consciousness is brought about by encouraging the patient to talk in free association and to talk about dreams. Another important element of psychoanalysis is a relative lack of direct involvement on the part of the analyst, which is meant to encourage the patient to project thoughts and feelings onto the analyst. Through this process, transference, the patient can reenact and resolve repressed conflicts, especially childhood conflicts with (or about) parents.

Perhaps the most significant contribution Freud made to Western thought was his argument for the existence of an unconscious mind. During the 19th century, the dominant trend in Western thought was positivism, which subscribed to the belief that people could ascertain real knowledge concerning themselves and their environment and judiciously exercise control over both. Freud, however, suggested that such declarations of free will are in fact delusions; that we are not entirely aware of what we think and often act for reasons that have little to do with our conscious thoughts.

The concept of the unconscious as proposed by Freud was considered by some to be groundbreaking in that he proposed that awareness existed in layers and that some thoughts occurred "below the surface." "It is difficult - or perhaps impossible - to find a nineteenth-century psychologist or psychiatrist who did not recognize unconscious cerebration as not only real but of the highest importance." Freud's advance was not, then, to uncover the unconscious but to devise a method for systematically studying it

Dreams, which he called the "royal road to the unconscious," provided the best access to our unconscious life and the best illustration of its "logic," which was different from the logic of conscious thought. Freud developed his first topology of the psyche in ***The Interpretation of Dreams*** (1899) in which he proposed the argument that the unconscious exists and described a method for gaining access to it. The preconscious was described as a layer between conscious and unconscious thought—that which we could access with a little effort. Thus for Freud, the ideals of the Enlightenment, positivism and rationalism, could be achieved through understanding, transforming, and mastering the unconscious, rather than through denying or repressing it.

Crucial to the operation of the unconscious is "repression." According to Freud, people often experience thoughts and feelings that are so painful that they cannot bear them. Such thoughts and feelings—and associated memories—could not, Freud argued, be banished from the mind, but could be banished from consciousness. Thus they come to constitute the unconscious. Although Freud later attempted to find patterns of repression among his patients in order to derive a general model of the mind, he also observed that individual patients repress different things. Moreover, Freud observed that the process of repression is itself a non-conscious act (in other words, it did not occur through people willing away certain thoughts or feelings). Freud supposed that what people repressed was in part determined by their unconscious. In other words, the unconscious was for Freud both a cause and effect of repression.

Eventually, Freud abandoned the idea of the system unconscious, replacing it with the concept of the Ego, super-ego, and id. Throughout his career, however, he retained the descriptive and dynamic conceptions of the unconscious.

Freud also believed that the libido developed in individuals by changing its object, a process codified by the concept of sublimation. He argued that humans are born "polymorphously perverse", meaning that any number of objects could be a source of pleasure. He further argued that, as humans develop, they become fixated on different and specific objects through their stages of development—first in the oral stage (exemplified by an infant's pleasure in nursing), then in the anal stage (exemplified by a toddler's pleasure in evacuating his or her bowels), then in the phallic stage. Freud argued that children then passed through a stage in which they fixated on the mother as a sexual object (known as the Oedipus Complex) but that the child eventually overcame and repressed this desire because of its taboo nature. (The lesser known Electra complex refers to such a fixation on the father.) The repressive or dormant latency stage of psychosexual development preceded the sexually mature genital stage of psychosexual development.

Freud's way of interpretation has been called phallocentric by many contemporary thinkers. This is because, for Freud, the unconscious always desires the phallus (penis). Males are afraid of castration - losing their phallus or masculinity to another male. Females always desire to have a phallus - an unfulfillable desire. Thus boys resent their fathers (fear of castration) and girls desire theirs. For Freud, desire is always defined in the negative term of lack - you always desire what you don't have or what you are not and it is very unlikely that you will fulfill this desire. Thus his psychoanalysis treatment is meant to teach the patient to cope with his or her insatiable desires.

In his later work, Freud proposed that the psyche could be divided into three parts: Ego, super-ego, and id. Freud discussed this structural model of the mind in the 1920 essay ***Beyond the Pleasure Principle***, and fully elaborated it in ***The Ego and the Id*** (1923), where he developed it as an alternative to his previous topographic schema (conscious, unconscious, preconscious).

Freud acknowledges that his use of the term Id (or the It) derives from the writings of Georg Grodeck. It is interesting to note that the term Id appears in the earliest writing of Boris Sidis, attributed to William James, as early as 1898.

Freud believed that humans were driven by two conflicting central desires: the life drive (libido) (survival, propagation, hunger, thirst, and sex) and the death drive (Thanatos). Freud's description of Cathexis, whose energy is known as libido, included all creative, life-producing drives. The death drive (or death instinct), whose energy is known as anticathexis, represented an urge inherent in all living things to return to a state of calm: in other words, an inorganic or dead state. He recognized Thanatos only in his later years and develops his theory on the death drive in ***Beyond the Pleasure Principle***.

While many enlightenment thinkers viewed rationality as both an unproblematic ideal and a defining feature of man, Freud's model of the mind drastically reduced the scope and power of reason. In Freud's view, reasoning occurs in the conscious mind--the ego--but this is only a small part of the whole. The mind also contains the hidden, irrational elements of id and superego, which lie outside of conscious control, drive behavior, and motivate conscious activities. As a result, these structures call into question humans' ability to act purely on the basis of reason, since lurking motives are also always at play. Moreover, this model of the mind makes rationality itself suspect, since it may be motivated by hidden urges or societal forces (e.g. defense mechanisms, where reasoning becomes "rationalizing").

Freud was one of the first to recognize the conflicted tripartite mind of contemporary theist.

36. Mohandas Gandhi
1869–1948

"When we see we have gone wrong, it is our duty to retrace our footsteps and proceed again by the right path."

"Cowards can never be moral."

"Fear has its use, but cowardice has none."

"The acquisition of the spirit of nonresistance is a matter of long training in self-denial and appreciation of the hidden forces within ourselves. It changes one's outlook on life . . . It is the greatest force because it is the highest expression of the soul."

"Truth never damages a cause that is just."

"A reformer has to sail not with the current. Very often he has to go against it even though it may cost him his life."

"The mind of a man who remains good under compulsion cannot improve; in fact, it worsens."

"To a true artist, only that face is beautiful which, quite apart from its exterior, shines with the truth within the soul."

"Honest differences are often a healthy sign of progress."

"The weak can never forgive. Forgiveness is the attribute of the strong."

"The greatest of man's spiritual needs is the need to be delivered from the evil and falsity that are in himself and in his society."

"Joy lives in the fight, in the attempt, in the suffering involved, not in the victory itself."

"It is beneath human dignity to lose one's individuality and become a mere cog in the machine."

"Where there is love, there is life; hatred leads to destruction."

"Every murder or other injury, no matter for what cause, committed or inflicted on another is a crime against humanity."

> *"Mankind has to get out of violence only through nonviolence."*
>
> *"True nonviolence is impossible without the possession of unadulterated fearlessness."*
>
> *"Silence becomes cowardice when occasion demands speaking out the whole truth and acting accordingly."*
>
> *"Rights that do not flow from duty well performed are not worth having."*
>
> *"Strength in numbers is the delight of the timid. The valiant in spirit glory in fighting alone."*
>
> *"I believe in God, not as a theory but as a fact more real than that of life itself."*

Gandhi was a major political and spiritual leader of India and the Indian independence movement. He was the pioneer of Satyagraha—resistance to tyranny through mass civil disobedience, firmly founded upon ahimsa or total non-violence—which led India to independence and inspired movements for civil rights and freedom across the world. Gandhi is commonly known in India and across the world as Mahatma Gandhi.

As a British-educated lawyer, Gandhi first employed his ideas of peaceful civil disobedience in the Indian community's struggle for civil rights in South Africa. Upon his return to India, he organized poor farmers and laborers to protest against oppressive taxation and widespread discrimination. Assuming leadership of the Indian National Congress, Gandhi led nationwide campaigns for the alleviation of poverty, for the liberation of women, for brotherhood amongst differing religions and ethnicities, for an end to untouchability and caste discrimination, and for the economic self-sufficiency of the nation, but above all for Swaraj—

the independence of India from foreign domination. Gandhi famously led Indians in the disobedience of the salt tax on the 400 kilometre (248 miles) Dandi Salt March in 1930, and in an open call for the British to Quit India in 1942. He was imprisoned for many years on numerous occasions in both South Africa and India.

Gandhi practiced and advocated non-violence and truth, even in the most extreme situations. A student of Hindu philosophy, he lived simply, organizing an ashram that was self-sufficient in its needs. Making his own clothes—the traditional Indian dhoti and shawl woven with a charkha—he lived on a simple vegetarian diet. He used rigorous fasts, for long periods, for both self-purification and protest.

Gandhi's first major achievements came in 1918 with the Champaran agitation. Suppressed by the militias of the landlords (mostly British), they were given very low compensation, leaving them mired in extreme poverty. The villages were kept extremely dirty and unhygienic; alcoholism and untouchability was rampant. In the throes of a devastating famine, the British levied an oppressive tax which they insisted on increasing. The situation was desperate. Gandhi organized scores of his veteran supporters and fresh volunteers. He organized a detailed study and survey of the villages, accounting for the atrocities and terrible episodes of suffering. Building on the confidence of villagers, he began leading the clean-up of villages, building schools and hospitals and encouraging the village leadership to undo and condemn many social evils.

But his main impact came when he was arrested by police on the charge of creating unrest and was ordered to leave the province. Hundreds of thousands of people protested and rallied outside the jail, police stations and courts demanding his release, which the court reluctantly granted. Gandhi led organized protests and strikes against the landlords, who with the guidance of the British government signed an agreement granting the poor farmers of the region more compensation and control over farming, and cancellation of revenue hikes and its collection until the famine ended. It was during this agitation, that Gandhi was addressed by the people as Bapu (Father) and Mahatma (Great Soul).

Non-cooperation and peaceful resistance were Gandhi's "weapons" in the fight against injustice. In Punjab, the Jallianwala Bagh massacre of civilians by British troops (also known as the Amritsar Massacre) caused deep trauma to the nation, leading to increased public anger and acts of violence. Gandhi criticized both the actions of the British Raj and the retaliatory violence of Indians. He authored the resolution offering condolences to British civilian victims and condemning the riots. After initial opposition in the party, his resolution was accepted following Gandhi's emotional speech advocating his principle that all violence was evil and could not be justified. But it was after the massacre and subsequent violence that Gandhi's mind focused upon obtaining complete self-government and control of all Indian government institutions, maturing soon into Swaraj or complete individual, spiritual, and political independence.

In December 1921, Gandhi was invested with executive authority on behalf of the Indian National Congress. Under his leadership, the Congress was reorganized with a new constitution, with the goal of Swaraj. Membership in the party was opened to anyone prepared to pay a token fee. A hierarchy of committees was set up to improve discipline, transforming the party from an elite organization to one of mass national appeal. Gandhi expanded his non-violence platform to include the swadeshi policy — the boycott of foreign-made goods, especially British goods. Linked to this was his advocacy that khadi (homespun cloth) be worn by all Indians instead of British-made textiles. Gandhi exhorted Indian men and women, rich or poor, to spend time each day spinning khadi in support of the independence movement. This was a strategy to inculcate discipline and dedication to weed out the unwilling and ambitious, and to include women in the movement at a time when many thought that such activities were not respectable activities for women. In addition to boycotting British products, Gandhi urged the people to boycott British educational institutions and law courts, to resign from government employment, and to forsake British titles and honors.

"Non-cooperation" enjoyed wide-spread appeal and success, increasing excitement and participation from all strata of Indian society. Yet, just as the movement reached its apex, it ended abruptly as a result of a violent clash in the town of Chauri Chaura in February 1922. Fearing that the movement was about to take a turn towards violence, and convinced that this would be the undoing of all his work, Gandhi called off the campaign of mass civil disobedience. Gandhi was arrested on March 10, 1922, tried for sedition, and sentenced to six years imprisonment. Beginning on March 18, 1922, he only served about two years of the sentence, being released in February 1924 after an operation for appendicitis.

Without Gandhi's uniting personality, the Indian National Congress began to splinter during his years in prison, splitting into two factions, one led by Chitta Ranjan Das and Motilal Nehru favouring party participation in the legislatures, and the other led by Chakravarti Rajagopalachari and Sardar Vallabhbhai Patel, opposing this move. Furthermore, cooperation among Hindus and Muslims, which had been strong at the height of the non-violence campaign, was breaking down. Gandhi attempted to bridge these differences through many means, including a three-week fast in the autumn of 1924, but with limited success.

Gandhi stayed out of the limelight for most of the 1920s, preferring to resolve the wedge between the Swaraj Party and the Indian National Congress, and expanding initiatives against untouchability, alcoholism, ignorance and poverty. He returned to the fore in 1928. The year before, the British government had appointed a new constitutional reform commission under Sir John Simon, with not a single Indian in its ranks. The result was a boycott of the commission by Indian political parties. Gandhi pushed through a resolution at the Calcutta Congress in December 1928 calling on the British government to grant India dominion status or face a new campaign of non-violence with complete independence for the country as its goal. Gandhi had not only moderated the views of younger men like Subhas Chandra Bose and Jawaharlal Nehru, who sought a demand for immediate independence, but also modified his own call to a one year wait, instead of two. The British did not respond. On December 31, 1929, the flag of India was unfurled in Lahore. January 26, 1930 was celebrated by the Indian National Congress, meeting in Lahore, as India's Independence Day. This day was commemorated by almost every other Indian organization. Making good on his word, he launched a new satyagraha against the tax on salt in March 1930, highlighted by the famous Salt March to Dandi from March 12 to April 6, marching 400 kilometres (248 miles) from Ahmedabad to Dandi, Gujarat to make salt himself. Thousands of Indians joined him on this march to the sea. This campaign was one of his most successful at upsetting British rule; Britain responded by imprisoning over 60,000 people.

The government, represented by Lord Edward Irwin, decided to negotiate with Gandhi. The Gandhi–Irwin Pact was signed in March 1931. The British Government agreed to set all political prisoners free in return for the suspension of the civil disobedience movement. Furthermore, Gandhi was invited to attend the Round Table Conference in London as the sole representative of the Indian National Congress. The conference was a disappointment to Gandhi and the nationalists, as it focused on the Indian princes and Indian minorities rather than the transfer of power. Furthermore, Lord Irwin's successor, Lord Willingdon, embarked on a new campaign of repression against the nationalists. Gandhi was again arrested, and the government attempted to destroy his influence by completely isolating him from his followers. This tactic was not successful. In 1932, through the campaigning of the Dalit leader B. R. Ambedkar, the government granted untouchables separate electorates under the new constitution. In protest, Gandhi embarked on a six-day fast in September 1932, successfully forcing the government to adopt a more equitable arrangement via negotiations mediated by the Dalit cricketer turned political leader Palwankar Baloo. This was the start of a new campaign by Gandhi to improve the lives of the untouchables, whom he named Harijans, the children of God. On May 8, 1933 Gandhi began a 21-day fast of self-purification to help the Harijan movement. In the summer of 1934, three unsuccessful attempts were made on his life.

When the Congress Party chose to contest elections and accept power under the Federation scheme, Gandhi decided to resign from party membership. He did not disagree with the party's move, but felt that if he resigned, his popularity with Indians would cease to stifle the party's membership, that actually varied from communists, socialists, trade unionists, students, religious conservatives, to those with pro-business convictions. Gandhi also did not want to prove a target for Raj propaganda by leading a party that had temporarily accepted political accommodation with the Raj.

Gandhi returned to the head in 1936, with the Nehru presidency and the Lucknow session of the Congress. Although Gandhi desired a total focus on the task of winning independence and not speculation about India's future, he did not restrain the Congress from adopting socialism as its goal. Gandhi had a clash with Subhas Bose, who had been elected to the presidency in 1938. Gandhi's main points of contention with Bose were his lack of commitment to democracy, and lack of faith in nonviolence. Bose won his second term despite Gandhi's criticism, but left the Congress when the All-India leaders resigned en masse in protest against his abandonment of the principles introduced by Gandhi.

World War II broke out in 1939 when Nazi Germany invaded Poland. Initially, Gandhi had favored offering "non-violent moral support" to the British effort, but other Congressional leaders were offended by the unilateral inclusion of India into the war, without consultation of the people's representatives. All Congressmen elected to resign from office en masse. After lengthy deliberations, Gandhi declared that India could not be party to a war ostensibly being fought for democratic freedom, while that freedom was denied to India itself. As the war progressed, Gandhi intensified his demand for independence, drafting a resolution calling for the British to Quit India. This was Gandhi's and the Congress Party's most definitive revolt aimed at securing the British exit from Indian shores.

Gandhi was criticized by some Congress party members and other Indian political groups, both pro-British and anti-British. Some felt that opposing Britain in its life or death struggle was immoral, and others felt that Gandhi wasn't doing enough. Quit India became the most forceful movement in the history of the struggle, with mass arrests and violence on an unprecedented scale. Thousands of freedom fighters were killed or injured by police gunfire, and hundreds of thousands were arrested. Gandhi and his supporters made it clear they would not support the war effort unless India were granted immediate independence. He even clarified that this time the movement would not be stopped if individual acts of violence were committed, saying that the "ordered anarchy" around him was "worse than real anarchy." He called on all Congressmen and Indians to maintain discipline via ahimsa, and Karo Ya Maro ("Do or Die") in the cause of ultimate freedom.

Gandhi and the entire Congress Working Committee were arrested in Bombay by the British on August 9, 1942. Gandhi was held for two years in the Aga Khan Palace in Pune. It was here that Gandhi suffered two terrible blows in his personal life. His 50-year old secretary Mahadev Desai died of a heart attack and 6 days later and his wife Kasturba died. He was released before the end of the war on May 6, 1944 because of his failing health and necessary surgery; the Raj did not want him to die in prison and enrage the nation. Although the Quit India movement had moderate success in its objective, the ruthless suppression of the movement brought order to India by the end of 1943.

At the end of the war, the British gave clear indications that power would be transferred to Indian hands. At this point Gandhi called off the struggle, and around 100,000 political prisoners were released, including the Congress's leadership.

Gandhi advised the Congress to reject the proposals the British Cabinet Mission offered in 1946, as he was deeply suspicious of the grouping proposed for Muslim-majority states—Gandhi viewed this as a precursor to partition. However, this became one of the few times the Congress broke from Gandhi's advice (though not his leadership), as Nehru and Patel knew that if the Congress did not approve the plan, the control of government would pass to the Muslim League. Between 1946 and 1948, over 5,000 people were killed in violence. Gandhi was vehemently opposed to any plan that partitioned India into two separate countries. An overwhelming majority of Muslims living in India, side by side with Hindus and Sikhs, were in favor of Partition. Additionally Muhammad Ali Jinnah, the leader of the Muslim League, commanded widespread support in West Punjab, Sindh, NWFP and East Bengal.

The partition plan was approved by the Congress leadership as the only way to prevent a wide-scale Hindu-Muslim civil war. Congress leaders knew that Gandhi would viscerally oppose partition, and it was impossible for the Congress to go ahead without his agreement, for Gandhi's support in the party and throughout India was strong. Gandhi's closest colleagues had accepted partition as the best way out, and Sardar Patel endeavored to convince Gandhi that it was the only way to avoid civil war. A devastated Gandhi gave his assent.

He conducted extensive dialogue with Muslim and Hindu community leaders, working to cool passions in northern India, as well as in Bengal. Despite the Indo- Pakistani War of 1947, he was troubled when the Government decided to deny Pakistan the Rs. 55 crores due as per agreements made by the Partition Council. Leaders like Sardar Patel feared that Pakistan would use the money to bankroll the war against India. Gandhi was also devastated when demands resurged for all Muslims to be deported to Pakistan, and when Muslim and Hindu leaders expressed frustration and an inability to come to terms with one another. He launched his last fast-unto-death in Delhi, asking that all communal violence be ended once and for all, and that the payment of Rs. 55 crores be made to Pakistan. Gandhi feared that instability and insecurity in Pakistan would increase their anger against India, and violence would spread across the borders. He further feared that Hindus and Muslims would renew their enmity and precipitate into an open civil war. After emotional debates with his life-long colleagues, Gandhi refused to budge, and the Government rescinded its policy and made the payment to Pakistan. Hindu, Muslim and Sikh community leaders, including the Rashtriya Swayamsevak Sangh and Hindu Mahasabha assured him that they would renounce violence and call for peace. Gandhi thus broke his fast by sipping orange juice.

On January 30, 1948, Gandhi was shot and killed while having his nightly public walk on the grounds of the Birla Bhavan (Birla House) in New Delhi. The assassin, Nathuram Godse, was a Hindu radical with links to the extremist Hindu Mahasabha, who held Gandhi responsible for weakening India by insisting upon a payment to Pakistan. Godse and his co-conspirator Narayan Apte were later tried and convicted; they were executed on 15 November 1949. Gandhi's memorial (or Samādhi) at Rāj Ghāt, New Delhi, bears the epigraph "He Ram", which may be translated as "Oh God". These are widely believed to be Gandhi's last words after he was shot, though the veracity of this statement has been disputed.

"An eye for an eye makes the whole world blind."

"There are many causes that I am prepared to die for but no causes that I am prepared to kill for."

For most of my life I have been fascinated with Gandhi. He has taught me a great deal and every time I read about him or read what he has written I learn more. There are far too few Gandhi's in the world.

37. Albert Einstein
1879--1955

Einstein received the 1921 Nobel Prize in Physics "for his services to Theoretical Physics, and especially for his discovery of the law of the photoelectric effect."

Einstein's many contributions to physics include:

1. The special theory of relativity, which reconciled mechanics with electromagnetism
2. The general theory of relativity, a new theory of gravitation obeying the equivalence principle
3. The founding of relativistic cosmology with a cosmological constant
4. The first post-Newtonian expansion, explaining the perihelion advance of Mercury
5. Prediction of the deflection of light by gravity and gravitational lensing
6. The first fluctuation dissipation theorem which explained the Brownian movement of molecules
7. The theory of density fluctuations in gasses and liquids, giving a criterion for critical opalescence
8. The photon theory and wave-particle duality derived from the thermodynamic properties of light
9. The quantum theory of atomic motion in solids
10. Zero-point energy concept
11. The semi classical version of the Schrödinger equation
12. Relations for atomic transition probabilities which predicted stimulated emission
13. The quantum theory of a monatomic gas which predicted Bose-Einstein condensation

15. A program for a unified field theory

16. The geometrization of fundamental physics

Einstein published more than 300 scientific works and more than 150 non-scientific works. He is often regarded as the father of Modern Physics and the greatest scientist of the 20th Century. In 1999 Time magazine named him the Person of the Century, beating contenders like Mahatma Gandhi and Franklin Roosevelt, and in the words of a biographer, "to the scientifically literate and the public at large, Einstein is synonymous with genius".

Einstein's early papers all come from attempts to demonstrate that atoms exist and have a finite nonzero size. At the time of his first paper in 1902, it was not yet completely accepted by physicists that atoms were real, even though chemists had good evidence ever since Antoine Lavoisier's work a century earlier. The reason physicists were skeptical was because no 19th century theory could fully explain the properties of matter from the properties of atoms.

Ludwig Boltzmann was a leading 19th century atomist physicist, who had struggled for years to gain acceptance for atoms. Boltzmann had given an interpretation of the laws of thermodynamics, suggesting that the law of entropy increase is statistical. In Boltzmann's way of thinking, the entropy is the logarithm of the number of ways a system could be configured inside. The reason the entropy goes up is only because it is more likely for a system to go from a special state with only a few possible internal configurations to a more generic state with many. While Boltzmann's statistical interpretation of entropy is universally accepted today, and Einstein believed it, at the turn of the 20th century it was a minority position.

The statistical idea was most successful in explaining the properties of gases. James Clerk Maxwell, another leading atomist, had found the distribution of velocities of atoms in a gas, and derived the surprising result that the viscosity of a gas should be independent of density. Intuitively, the friction in a gas would seem to go to zero as the density goes to zero, but this is not so, because the mean free path of atoms becomes large at low densities. A subsequent experiment by Maxwell and his wife confirmed this surprising prediction. Other experiments on gases and vacuum, using a rotating slitted drum, showed that atoms in a gas had velocities distributed according to Maxwell's distribution law.

In addition to these successes, there were also inconsistencies. Maxwell noted that at cold temperatures, atomic theory predicted specific heats that are too large. In classical statistical mechanics, every spring-like motion has thermal energy $k_B T$ on average at temperature T, so that the specific heat of every spring is Boltzmann's constant k_B. A monatomic solid with N atoms can be thought of as N little balls representing N atoms attached to each other in a box grid with 3N springs, so the specific heat of every solid is $3Nk_B$, a result which became known as the Dulong–Petit law. This law is true at room temperature, but not for colder temperatures. At temperatures near zero, the specific heat goes to zero.

Similarly, a gas made up of two atoms can be thought of as two balls on a spring. This spring has energy $k_B T$ at high temperatures, and should contribute an extra k_B to the specific heat. It does at room temperature, but at low temperature, this contribution disappears. At zero temperature, all other contributions to the specific heat from rotations and vibrations also disappear. This behavior was inconsistent with classical physics.

The most glaring inconsistency was in the theory of light waves. Continuous waves in a box can be thought of as infinitely many spring-like motions, one for each possible standing wave. Each standing wave has a specific heat of k_B, so the total specific heat of a continuous wave like light should be infinite in classical mechanics. This is obviously wrong, because it would mean that all energy in the universe would be instantly sucked up into light waves, and everything would slow down and stop.

These inconsistencies led some people to say that atoms were not physical, but mathematical. Notable among the skeptics was Ernst Mach, whose logical positivist philosophy led him to demand that if atoms are real, it should be possible to see them directly. Mach believed that atoms were a useful fiction, that in reality they could be assumed to be infinitesimally small, that Avogadro's number was infinite, or so large that it might as well be infinite, and k_B was infinitesimally small. Certain experiments could then be explained by atomic theory, but other experiments could not, and this is the way it will always be.

Einstein opposed this position. Throughout his career, he was a realist. He believed that a single consistent theory should explain all observation, and that this theory would be a description what was really going on, underneath it all. So he set out to show that the atomic point of view was correct. This led him first to thermodynamics, then to statistical physics, and to the theory of specific heats of solids.

In 1905, while he was working in the patent office, the leading German language physics journal ***Annalen der Physik*** published four of Einstein's papers. The four papers eventually were recognized as revolutionary, and 1905 became known as Einstein's "Miracle Year", and the papers, as the ***Annus Mirabilis Papers***.

Main article: ***Annus Mirabilis Papers***

Einstein's earliest papers were concerned with thermodynamics. He wrote a paper establishing a thermodynamic identity in 1902, and a few other papers which attempted to interpret phenomena from a statistical atomic point of view.

His research in 1903 and 1904 was mainly concerned with the effect of finite atomic size on diffusion phenomena. As in Maxwell's work, the finite nonzero size of atoms leads to effects which can be observed. This research, and the thermodynamic identity, was well within the mainstream of physics in his time. They would eventually form the content of his PhD thesis.

His first major result in this field was the theory of thermodynamic fluctuations. When in equilibrium, a system has maximum entropy. According to the statistical interpretation, the entropy can fluctuate a little bit. Einstein pointed out that the statistical fluctuations of a macroscopic object, like a mirror suspended on spring, would be completely determined by the second derivative of the entropy with respect to the position of the mirror. This makes a connection between microscopic and macroscopic objects.

Searching for ways to test this relation, his great breakthrough came in 1905. The theory of fluctuations, he realized, would have a visible effect for an object which could move around freely. Such an object would have a velocity which is random, and would move around randomly, just like an individual atom. The average kinetic energy of the object would be kT, and the time decay of the fluctuations would be entirely determined by the law of friction.

The law of friction for a small ball in a viscous fluid like water was discovered by George Stokes. He showed that for small velocities, the friction force would be proportional to the velocity, and to the radius of the particle (see Stokes' law). This relation could be used to calculate how far a small ball in water would travel due to its random thermal motion, and Einstein noted that such a ball, of size about a micron, would travel about a few microns per second. This motion could be easily observed with a microscope. Such a motion had already been observed with a microscope by a Botanist named Brown, and had been called Brownian motion. Einstein was able to identify this motion with the motion predicted by his theory. Since the fluctuations which give rise to Brownian motion are just the same as the fluctuations of the velocities of atoms, measuring the precise amount of Brownian motion using Einstein's theory would show that Boltzmann's constant is nonzero. It would measure Avogadro's number.

These experiments were carried out a few years later, and gave a rough estimate of Avogadro's number consistent with the more accurate estimates due to Max Planck's theory of blackbody light, and Robert Millikan's measurement of the charge of the electron.[32] Unlike the other methods, Einstein's required very few theoretical assumptions or new physics, since it was directly measuring atomic motion on visible grains.

Einstein's theory of Brownian motion was the first paper in the field of statistical physics. It established that thermodynamic fluctuations were related to dissipation. This was shown by Einstein to be true for time-independent fluctuations, but in the Brownian motion paper he showed that dynamical relaxation rates calculated from classical mechanics could be used as statistical relaxation rates to derive dynamical diffusion laws. These relations are known as Einstein relations.

The theory of Brownian motion was the least revolutionary of Einstein's ***Annus mirabilis*** papers, but it had an important role in securing the acceptance of the atomic theory by physicists.

A-priori principles

Einstein's thinking underwent a transformation in 1905. He had come to understand that quantum properties of light mean that Maxwell's equations were only an approximation. He knew that new laws would have to replace these, but he did not know how to go about finding those laws. He felt that guessing formal relations would not go anywhere. So he decided to focus on a-priori principles instead, which are statements about physical laws which can be understood to hold in a very broad sense even in domains where they have not yet been shown to apply. A well accepted example of an a-priori principle is rotational invariance. If a new force is discovered in physics, it is assumed to be rotationally invariant almost automatically, without thought. Einstein sought new principles of this sort, to guide the production of physical ideas. Once enough principles are found, then the new physics will be the simplest theory consistent with the principles and with previously known laws.

The first general a-priori principle he found was the principle of relativity, that uniform motion is indistinguishable from rest. This was understood by Hermann Minkowski to be a generalization of rotational invariance from space to space-time. Other principles postulated by Einstein and later vindicated, are the principle of equivalence and the principle of adiabatic invariance of the quantum number. Another of Einstein's general principles, Mach's principle is fiercely debated, and whether it holds in our world or not is still not definitively established.

The use of a-priori principles is a distinctive unique signature of Einstein's early work, which has become a standard tool in modern theoretical physics.

Special relativity

Einstein's 1905 paper on the electrodynamics of moving bodies introduced the radical theory of special relativity, which showed that the observed independence of the speed of light on the observer's state of motion required fundamental changes to the notion of simultaneity. Consequences of this include the time-space frame of a moving body slowing down and contracting (in the direction of motion) relative to the frame of the observer. This paper also argued that the idea of a luminiferous aether—one of the leading theoretical entities in physics at the time—was superfluous. In his paper on mass–energy equivalence, which had previously considered to be distinct concepts, Einstein deduced from his equations of special relativity what has been called the twentieth century's best-known equation: $E = mc^2$. This equation suggests that tiny amounts of mass could be converted into huge amounts of energy and presaged the development of nuclear power. Einstein's 1905 work on relativity remained controversial for many years, but was accepted by leading physicists, starting with Max Planck.

Photons

In a 1905 paper, Einstein postulated that light itself consists of localized particles (*quanta*). Einstein's light quanta were nearly universally rejected by all physicists, including Max Planck and Niels Bohr. This idea only became universally accepted in 1919, with Robert Millikan's detailed experiments on the photoelectric effect, and with the measurement of Compton scattering.

Einstein's paper on the light particles was almost entirely motivated by thermodynamic considerations. He was not at all motivated by the detailed experiments on the photoelectric effect, which did not confirm his theory until fifteen years later. Einstein considers the entropy of light at temperature T, and decomposes it into a low-frequency part and a high-frequency part. The high-frequency part, where the light is described by Wien's law, has an entropy which looks exactly the same as the entropy of a gas of classical particles.

Since the entropy is the logarithm of the number of possible states, Einstein concludes that the number of states of short wavelength light waves in a box with volume V is equal to the number of states of a group of localizable particles in the same box. Since (unlike others) he was comfortable with the statistical interpretation, he confidently postulates that the light itself is made up of localized particles, as this is the only reasonable interpretation of the entropy.

This leads him to conclude that each wave of frequency f is associated with a collection of photons with energy hf each, where h is Planck's constant. He does not say much more, because he is not sure how the particles are related to the wave. But he does suggest that this idea would explain certain experimental results, notably the photoelectric effect.

Quantized atomic vibrations

Einstein continued his work on quantum mechanics in 1906, by explaining the specific heat anomaly in solids. This was the first application of quantum theory to a mechanical system. Since Planck's distribution for light oscillators had no problem with infinite specific heats, the same idea could be applied to solids to fix the specific heat problem there. Einstein showed in a simple model that the hypothesis that solid motion is quantized explains why the specific heat of a solid goes to zero at zero temperature.

Einstein's model treats each atom as connected to a single spring. Instead of connecting all the atoms to each other, which leads to standing waves with all sorts of different frequencies, Einstein imagined that each atom was attached to a fixed point in space by a spring. This is not physically correct, but it still predicts that the specific heat is $3Nk_B$, since the number of independent oscillations stays the same.

Einstein then assumes that the motion in this model are quantized, according to the Planck law, so that each independent spring motion has energy which is an integer multiple of hf, where f is the frequency of oscillation. With this assumption, he applied Boltzmann's statistical method to calculate the average energy of the spring. The result was the same as the one that Planck had derived for light: for temperatures where $k_B T$ is much smaller than *hf*, the motion is frozen, and the specific heat goes to zero.

So Einstein concluded that quantum mechanics would solve the main problem of classical physics, the specific heat anomaly. The particles of sound implied by this formulation are now called phonons. Because all of Einstein's springs have the same stiffness, they all freeze out at the same temperature, and this leads to a prediction that the specific heat should go to zero exponentially fast when the temperature is low. The solution to this problem is to solve for the independent normal modes individually, and to quantize those. Then each normal mode has a different frequency, and long wavelength vibration modes freeze out at colder temperatures than short wavelength ones. This was done by Debye, and after this modification, Einstein's quantization method reproduced quantitatively the behavior of the specific heats of solids at low temperatures.

This work was the foundation of condensed matter physics.

Adiabatic principle and action-angle variables

Throughout the 1910s, quantum mechanics expanded in scope to cover many different systems. After Ernest Rutherford discovered the nucleus and proposed that electrons orbit like planets, Niels Bohr was able to show that the same quantum mechanical postulates introduced by Planck and developed by Einstein would explain the discrete motion of electrons in atoms, and the periodic table of the elements.

Einstein contributed to these developments by linking them with the 1898 arguments Wilhelm Wien had made. Wien had shown that the hypothesis of adiabatic invariance of a thermal equilibrium state allows all the blackbody curves at different temperature to be derived from one another by a simple shifting process. Einstein noted in 1911 that the same adiabatic principle shows that the quantity which is quantized in any mechanical motion must be an adiabatic invariant. Arnold Sommerfeld identified this adiabatic invariant as the action variable of classical mechanics. The law that the action variable is quantized was the basic principle of the quantum theory as it was known between 1900 and 1925.

Wave–particle duality

Although the patent office promoted Einstein to Technical Examiner Second Class in 1906, he had not given up on *academia*. In 1908, he became a *privatdozent* at the University of Bern. In "über die Entwicklung unserer Anschauungen über das Wesen und die Konstitution der Strahlung" ("The Development of Our Views on the Composition and Essence of Radiation"), on the quantization of light, and in an earlier 1909 paper, Einstein showed that Max Planck's energy quanta must have well-defined momenta and act in some respects as independent, point-like particles. This paper introduced the *photon* concept (although the name *photon* was introduced later by Gilbert N. Lewis in 1926) and inspired the notion of wave–particle duality in quantum mechanics.

Theory of Critical Opalescence

Einstein returned to the problem of thermodynamic fluctuations, giving a treatment of the density variations in a fluid at its critical point. Ordinarily the density fluctuations are controlled by the second derivative of the free energy with respect to the density. At the critical point, this derivative is zero, leading to large fluctuations. The effect of density fluctuations is that light of all wavelengths is scattered, making the fluid look milky white. Einstein relates this to Raleigh scattering, which is what happens when the fluctuation size is much smaller than the wavelength, and which explains why the sky is blue.

Zero-point energy

Einstein's physical intuition led him to note that Planck's oscillator energies had an incorrect zero point. He modified Planck's hypothesis by stating that the lowest energy state of an oscillator is equal to $\frac{1}{2}hf$, to half the energy spacing between levels. This argument, which was made in 1913 in collaboration with Otto Stern, was based on the thermodynamics of a diatomic molecule which can split apart into two free atoms.

Principle of equivalence

In 1907, while still working at the patent office, Einstein had what he would call his "happiest thought". He realized that the principle of relativity could be extended to gravitational fields. He thought about the case of a uniformly accelerated box not in a gravitational field, and noted that it would be indistinguishable from a box sitting still in an unchanging gravitational field. He used special relativity to see that the rate of clocks at the top of a box accelerating upward would be faster than the rate of clocks at the bottom. He concludes that the rates of clocks depend on their position in a gravitational field, and that the difference in rate is proportional to the gravitational potential to first approximation.

Although this approximation is crude, it allowed him to calculate the deflection of light by gravity, and show that it is nonzero. This gave him confidence that the scalar theory of gravity proposed by Gunnar Nordstrom was incorrect. But the actual value for the deflection that he calculated was too small by a factor of two, because the approximation he used doesn't work well for things moving at near the speed of light. When Einstein finished the full theory of general relativity, he would rectify this error, and predict the correct amount of light deflection by the sun.

From Prague, Einstein published a paper about the effects of gravity on light, specifically the gravitational redshift and the gravitational deflection of light. The paper challenged astronomers to detect the deflection during a solar eclipse. German astronomer Erwin Finlay-Freundlich publicized Einstein's challenge to scientists around the world.

Einstein thought about the nature of the gravitational field in the years 1909-1912, studying its properties by means of simple thought experiments. A notable one is the rotating disk. Einstein imagined an observer making experiments on a rotating turntable. He noted that such an observer would find a different value for the mathematical constant pi than the one predicted by Euclidean geometry. The reason is that the radius of a circle would be measured with an uncontracted ruler, but according to special relativity, the circumference would seem to be longer, because the ruler would be contracted.

Since Einstein believed that the laws of physics were local, described by local fields, he concluded from this that spacetime could be locally curved. This led him to study Riemannian geometry, and to formulate general relativity in this language.

Hole argument and Entwurf theory

While developing general relativity, Einstein became confused about the gauge invariance in the theory. He formulated an argument that led him to conclude that a general relativistic field theory is impossible. He gave up looking for fully generally covariant tensor equations, and searched for equations that would be invariant under general linear transformations only.

The Entwurf theory was the result of these investigations. As it name suggests, it was a sketch of a theory, with the equations of motion supplemented by additional gauge fixing conditions. Simultaneously less elegant and more difficult than general relativity, Einstein abandoned the theory after realizing that the hole argument was mistaken.

General relativity

In 1912, Einstein returned to Switzerland to accept a professorship at his *alma mater,* the Eidgenössische Technische Hochschule. Once back in Zurich, he immediately visited his old classmate Marcel Grossmann, now a professor of mathematics, who introduced him to Riemannian geometry and, more generally, to differential geometry. On the recommendation of Italian mathematician Tullio Levi-Civita, Einstein began exploring the usefulness of general covariance (essentially the use of tensors) for his gravitational theory. For a while Einstein thought that there were problems with the approach, but he later returned to it and, by late 1915, had published his general theory of relativity in the form in which it is used today. This theory explains gravitation as distortion of the structure of spacetime by matter, affecting the inertial motion of other matter. During World War I, the work of Central Powers scientists was available only to Central Powers academics, for national security reasons. Some of Einstein's work did reach the United Kingdom and the United States through the efforts of the Austrian Paul Ehrenfest and physicists in the Netherlands, especially 1902 Nobel Prize-winner Hendrik Lorentz and Willem de Sitter of Leiden University. After the war ended, Einstein maintained his relationship with Leiden University, accepting a contract as an *Extraordinary Professor*; for ten years, from 1920 to 1930, he travelled to Holland regularly to lecture.

In 1917, several astronomers accepted Einstein 's 1911 challenge from Prague. The Mount Wilson Observatory in California, U.S., published a solar spectroscopic analysis that showed no gravitational redshift. In 1918, the Lick Observatory, also in California, announced that it too had disproved Einstein's prediction, although its findings were not published.

However, in May 1919, a team led by the British astronomer Arthur Stanley Eddington claimed to have confirmed Einstein's prediction of gravitational deflection of starlight by the Sun while photographing a solar eclipse with dual expeditions in Sobral, northern Brazil, and Príncipe, a west African island. Nobel laureate Max Born praised general relativity as the "greatest feat of human thinking about nature"; fellow laureate Paul Dirac was quoted saying it was "probably the greatest scientific discovery ever made". The international media guaranteed Einstein's global renown. There have been later claims that scrutiny of the specific photographs taken on the Eddington expedition showed the experimental uncertainty to be comparable to the same magnitude as the effect Eddington claimed to have demonstrated, and that a 1962 British expedition concluded that the method was inherently unreliable. The deflection of light during a solar eclipse was confirmed by later, more accurate observations. Some resented the newcomer's fame, notably among some German physicists, who later started the *Deutsche Physik* (German Physics) movement.

Cosmology

In 1917, Einstein applied the General theory of relativity to model the structure of the universe as a whole. He wanted the universe to be eternal and unchanging, but this type of universe is not consistent with relativity. To fix this, Einstein modified the general theory by introducing a new notion, the cosmological constant. With a positive cosmological constant, the universe could be an eternal static sphere.

Einstein believed a spherical static universe is philosophically preferred, because it would obey Mach's principle. He had shown that general relativity incorporates Mach's principle to a certain extent in frame dragging by gravitomagnetic fields, but he knew that Mach's idea would not work if space goes on forever. In a closed universe, he believed that Mach's principle would hold.

Modern quantum theory

In 1917, at the height of his work on relativity, Einstein published an article in *Physikalische Zeitschrift* that proposed the possibility of stimulated emission, the physical process that makes possible the maser and the laser. This article showed that the statistics of absorption and emission of light would only be consistent with Planck's distribution law if the emission of light into a mode with n photons would be enhanced statistically compared to the emission of light into an empty mode. This paper was enormously influential in the later development of quantum mechanics, because it was the first paper to show that the statistics of atomic transitions had simple laws. Einstein discovered Louis de Broglie's work, and supported his ideas, which were received skeptically at first. In another major paper from this era, Einstein gave a wave equation for de Broglie waves, which Einstein suggested was the Hamilton–Jacobi equation of mechanics. This paper would inspire Schrödinger's work of 1926.

Bose–Einstein statistics

In 1924, Einstein received a description of a statistical model from Indian physicist Satyendra Nath Bose, based on a counting method that assumed that light could be understood as a gas of indistinguishable particles. Einstein noted that Bose's statistics applied to some atoms as well as to the proposed light particles, and submitted his translation of Bose's paper to the *Zeitschrift für Physik*. Einstein also published his own articles describing the model and its implications, among them the Bose–Einstein condensate phenomenon that some particulates should appear at very low temperatures. It was not until 1995 that the first such condensate was produced experimentally by Eric Allin Cornell and Carl Wieman using ultra-cooling equipment built at the NIST-JILA laboratory at the University of Colorado at Boulder. Bose–Einstein statistics are now used to describe the behaviors of any assembly of bosons. Einstein's sketches for this project may be seen in the Einstein Archive in the library of the Leiden University.

Energy momentum pseudotensor

General relativity includes a dynamical spacetime, so it is difficult to see how to identify the conserved energy and momentum. Noether's theorem allows these quantities to be determined from a Lagrangian with translation invariance, but general covariance makes translation invariance into something of a gauge symmetry. The energy and momentum derived within general relativity by Noether's prescriptions do not make a real tensor for this reason.

Einstein argued that this is true for fundamental reasons, because the gravitational field could be made to vanish by a choice of coordinates. He maintained that the noncovariante energy momentum pseudotensor was in fact the best description of the energy momentum distribution in a gravitational field. This approach has been echoed by Lev Landau and Evgeny Lifshitz, and others, and has become standard.

Unified field theory

Following his research on general relativity, Einstein entered into a series of attempts to generalize his geometric theory of gravitation, which would allow the explanation of electromagnetism. In 1950, he described his "unified field theory" in a *Scientific American* article entitled "On the Generalized Theory of Gravitation." Although he continued to be lauded for his work, Einstein became increasingly isolated in his research, and his efforts were ultimately unsuccessful. In his pursuit of a unification of the fundamental forces, Einstein ignored some mainstream developments in physics, most notably the strong and weak nuclear forces, which were not well understood until many years after his death. Mainstream physics, in turn, largely ignored Einstein's approaches to unification. Einstein's dream of unifying other laws of physics with gravity motivates modern quests for a theory of everything and in particular string theory, where geometrical fields emerge in a unified quantum-mechanical setting.

Equations of motion

The theory of general relativity has two fundamental laws—the Einstein equations which describe how space curves, and the geodesic equation which describes how particles move. Since the equations of general relativity are non-linear, a lump of energy made out of pure gravitational fields, like a black hole, would move on a trajectory which is determined by the Einstein equations themselves, not by a new law. So Einstein proposed that the path of a singular solution, like a black hole, would be determined to be a geodesic from general relativity itself. This was established by Einstein, Infeld and Hoffmann for pointlike objects without angular momentum, and by Roy Kerr for spinning objects.

Einstein and the Atom Bomb

Albert Einstein did not directly participate in the invention of the atomic bomb. He was instrumental in facilitating its development. But bombs were not what Einstein had in mind when he published $E=MC^2$. Indeed, he considered himself to be a pacifist. In 1929, he publicly declared that if a war broke out he would "unconditionally refuse to do war service, direct or indirect... regardless of how the cause of the war should be judged."

His position changed in 1933, as the result of Adolf Hitler's ascent to power in Germany. While still promoting peace, Einstein no longer fit his previous self-description of being an "absolute pacifist".

Einstein's greatest role in the invention of the atomic bomb was signing a letter to President Franklin Roosevelt urging that the bomb be built. The splitting of the uranium atom in Germany in December 1938 plus continued German aggression led some physicists to fear that Germany might be working on an atomic bomb. Among those concerned were physicists Leo Szilard and Eugene Wigner. But Szilard and Wigner had no influence with those in power. So in July 1939 they explained the problem to someone who did: Albert Einstein. According to Szilard, Einstein said the possibility of a chain reaction "never occurred to me", although Einstein was quick to understand the concept . After consulting with Einstein, in August 1939 Szilard wrote a letter to President Roosevelt with Einstein's signature on it. The letter was delivered to Roosevelt in October 1939 by Alexander Sachs, a friend of the President. Germany had invaded Poland the previous month; the time was ripe for action. That October the Briggs Committee was appointed to study uranium chain reactions.

But the Briggs Committee moved very slowly, prompting Einstein, Szilard, and Sachs to write to FDR in March 1940, pointing again to German progress in uranium research. In April 1940 an Einstein letter, ghost-written by Szilard, pressed Briggs Committee chairman Lyman Briggs on the need for "greater speed"

Research still proceeded slowly, because the invention of the atomic bomb seemed distant and unlikely, rather than a weapon that might be used in the current war. It was not until after the British MAUD Report was presented to FDR in October 1941 that a more accelerated pace was taken. This British document stated that an atomic bomb **could** be built and that it might be ready for use by late 1943, in time for use during the war.

The atomic bomb related work that Einstein did was very limited and he completed it in two days during December 1941. Vannevar Bush, who was coordinating the scientific work on the a-bomb at that time, asked Einstein's advice on a theoretical problem involved in separating fissionable material by gaseous diffusion. But Bush and other leaders in the atomic bomb project excluded Einstein from any other a-bomb related work. Bush didn't trust Einstein to keep the project a secret.

As the realization of nuclear weapons grew near, Einstein looked beyond the current war to future problems that such weapons could bring. He wrote to physicist Niels Bohr in December 1944.

"when the war is over, then there will be in all countries a pursuit of secret war preparations with technological means which will lead inevitably to preventative wars and to destruction even more terrible than the present destruction of life."

The atomic bombings of Japan occurred three months after the surrender of Germany, whose potential for creating a Nazi a-bomb had led Einstein to push for the development of an a-bomb for the Allies. Einstein withheld public comment on the atomic bombing of Japan until a year afterward. A short article on the front page of the New York Times contained his view: "Prof. Albert Einstein... said that he was sure that Roosevelt would have forbidden the atomic bombing of Hiroshima had he been alive and that it was probably carried out to end the Pacific war before Russia could participate." Einstein later wrote, "I have always condemned the use of the atomic bomb against Japan."

In November 1954, five months before his death, Einstein summarized his feelings about his role in the creation of the atomic bomb: "I made one great mistake in my life... when I signed the letter to President Roosevelt recommending that atom bombs be made; but there was some justification - the danger that the Germans would make them."

Einstein's Religious Views

The question of scientific determinism gave rise to questions about Einstein's position on theological determinism, and whether or not he believed in God, or in a God. In 1929, Einstein told Rabbi Herbert S. Goldstein "I believe in Spinoza's God, who reveals Himself in the lawful harmony of the world, not in a God who concerns Himself with the fate and the doings of mankind."

38. Ludwig von Mises
1881-1973

"It is certainly true that our age is full of conflicts which generate war. However, these conflicts do not spring from the operation of the unhampered market society. It may be permissible to call them economic conflicts because they concern that sphere of human life which is, in common speech, known as the sphere of economic activities. But it is a serious blunder to infer from this appellation that the source of these conflicts is conditions which develop within the frame of a market society. It is not capitalism that produces them, but precisely the anti-capitalistic policies designed to check the functioning of capitalism. They are an outgrowth of the various governments' interference with business, of trade and migration barriers and discrimination against foreign labor, foreign products, and foreign capital."

To avoid the influence of Nazis in his Austrian homeland, and fearing repression due to his Jewish ancestry, in 1934 Mises left for Geneva, Switzerland, where he was a professor at the Graduate Institute of International Studies until 1940. In 1940, he immigrated to New York City. He was a visiting professor at New York University from 1945 until he retired in 1969, though he was not salaried by the university. Instead, he earned his living from funding by businessmen such as Lawrence Fertig. For part of this period Mises worked on currency issues. He also wrote and lectured extensively on behalf of classical liberalism and is seen as one of the leaders of the Austrian School of economics. In his treatise on economics, **_Human Action,_** Mises introduced praxeology as the conceptual foundation of the science of human action, establishing economic laws of apodictic certainty rejecting positivism and material causality.

Mises argued that money is demanded for its usefulness in purchasing other goods, rather than for its own sake and that any unsound credit expansion causes business cycles. His other notable contribution was his argument that socialism must fail economically because of the economic calculation problem — the impossibility of a socialist government being able to make the economic calculations required to organize a complex economy. Mises projected that without a market economy there would be no functional price system, which he held essential for achieving rational allocation of capital goods to their most productive uses. Socialism would fail as demand cannot be known without prices, according to Von Mises. Mises' criticism of socialist paths of economic development is well-known:

"The only certain fact about Russian affairs under the Soviet regime with regard to which all people agree is: that the standard of living of the Russian masses is much lower than that of the masses in the country which is universally considered as the paragon of capitalism, the United States of America. If we were to regard the Soviet regime as an experiment, we would have to say that the experiment has clearly demonstrated the superiority of capitalism and the inferiority of socialism."

These arguments were elaborated on by subsequent Austrian economists such as Friedrich Hayek.

In ***Interventionism, An Economic Analysis*** (1940), Mises wrote:

> "The usual terminology of political language is stupid. What is 'left' and what is 'right'? Why should Hitler be 'right' and Stalin, his temporary friend, be 'left'? Who is 'reactionary' and who is 'progressive'? Reaction against an unwise policy is not to be condemned. And progress towards chaos is not to be commended. Nothing should find acceptance just because it is new, radical, and

fashionable. 'Orthodoxy' is not an evil if the doctrine on which the 'orthodox' stand is sound. Who is anti-labor, those who want to lower labor to the Russian level, or those who want for labor the capitalistic standard of the United States? Who is 'nationalist,' those who want to bring their nation under the heel of the Nazis, or those who want to preserve its independence?"

Bibliography

The Development of the Relationship Between Peasant and Lord of the Manor in Galicia, 1772-1848 (1902, never translated into English)

The Theory of Money and Credit (1912, enlarged US edition 1953)

Nation, State, and Economy (1919)

Socialism: An Economic and Sociological Analysis (online version) (1922, 1932, 1951)

Liberalism (1927, 1962)

Critique of Interventionism (1929)

Epistemological Problems of Economics (1933, 1960)

Omnipotent Government: The Rise of Total State and Total War (1944) Preview

Bureaucracy (1944)

Planned Chaos (1947, added to 1951 edition of Socialism)

Human Action: A Treatise on Economics (1949, 1963, 1966, 1996)

Planning for Freedom (1952, enlarged editions in 1962, 1974, and 1980)

The Anti-Capitalistic Mentality (1956)

Theory and History: An Interpretation of Social and Economic Evolution (1957)

The Ultimate Foundation of Economic (1962)

Notes and Recollections (1978)

Economic Policy: Thoughts for Today and Tomorrow (1979, lectures given in 1959)

Interventionism: An Economic Analysis (1998)

Von Mises is to economics what Einstein is to physics. And yet most people have never heard of him. Compare the state of physics with the state of economics and you know why everyone should know and study him.

39. Ayn Rand
Alisa Zinov'yevna Rosenbaum
1905 -1982

"A government is the most dangerous threat to man's rights: it holds a legal monopoly on the use of physical force against legally disarmed victims."

"Achievement of your happiness is the only moral purpose of your life, and that happiness, not pain or mindless self-indulgence, is the proof of your moral integrity, since it is the proof and the result of your loyalty to the achievement of your values."

"Achieving life is not the equivalent of avoiding death."

> "Ask yourself whether the dream of heaven and greatness should be waiting for us in our graves - or whether it should be ours here and now and on this earth."

> "Civilization is the progress toward a society of privacy. The savage's whole existence is public, ruled by the laws of his tribe. Civilization is the process of setting man free from men."

> "Do not ever say that the desire to "do good" by force is a good motive. Neither power-lust nor stupidity are good motives."

> "Every aspect of Western culture needs a new code of ethics - a rational ethics - as a precondition of rebirth."

> "Force and mind are opposites; morality ends where a gun begins."

> "It only stands to reason that where there's sacrifice, there's someone collecting the sacrificial offerings. Where there's service, there is someone being served. The man who speaks to you of sacrifice is speaking of slaves and masters, and intends to be the master."

Rand was a Russian-born American novelist and philosopher. She is widely known for her best-selling novels ***The Fountainhead*** and ***Atlas Shrugged***, and for developing a philosophical system she called Objectivism.

She was an uncompromising advocate of rational individualism and laissez-faire capitalism, and vociferously opposed socialism, altruism, and other contemporary philosophical trends.

Rand's writing (both fiction and non-fiction) emphasizes the philosophic concepts of objective reality in metaphysics, reason in epistemology, and rational egoism in ethics. In politics she was a proponent of laissez-faire capitalism and a staunch defender of individual rights, believing that the sole function of a proper government is protection of individual rights (including property rights).

She believed that individuals must choose their values and actions solely by reason, and that;

"Man — every man — is an end in himself, not the means to the ends of others."

According to Rand, the individual;

"must exist for his own sake, neither sacrificing himself to others nor sacrificing others to himself. The pursuit of his own rational self-interest and of his own happiness is the highest moral purpose of his life."

Rand decried the initiation of force and fraud, and held that government action should consist only in protecting citizens from criminal aggression (via the police) and foreign aggression (via the military) and in maintaining a system of courts to decide guilt or innocence for objectively defined crimes and to resolve disputes. Her politics are generally described as minarchist and libertarian, though she did not use the first term and disavowed any connection to the second.

Rand's magnum opus, ***Atlas Shrugged***, was published in 1957. The book went on to become an international bestseller. Although the frequent claim that Atlas Shrugged became the "second most influential book in America, after ***The Bible***," may be an exaggeration of the findings of a 1991 survey, ***Atlas Shrugged*** has been cited in numerous interviews as the book that most influenced the subject.

Atlas Shrugged is often seen as Rand's most extensive statement of Objectivism in any of her works of fiction. In its appendix, she offered this summary:

"My philosophy, in essence, is the concept of man as a heroic being, with his own happiness as the moral purpose of his life, with productive achievement as his noblest activity, and reason as his only absolute."

The theme of **_Atlas Shrugged_** is "The role of man's mind in society." Rand upheld the industrialist as one of the most admirable members of any society and fiercely opposed the popular resentment accorded to industrialists. This led her to envision a novel wherein the industrialists of America go on strike and retreat to a mountainous hideaway, where they build an independent free economy with gold currency. The American economy and its society in general, deprived of its most productive members, slowly start to collapse. The government responds by increasing the already stifling controls on industry.

The novel, which includes elements of mystery and science fiction, deals with other diverse issues as wide-ranging as sex, music, medicine, politics, philosophy, industry, and human ability.

Rand's philosophical system, Objectivism, encompasses positions on metaphysics, epistemology, ethics, politics and aesthetics. While there have been "objectivist" theories in the past, Rand's Objectivism uses the term in a new way: it treats knowledge and values as neither subjective, nor intrinsic in existence (the traditional meaning of "objective") but rather as the factual identification, by Man's mind, of what exists.

Rand's greatest influence was Aristotle, especially **_Organon_** ("Logic"); she considered Aristotle the greatest philosopher. In particular, her philosophy reflects an Aristotelian epistemology and metaphysics – both Aristotle and Rand argued that "there exists an objective reality that is independent of mind and that is capable of being known." Although Rand was ultimately critical of Aristotle's ethics, others have noted her egoistic ethics "is a system of guidelines required by human beings to live their lives successfully, to flourish, and to survive as 'man qua man

Rand recognized the evil of collectivism and self-sacrifice. My admiration and indebtedness to Rand is immense. This book would not have possible without the insights Rand provided. Imagine making the claim that every person, every member of the masses, has a sovereign right to their own life! Her influence upon civilization will be enormously positive and it is just beginning. Everyone concerned with the abolition of slavery and with freedom must read and understand Rand!

40. Martin Luther King, Jr.
1929 – 1968

"I look to a day when people will not be judged by the color of their skin, but by the content of their character."

"I refuse to accept the view that mankind is so tragically bound to the starless midnight of racism and war that the bright daybreak of peace and brotherhood can never become a reality... I believe that unarmed truth and unconditional love will have the final word."

"Means we use must be as pure as the ends we seek."

"Never forget that everything Hitler did in Germany was legal."

"Our scientific power has outrun our spiritual power. We have guided missiles and misguided men."

Martin Luther King, Jr.

Martin Luther King, Jr. was an American clergyman, activist and prominent leader in the African-American civil rights movement. His main legacy was to secure progress on civil rights in the United States, and he has become a human rights icon: King is recognized as a martyr by two Christian churches. A Baptist minister, King became a civil rights activist early in his career. He led the 1955 Montgomery Bus Boycott and helped found the Southern Christian Leadership Conference in 1957, serving as its first president. King's efforts led to the 1963 March on Washington, where King delivered his "I Have a Dream" speech. There, he raised public consciousness of the civil rights movement and established himself as one of the greatest orators in U.S. history.

In 1964, King became the youngest person to receive the Nobel Peace Prize for his work to end racial segregation and racial discrimination through civil disobedience and other non-violent means. By the time of his death in 1968, he had refocused his efforts on ending poverty and opposing the Vietnam War, both from a religious perspective. King was assassinated on April 4, 1968, in Memphis, Tennessee. He was posthumously awarded the Presidential Medal of Freedom in 1977 and Congressional Gold Medal in 2004; Martin Luther King, Jr. Day was established as a U.S. national holiday in 1986.

Martin Luther King, Jr., was born on January 15, 1929, in Atlanta, Georgia. He was the son of the Reverend Martin Luther King, Sr. and Alberta Williams King. King's father was born "Michael King," and Martin Luther King, Jr., was originally named "Michael King, Jr.," until the family traveled to Europe in 1934 and visited Germany. His father soon changed both of their names to Martin Luther in honor of the German Protestant leader Martin Luther. He had an older sister, Willie Christine King, and a younger brother Alfred Daniel Williams King. King sang with his church choir at the 1939 Atlanta premiere of the movie *Gone with the Wind.*

Growing up in Atlanta, King attended Booker T. Washington High School. He skipped ninth and twelfth grade, and entered Morehouse College at age fifteen without formally graduating from high school. In 1948, he graduated from Morehouse with a Bachelor of Arts degree in sociology, and enrolled in Crozer Theological Seminary in Chester, Pennsylvania, from which he graduated with a Bachelor of Divinity degree in 1951. King then began doctoral studies in systematic theology at Boston University and received his Doctor of Philosophy on June 5, 1955. A 1980s inquiry concluded portions of his dissertation had been plagiarized and he had acted improperly but that his dissertation still "makes an intelligent contribution to scholarship."

King married Coretta Scott, on June 18, 1953, on the lawn of her parents' house in her hometown of Heiberger, Alabama. King and Scott had four children; Yolanda King, Martin Luther King III, Dexter Scott King, and Bernice King. King became pastor of the Dexter Avenue Baptist Church in Montgomery, Alabama when he was twenty-five years old in 1954.

Populist tradition and Black populism

Harry C. Boyte, a self-proclaimed populist, field secretary of the Southern Christian Leadership Conference and white civil rights activist describes an episode in his life that gives insight on some of King's influences:

My first encounter with deeper meanings of populism came when I was nineteen, working as a field secretary for the Southern Christian Leadership Conference (SCLC) in St. Augustine, Florida in 1964. One day I was caught by five men and a woman who were members of the Ku Klux Klan. They accused me of being a "communist and a Yankee." I replied, "I'm no Yankee – my family has been in the South since before the Revolution. And I'm not a communist. I'm a populist. I believe that blacks and poor whites should join to do something about the big shots who keep us divided." For a few minutes we talked about what such a movement might look like. Then they let me go.

When he learned of the incident, Martin Luther King, head of SCLC, told me that he identified with the populist tradition and assigned to organize poor whites.

Thurman

Civil rights leader, theologian, and educator Howard Thurman was an early influence on King. A classmate of King's father at Morehouse College, Thurman mentored the young King and his friends. Thurman's missionary work had taken him abroad where he had met and conferred with Mahatma Gandhi. When he was a student at Boston University, King often visited Thurman, who was the dean of Marsh Chapel. Walter Fluker, who has studied Thurman's writings, has stated, "I don't believe you'd get a Martin Luther King, Jr. without a Howard Thurman".

Gandhi and Rustin

Inspired by Gandhi's success with non-violent activism, King visited the Gandhi family in India in 1959, with assistance from the Quaker group the American Friends Service Committee. The trip to India affected King in a profound way, deepening his understanding of non-violent resistance and his commitment to America's struggle for civil rights. In a radio address made during his final evening in India, King reflected, "Since being in India, I am more convinced than ever before that the method of nonviolent resistance is the most potent weapon available to oppressed people in their struggle for justice and human dignity. In a real sense, Mahatma Gandhi embodied in his life certain universal principles that are inherent in the moral structure of the universe, and these principles are as inescapable as the law of gravitation." African American civil rights activist Bayard Rustin, who had studied Gandhi's teachings, counseled King to dedicate himself to the principles of non-violence, served as King's main advisor and mentor throughout his early activism, and was the main organizer of the 1963 March on Washington Rustin's open homosexuality, support of democratic socialism, and his former ties to the Communist Party USA caused many white and African-American leaders to demand King distance himself from Rustin.

Main article: Sermons and speeches of Martin Luther King, Jr.

Throughout his career of service, King wrote and spoke frequently, drawing on his experience as a preacher. His "Letter from Birmingham Jail", written in 1963, is a "passionate" statement of his crusade for justice. On October 14, 1964, King became the youngest recipient of the Nobel Peace Prize, which was awarded to him for leading non-violent resistance to end racial prejudice in the United States.

Montgomery Bus Boycott, 1955

In March 1955, a fifteen-year-old school girl, Claudette Colvin, refused to give up her bus seat to a white man in compliance with the Jim Crow laws. King was on the committee from the Birmingham African-American community that looked into the case; Edgar Nixon and Clifford Durr decided to wait for a better case to pursue. On December 1, 1955, Rosa Parks was arrested for refusing to give up her seat. The Montgomery Bus Boycott, urged and planned by Nixon and led by King, soon followed. The boycott lasted for 385 days, and the situation became so tense that King's house was bombed. King was arrested during this campaign, which ended with a United States District Court ruling in *Browder v. Gayle* that ended racial segregation on all Montgomery public buses.

SCLC
Southern Christian Leadership Conference

In 1957, King, Ralph Abernathy, and other civil rights activists founded the Southern Christian Leadership Conference (SCLC). The group was created to harness the moral authority and organizing power of black churches to conduct non-violent protests in the service of civil rights reform. King led the SCLC until his death. In 1958, while signing copies of his book *Stride Toward Freedom* in a Harlem department store, he was stabbed in the chest by Izola Curry, a deranged black woman with a letter opener, and narrowly escaped death.

Gandhi's nonviolent techniques were useful to King's campaign to correct the civil rights laws implemented in Alabama. King applied non-violent philosophy to the protests organized by the SCLC. In 1959, he wrote *The Measure of A Man*, from which the piece *What is Man?*, an attempt to sketch the optimal political, social, and economic structure of society, is derived. His SCLC secretary and personal assistant in this period was Dora McDonald.

The FBI, under written directive from Attorney General Robert F. Kennedy, began telephone tapping King in the Fall of 1963. Concerned that allegations (of Communists in the SCLC), if made public, would derail the Administration's civil rights initiatives, Kennedy warned King to discontinue the suspect associations, and later felt compelled to issue the written directive authorizing the FBI to wiretap King and other leaders of the Southern Christian Leadership Conference. J. Edgar Hoover feared Communists were trying to infiltrate the Civil Rights Movement, but when no such evidence emerged, the bureau used the incidental details caught on tape over the next five years in attempts to force King out of the preeminent leadership position.

King believed that organized, nonviolent protest against the system of southern segregation known as Jim Crow laws would lead to extensive media coverage of the struggle for black equality and voting rights. Journalistic accounts and televised footage of the daily deprivation and indignities suffered by southern blacks, and of segregationist violence and harassment of civil rights workers and marchers, produced a wave of sympathetic public opinion that convinced the majority of Americans that the Civil Rights Movement was the most important issue in American politics in the early 1960s.

King organized and led marches for blacks' right to vote, desegregation, labor rights and other basic civil rights. Most of these rights were successfully enacted into the law of the United

States with the passage of the Civil Rights Act of 1964 and the 1965 Voting Rights Act.

King and the SCLC applied the principles of nonviolent protest with great success by strategically choosing the method of protest and the places in which protests were carried out. There were often dramatic stand-offs with segregationist authorities. Sometimes these confrontations turned violent.

Albany movement

The Albany Movement was a desegregation coalition formed in Albany, Georgia in November, 1961. In December King and the SCLC became involved. The movement mobilized thousands of citizens for a broad-front nonviolent attack on every aspect of segregation within the city and attracted nationwide attention. When King first visited on December 15, 1961, he "had planned to stay a day or so and return home after giving counsel." But the following day he was swept up in a mass arrest of peaceful demonstrators, and he declined bail until the city made concessions. "Those agreements", said King, "were dishonored and violated by the city," as soon as he left town. King returned in July 1962, and was sentenced to forty-five days in jail or a $178 fine. He chose jail. Three days into his sentence, Chief Pritchett discreetly arranged for King's fine to be paid and ordered his release. "We had witnessed persons being kicked off lunch counter stools ... ejected from churches ... and thrown into jail ... But for the first time, we witnessed being kicked out of jail."

After nearly a year of intense activism with few tangible results, the movement began to deteriorate. King requested a halt to all demonstrations and a "Day of Penance" to promote nonviolence and maintain the moral high ground. Divisions within the black community and the canny, low-key response by local government defeated efforts. However, it was credited as a key lesson in tactics for the national civil rights movement.

Birmingham campaign

The Birmingham campaign was a strategic effort by the SCLC to promote civil rights for African Americans. Many of its tactics of "Project C" were developed by Rev. Wyatt Tee Walker, Executive Director of SCLC from 1960–1964. Based on actions in Birmingham, Alabama, its goal was to end the city's segregated civil and discriminatory economic policies. The campaign lasted for more than two months in the spring of 1963. To provoke the police into filling the city's jails to overflowing, King and black citizens of Birmingham employed nonviolent tactics to flout laws they considered unfair. King summarized the philosophy of the Birmingham campaign when he said, "The purpose of ... direct action is to create a situation so crisis-packed that it will inevitably open the door to negotiation".

Protests in Birmingham began with a boycott to pressure businesses to sales jobs and other employment to people of all races, as well as to end segregated facilities in the stores. When business leaders resisted the boycott, King and the SCLC began what they termed Project C, a series of sit-ins and marches intended to provoke arrest. After the campaign ran low on adult volunteers, it recruited children for what became known as the "Children's Crusade". During the protests, the Birmingham Police Department, led by Eugene "Bull" Connor, used high-pressure water jets and police dogs to control protesters, including children. Not all of the demonstrators were peaceful, despite the avowed intentions of the SCLC. In some cases, bystanders attacked the police, who responded with force. King and the SCLC were criticized for putting children in harm's way. By the end of the campaign, King's reputation improved immensely, Connor lost his job, the "Jim Crow" signs in Birmingham came down, and public places became more open to blacks.

Augustine and Selma

King and SCLC were also driving forces behind the protest in St. Augustine, Florida, in 1964. The movement engaged in nightly marches in the city met by white segregationists who violently assaulted them. Hundreds of the marchers were arrested and jailed.

King and the SCLC joined forces with the Student Nonviolent Coordinating Committee (SNCC) in Selma, Alabama, in December 1964, where SNCC had been working on voter registration for several months. A sweeping injunction issued by a local judge barred any gathering of 3 or more people under sponsorship of SNCC, SCLC, or DCVL, or with the involvement of 41 named civil rights leaders. This injunction temporarily halted civil rights activity until King defied it by speaking at Brown Chapel on January 2 1965.

March on Washington, 1963

King, representing SCLC, was among the leaders of the so-called "Big Six" civil rights organizations who were instrumental in the organization of the March on Washington for Jobs and Freedom in 1963. The other leaders and organizations comprising the Big Six were: Roy Wilkins from the National Association for the Advancement of Colored People; Whitney Young, National Urban League; A. Philip Randolph, Brotherhood of Sleeping Car Porters; John Lewis, SNCC; and James L. Farmer, Jr. of the Congress of Racial Equality. The primary logistical and strategic organizer was King's colleague Bayard Rustin. For King, this role was another which courted controversy, since he was one of the key figures who acceded to the wishes of President John F. Kennedy in changing the focus of the march Kennedy initially opposed the march outright, because he was concerned it would negatively impact the drive for passage of civil rights legislation, but the organizers were firm that the march would proceed.

The march originally was conceived as an event to dramatize the desperate condition of blacks in the southern United States and a very public opportunity to place organizers' concerns and grievances squarely before the seat of power in the nation's capital. Organizers intended to excoriate and then challenge the federal government for its failure to safeguard the civil rights and physical safety of civil rights workers and blacks, generally, in the South. However, the group acquiesced to presidential pressure and influence, and the event ultimately took on a far less strident tone. As a result, some civil rights activists felt it presented an inaccurate, sanitized pageant of racial harmony; Malcolm X called it the "Farce on Washington," and members of the Nation of Islam were not permitted to attend the march.

King is perhaps most famous for his "I Have a Dream" speech, given in front of the Lincoln Memorial during the 1963 March on Washington for Jobs and Freedom. The march did, however, make specific demands: an end to racial segregation in public school; meaningful civil rights legislation, including a law prohibiting racial discrimination in employment; protection of civil rights workers from police brutality; a $2 minimum wage for all workers; and self-government for Washington, D.C., then governed by congressional committee. Despite tensions, the march was a resounding success. More than a quarter million people of diverse ethnicities attended the event, sprawling from the steps of the Lincoln Memorial onto the National Mall and around the reflecting pool. At the time, it was the largest gathering of protesters in Washington's history. King's "I Have a Dream" speech electrified the crowd. It is regarded, along with Abraham Lincoln's Gettysburg Address and Franklin D. Roosevelt's Infamy Speech, as one of the finest speeches in the history of American oratory.

Stance on compensation

Martin Luther King Jr. expressed a view that black Americans, as well as other disadvantaged Americans, should be compensated for historical wrongs. In an interview conducted for *Playboy* in 1965, he said that granting black Americans only equality could not realistically close the economic gap between them and whites. King said that he did not seek a full restitution of wages lost to slavery, which he believed impossible, but proposed a government compensatory program of US$50 billion over ten years to all disadvantaged groups. He posited that "the money spent would be more than amply justified by the benefits that would accrue to the nation through a spectacular decline in school dropouts, family breakups, crime rates, illegitimacy, swollen relief rolls, rioting and other social evils". He presented this idea as an application of the common law regarding settlement of unpaid labor but clarified that he felt that the money should not be spent exclusively on blacks. He stated, "It should benefit the disadvantaged of *all* races".

"Bloody Sunday", 1965

King and SCLC, in partial collaboration with SNCC, attempted to organize a march from Selma to the state capital of Montgomery, for March 7, 1965. The first attempt to march on March 7 was aborted because of mob and police violence against the demonstrators. This day has since become known as Bloody Sunday. Bloody Sunday was a major turning point in the effort to gain public support for the Civil Rights Movement, the clearest demonstration up to that time of the dramatic potential of King's nonviolence strategy. King, however, was not present. After meeting with President Lyndon B. Johnson, he decided not to endorse the march, but it was carried out against his wishes and without his presence on March 7 by local civil rights leaders. Footage of police brutality against the protesters was broadcast extensively and aroused national public outrage.

King next attempted to organize a march for March 9. The SCLC petitioned for an injunction in federal court against the State of Alabama; this was denied and the judge issued an order blocking the march until after a hearing. Nonetheless, King led marchers on March 9 to the Edmund Pettus bridge, then held a short prayer session before turning the marchers around and asking them to disperse so as not to violate the court order. The unexpected ending of this second march aroused the surprise and anger of many within the local movement. The march finally went ahead fully on March 25. At the conclusion of the march and on the steps of the state capitol, King delivered a speech that has become known as "How Long, Not Long".

Chicago, 1966

In 1966, after several successes in the South, King and others in the civil rights organizations tried to spread the movement to the North, with Chicago as its first destination. King and Ralph Abernathy, both from the middle classes, moved into the slums of North Lawndale on the west side of Chicago as an educational experience and to demonstrate their support and empathy for the poor.

The SCLC formed a coalition with CCCO, Coordinating Council of Community Organizations, an organization founded by Albert Raby, and the combined organizations' efforts were fostered under the aegis of The Chicago Freedom Movement. During that spring, several dual white couple/black couple tests on real estate offices uncovered the practice (now banned in the U.S.) of racial steering. These tests revealed the racially selective processing of housing requests by couples who were exact matches in income, background, number of children, and other attributes, with the only difference being their race.

The needs of the movement for radical change grew, and several larger marches were planned and executed, including those in the following neighborhoods: Bogan, Belmont Cragin, Jefferson Park, Evergreen Park (a suburb southwest of Chicago), Gage Park and Marquette Park, among others.

In Chicago, Abernathy later wrote that they received a worse reception than they had in the South. Their marches were met by thrown bottles and screaming throngs, and they were truly afraid of starting a riot. King's beliefs mitigated against his staging a violent event, and he negotiated an agreement with Mayor Richard J. Daley to cancel a march in order to avoid the violence that he feared would result from the demonstration. King, who received death threats throughout his involvement in the civil rights movement, was hit by a brick during one march but continued to lead marches in the face of personal danger.

When King and his allies returned to the south, they left Jesse Jackson, a seminary student who had previously joined the movement in the South, in charge of their organization. Jackson continued their struggle for civil rights by organizing the Operation Breadbasket movement that targeted chain stores that did not deal fairly with blacks.

Opposition to the Vietnam War

Starting in 1965, King began to express doubts about the United States' role in the Vietnam War. In an April 4, 1967 appearance at the New York City Riverside Church—exactly one year before his death—King delivered a speech titled "Beyond Vietnam". In the speech, he spoke strongly against the U.S.'s role in the war, insisting that the U.S. was in Vietnam "to occupy it as an American colony" and calling the U.S. government "the greatest purveyor of violence in the world today". He also argued that the country needed larger and broader moral changes:

A true revolution of values will soon look uneasily on the glaring contrast of poverty and wealth. With righteous indignation, it will look across the seas and see individual capitalists of the West investing huge sums of money in Asia, Africa and South America, only to take the profits out with no concern for the social betterment of the countries, and say: "This is not just."

King also was opposed to the Vietnam War on the grounds that the war took money and resources that could have been spent on social welfare services like the War on Poverty. The United States Congress was spending more and more on the military and less and less on anti-poverty programs at the same time. He summed up this aspect by saying, "A nation that continues year after year to spend more money on military defense than on programs of social uplift is approaching spiritual death".

Many white southern segregationists vilified King; moreover, this speech soured his relationship with many members of the mainstream media. *Life* magazine called the speech "demagogic slander that sounded like a script for Radio Hanoi", and *The Washington Post* declared that King had "diminished his usefulness to his cause, his country, his people."

King stated that North Vietnam "did not begin to send in any large number of supplies or men until American forces had arrived in the tens of thousands". King also criticized the United States' resistance to North Vietnam's land reforms. He accused the United States of having killed a million Vietnamese, "mostly children."

The speech was a reflection of King's evolving political advocacy in his later years, which paralleled the teachings of the progressive Highlander Research and Education Center, with whom King was affiliated. King began to speak of the need for fundamental changes in the political and economic life of the nation. Toward the end of his life, King more frequently expressed his opposition to the war and his desire to see a redistribution of resources to correct racial and economic injustice. Though his public language was guarded, so as to avoid being linked to communism by his political enemies, in private he sometimes spoke of his support for democratic socialism. In one speech, he stated that "something is wrong with capitalism" and claimed, "There must be a better distribution of wealth, and maybe America must move toward a democratic socialism."

King had read Marx while at Morehouse, but while he rejected "traditional capitalism," he also rejected Communism because of its "materialistic interpretation of history" that denied religion, its "ethical relativism," and its "political totalitarianism."

King also stated in his "Beyond Vietnam" speech that "true compassion is more than flinging a coin to a beggar....it comes to see that an edifice which produces beggars needs restructuring". King quoted a United States official, who said that, from Vietnam to South America to Latin America, the country was "on the wrong side of a world revolution" King condemned America's "alliance with the landed gentry of Latin America," and said that the United States should support "the shirtless and barefoot people" in the Third World rather than suppressing their attempts at revolution.

King spoke at an Anti-Vietnam demonstration where he also brought up issues of civil rights and the draft.

"I have not urged a mechanical fusion of the civil rights and peace movements. There are people who have come to see the moral imperative of equality, but who cannot yet see the moral imperative of world brotherhood. I would like to see the fervor of the civil-rights movement imbued into the peace movement to instill it with greater strength. And I believe everyone has a duty to be in both the civil-rights and peace movements. But for those who presently choose but one, I would hope they will finally come to see the moral roots common to both."

Poor People's Campaign, 1968

In 1968, King and the SCLC organized the "Poor People's Campaign" to address issues of economic justice. The campaign culminated in a march on Washington, D.C. demanding economic aid to the poorest communities of the United States. King traveled the country to assemble "a multiracial army of the poor" that would march on Washington to engage in nonviolent civil disobedience at the Capitol until Congress created a bill of rights for poor Americans.

However, the campaign was not unanimously supported by other leaders of the Civil Rights Movement. Rustin resigned from the march stating that the goals of the campaign were too broad, the demands unrealizable, and thought these campaigns would accelerate the backlash and repression on the poor and the black. Throughout his participation in the civil rights movement, King was criticized by many groups. This included opposition by more militant blacks and such prominent critics as Nation of Islam member Malcolm X. Stokely Carmichael was a separatist and disagreed with King's plea for racial integration because he considered it an insult to a uniquely African-American culture. Omali Yeshitela urged Africans to remember the history of violent European colonization and how power was not secured by Europeans through integration, but by violence and force.

King and the SCLC called on the government to invest in rebuilding America's cities. He felt that Congress had shown "hostility to the poor" by spending "military funds with alacrity and generosity". He contrasted this with the situation faced by poor Americans, claiming that Congress had merely provided "poverty funds with miserliness". His vision was for change that was more revolutionary than mere reform: he cited systematic flaws of "racism, poverty, militarism and materialism", and argued that "reconstruction of society itself is the real issue to be faced".

Assassination

On March 29, 1968, King went to Memphis, Tennessee in support of the black sanitary public works employees, represented by AFSCME Local 1733, who had been on strike since March 12 for higher wages and better treatment. In one incident, black street repairmen received pay for two hours when they were sent home because of bad weather, but white employees were paid for the full day.

On April 3, King addressed a rally and delivered his "I've Been to the Mountaintop" address at Mason Temple, the world headquarters of the Church of God in Christ. King's flight to Memphis had been delayed by a bomb threat against his plane. In the close of the last speech of his career, in reference to the bomb threat, King said the following:

And then I got to Memphis. And some began to say the threats, or talk about the threats that were out. What would happen to me from some of our sick white brothers? Well, I don't know what will happen now. We've got some difficult days ahead. But it doesn't matter with me now. Because I've been to the mountaintop. And I don't mind. Like anybody, I would like to live a long life. Longevity has its place. But I'm not concerned about that now. I just want to do God's will. And He's allowed me to go up to the mountain. And I've looked over. And I've seen the Promised Land. I may not get there with you. But I want you to know tonight, that we, as a people, will get to the Promised Land. And I'm happy, tonight. I'm not worried about anything. I'm not fearing any man. Mine eyes have seen the glory of the coming of the Lord.

King was booked in room 306 at the Lorraine Motel, owned by Walter Bailey, in Memphis. The Reverend Ralph Abernathy, King's close friend and colleague who was present at the assassination, swore under oath to the United States House Select Committee on Assassinations that King and his entourage stayed at room 306 at the Lorraine Motel so often it was known as the 'King-Abernathy suite.' King was shot at 6:01 p.m. April 4, 1968 while he was standing on the motel's second floor balcony. The bullet entered through his right cheek smashing his jaw and then traveled down his spinal cord before lodging in his shoulder. According to Jesse Jackson, who was present, King's last words on the balcony were to musician Ben Branch, who was scheduled to perform that night at an event King was attending: "Ben, make sure you play "Take My Hand, Precious Lord" in the meeting tonight. Play it real pretty." Abernathy heard the shot from inside the motel room and ran to the balcony to find King on the floor. The events following the shooting have been disputed, as some people have accused Jackson of exaggerating his response.

After emergency chest surgery, King was pronounced dead at St. Joseph's Hospital at 7:05 p.m. According to biographer Taylor Branch, King's autopsy revealed that though only thirty-nine years old, he had the heart of a sixty-year-old man, perhaps a result of the stress of thirteen years in the civil rights movement.

The assassination led to a nationwide wave of riots in more than 100 cities. Presidential nominee Robert Kennedy was on his way to Indianapolis for a campaign rally when he was informed of King's death. He gave a short speech to the gathering of supporters informing them of the tragedy and asking them to continue King's idea of non-violence. President Lyndon B. Johnson declared April 7 a national day of mourning for the civil rights leader. Vice-President Hubert Humphrey attended King's funeral on behalf of Lyndon B. Johnson, as there were fears that Johnson's presence might incite protests and perhaps violence. At his widow's request, King's last sermon at Ebenezer Baptist Church was played at the funeral. It was a recording of his "Drum Major" sermon, given on February 4, 1968. In that sermon, King made a request that at his funeral no mention of his awards and honors be made, but that it be said that he tried to "feed the hungry", "clothe the naked", "be right on the [Vietnam] war question", and "love and serve humanity". His good friend Mahalia Jackson sang his favorite hymn, "Take My Hand, Precious Lord", at the funeral. The city of Memphis quickly settled the strike on terms favorable to the sanitation workers.

Two months after King's death, escaped convict James Earl Ray was captured at London Heathrow Airport while trying to leave the United Kingdom on a false Canadian passport in the name of Ramon George Sneyd on his way to white-ruled Rhodesia. Ray was quickly extradited to Tennessee and charged with King's murder. He confessed to the assassination on March 10, 1969, though he recanted this confession three days later. On the advice of his attorney Percy Foreman, Ray pleaded guilty to avoid a trial conviction and thus the possibility of receiving the death penalty. Ray was sentenced to a 99-year prison term. Ray fired Foreman as his attorney, from then on derisively calling him "Percy Fourflusher". He claimed a man he met in Montreal, Quebec with the alias "Raoul" was involved and that the assassination was the result of a conspiracy. He spent the remainder of his life attempting (unsuccessfully) to withdraw his guilty plea and secure the trial he never had. On June 10, 1977, shortly after Ray had testified to the House Select Committee on Assassinations that he did not shoot King, he and six other convicts escaped from Brushy Mountain State Penitentiary in Petros, Tennessee. They were recaptured on June 13 and returned to prison.

Allegations of conspiracy

Ray's lawyers maintained he was a scapegoat similar to the way that alleged John F. Kennedy assassin Lee Harvey Oswald is seen by conspiracy theorists. One of the claims used to support this assertion is that Ray's confession was given under pressure, and he had been threatened with the death penalty. Ray was a thief and burglar, but he had no record of committing violent crimes with a weapon.

Those suspecting a conspiracy in the assassination point out the two separate ballistics tests conducted on the Remington Gamemaster recovered by police had neither conclusively proved Ray had been the killer nor that it had even been the murder weapon. Moreover, witnesses surrounding King at the moment of his death say the shot came from another location, from behind thick shrubbery near the rooming house – which had been inexplicably cut away in the days following the assassination – and not from the rooming house window.

Bibliography

Stride toward freedom; the Montgomery story (1958)

The Measure of a Man (1959)

Strength to Love (1963)

Why We Can't Wait (1964)

Where do we go from here: Chaos or community? (1967)

The Trumpet of Conscience (1968)

A Testament of Hope : The Essential Writings and Speeches of Martin Luther King, Jr. (1986)

The Autobiography of Martin Luther King, Jr. (1998), ed. Clayborne Carson

Martin Luther King, Jr. accomplished a great deal for civil rights. He was murdered at the age of 39! Had he lived a long life what additional great good may he have accomplished? His life and efforts to free the slaves is typical of heretics and the forces arraigned against them. There can never be too many Martin Luther King, Jr.s.

41. Murray Rothbard
1926–1995

Rothbard was an influential American historian, numismatist, natural law theorist, Aristotelian and economist of the Austrian School who helped define modern libertarianism and founded a form of free-market anarchism he termed "anarcho-capitalism". Rothbard took the Austrian School's emphasis on spontaneous order and condemnation of central planning to an individualist anarchist conclusion. In addition to his work on economics and political theory, Rothbard also wrote on economic history. He is one of the few economic authors who studied the pre-Adam Smith economic schools, such as the Scholastics and the Physiocrats. These are discussed in his unfinished, multi-volume work, **_An Austrian Perspective on the History of Economic Thought._**

Rothbard opposed what he considered the overspecialization of the academy and sought to fuse the disciplines of economics, history, ethics, and political science to create a "science of liberty," as reflected in his many books and articles. His approach was influenced by the arguments of Ludwig von Mises in such books as **_Human Action_** and **_Theory and History_** that the foundations of the social sciences are in logic of human action that can be known prior to empirical investigation. Rothbard sought to use such insights to guide historical research, especially in his work on economic history, but also in his four-volume history of the **_American Revolution, Conceived in Liberty._** Rothbard argued that the entire Austrian economic theory is the working out of the logical implications of the fact that humans engage in purposeful action.

It was in 1949 that Rothbard first concluded that the free market could provide all services, including police, courts, and defense services, better than could the State. Prior to this it was advocated by nineteenth century individualist anarchists such as Benjamin Tucker, whose writings were an influence on Rothbard. Prior to this it was advocated by Gustave de Molinari who Rothbard calls the first anarcho-capitalist. Rothbard described the moral basis for his anarcho-capitalist position in two of his books, **_For a New Liberty_**, published in 1972, and **_The Ethics of Liberty_**, published in 1982. He described how a stateless economy would function in his book **_Power and Market_**. According to Rothbard, the difference between a state and voluntary defense is that a state taxes and it enforces a territorial monopoly, over property that it does not own (private property), on the use of defense and punitive force. Private defense relies on voluntary payments and it does not forcefully prevent other private defenders from competing for business. For example, if someone subscribed to a private police agency, and someone had broken into that person's home, then that individual could call the private police to come to the home and arrest the intruder and take him to a private jail and private court. A state claims a monopoly over such force on property that anarcho-capitalists do not believe that the state owns (e.g. the person's home); it does not permit this kind of competition, by definition.

In **_The Ethics of Liberty_**, Rothbard asserted the right of 100 percent self-ownership, as the only principle compatible with a moral code that applies to every person — a "universal ethic" — and that it is a natural law by being what is naturally best for man. He believed that, as a result, individuals owned the fruits of their labor. Accordingly, each person had the right to exchange his property with others. He believed that if an individual mixes his labor with unowned land then he is the proper owner, and from that point on it is private property that may only exchange hands by trade or gift. He also argued that such land would tend not to remain unused unless it makes economic sense to not put it to use. Rothbard defined the libertarian position through what is called the non-aggression principle that "No person may aggress against anybody else." Rothbard attacked taxation as theft, because it was taking someone else's property without his consent. Further, conscription was slavery, and war was murder. Rothbard also opposed compulsory jury service and involuntary mental hospitalization.

Rothbard was an ardent critic of the influential economist John Maynard Keynes and Keynesian economic thought. His essay ***Keynes, the Man***, is an attack upon Keynes' economic ideas and personage.

Rothbard was also severely critical of, among others, utilitarian philosopher Jeremy Bentham in his essay, "***Jeremy Bentham: The Utilitarian as Big Brother***" published in his work, ***Classical Economics***.

Murray Rothbard devotes a chapter of ***Power and Market*** to the traditional role of the economist in public life. Rothbard notes that the functions of the economist on the free market differ strongly from those of the economist on the hampered market. "What can the economist do on the purely free market?" Rothbard asks. "He can explain the workings of the market economy (a vital task, especially since the untutored person tends to regard the market economy as sheer chaos), but he can do little else."

Rothbard was highly active in libertarian, Libertarian Radical Caucus, Libertarian International Organization, and especially U.S. Libertarian Party causes. He wrote many of the party's initial manifestos, and, despite initial skepticism on its timeliness, was its first Presidential nominee in a preliminary vote (he declined). He was heavily involved in its first strategic plan, and in using it with the Party as a springboard for what he called 'intellectual entrepreneurship' in building a Libertarian intellectual superstructure. These included activities such as the CATO Institute, the Center for Libertarian Studies, plus a variety of magazines, scholarly dialogues, colloquia, scholar and candidate mentoring initiatives, and journals. He defined Libertarianism as "the interdisciplinary study of individual rights," and wrote several cross-discipline histories detailing many long-forgotten libertarian-interest scientists or events, and co-organized the first international conferences on the subject.

Works

Man, Economy, and State (Full Text; ISBN 0-945466-30-7) (1962)

The Panic of 1819. 1962, 2006 edition: ISBN 1-933550-08-2.

America's Great Depression. ISBN 0-945466-05-6. (1963, 1972, 1975, 1983, 2000)

What Has Government Done to Our Money? (Full Text / Audio Book) ISBN 0-945466-44-7. (1963)

Economic Depressions: Causes and Cures (1969)

Power and Market. ISBN 1-933550-05-8. (1970) (restored to ***Man, Economy, and State*** ISBN 0-945466-30-7, 2004)

Education: Free and Compulsory. ISBN 0-945466-22-6. (1972)

Left and Right, Selected Essays 1954-65 (1972)

For a New Liberty: The Libertarian Manifesto (Full text / Audio book) ISBN 0-945466-47-1. (1973, 1978)

The Essential von Mises (1973)

The Case for the 100 Percent Gold Dollar. ISBN 0-945466-34-X. (Full Text / Audio Book) (1974)

Egalitarianism as a Revolt Against Nature and Other Essays ISBN 0-945466-23-4. (1974)

Conceived in Liberty (4 vol.) ISBN 0-945466-26-9. (1975-79)

Individualism and the Philosophy of the Social Sciences. ISBN 0-932790-03-8. (1979)

The Ethics of Liberty (Full Text / Audio Book) ISBN 0-8147-7559-4. (1982)

The Mystery of Banking (Full text). ISBN 0-943940-04-4. (1983)

Ludwig von Mises: Scholar, Creator, Hero. OCLC 20856420. (1988)

Freedom, Inequality, Primitivism, and the Division of Labor. Full text (included as Chapter 16 in Egalitarianism above) (1991)

The Case Against the Fed. ISBN 0-945466-17-X. (1994)

An Austrian Perspective on the History of Economic Thought (2 vol.) ISBN 0-945466-48-X. (1995)

Wall Street, Banks, and American Foreign Policy. (Full Text) with an introduction by Justin Raimondo. (1995)

Making Economic Sense. ISBN 0-945466-18-8. (1995, 2006)

Logic of Action (2 vol.) ISBN 1-85898-015-1 and ISBN 1-85898-570-6. (1997)

The Austrian Theory of the Trade Cycle and Other Essays. ISBN 0-945466-21-8. (also by Mises, Hayek, & Haberler)

Irrepressible Rothbard: The Rothbard-Rockwell Report Essays of Murray N. Rothbard. (Full Text.) ISBN 1-883959-02-0. (2000)

History of Money and Banking in the United States. ISBN 0-945466-33-1. (2005)

The Complete Libertarian Forum (2 vol.) (Full Text) ISBN 1-933550-02-3. (2006)

Economic Controversies (to be published 2007)

The Betrayal of the American Right ISBN 978-1-933550-13-8 (2007)

I know of no one who has more brilliantly explained the delusions upon which the State is based. Murray Rothbard is a towering hero of our time. He is recognized by few, but his stature will grow as more and more people appreciate his tremendous contribution.

42. Robert E. Kahn
1938

Robert Elliot Kahn, is an engineer who, along with Vinton G. Cerf, invented the Transmission Control Protocol (TCP) and the Internet Protocol (IP), the technologies used to transmit information on the Internet. After receiving a B.E.E. degree from the City College of New York in 1960, Kahn earned M.A. and Ph.D. degrees from Princeton University in 1962 and 1964 respectively. In 1972 he moved to ARPA (now known as DARPA), and in October of that year, he demonstrated the ARPANET by connecting 40 different computers at the International Computer Communication Conference, publicizing the network to the general public for the first time. After he became Director of DARPA's Information Processing Techniques Office (IPTO), he started the United States government's billion dollar Strategic Computing Initiative, the largest computer research and development program ever undertaken by the U.S. federal government.

After thirteen years with DARPA, he left to found the Corporation for National Research Initiatives (CNRI) in 1986, and as of 2009 is the Chairman, CEO and President. CNRI is a nonprofit organization which is intended to provide leadership and funding for research and development of the National Information Infrastructure.

43. Vinton Gray "Vint" Cerf
1943

Cerf is an American computer scientist who is the "person most often called 'the father of the Internet'." His contributions have been recognized repeatedly, with honorary degrees and awards that include the National Medal of Technology, the Turing Award, and the Presidential Medal of Freedom.

In the early days, Cerf was a DOD DARPA program manager funding various groups to develop TCP/IP technology. When the Internet began to transition to a commercial opportunity, Cerf moved to MCI where he was instrumental in the development of the first commercial email system (MCI Mail) connected to the Internet.

Vinton Cerf was instrumental in the funding and formation of ICANN from the start. Cerf went to the same high school as Jon Postel and Steve Crocker. Cerf waited in the wings for a year before he stepped forward to join the ICANN Board. Eventually he became the Chairman of ICANN.

Cerf has worked for Google as its Vice President and Chief Internet Evangelist since September 2005. In this function he has become well known for his predictions on how technology will affect future society, encompassing such areas as artificial intelligence, environmentalism, the advent of IPv6 and the transformation of the television industry and its delivery model.

44. Timothy John "Tim" Berners-Lee
1955

Berners-Lee is a British engineer and computer scientist and MIT professor credited with inventing the World Wide Web, making the first proposal for it in March 1989. On 25 December 1990, with the help of Robert Cailliau and a young student staff at CERN, he implemented the first successful communication between an HTTP client and server via the Internet. Berners-Lee is the director of the World Wide Web Consortium (W3C), which oversees the Web's continued development. He is also the founder of the World Wide Web Foundation, and is a senior researcher and holder of the 3Com Founders Chair at the MIT Computer Science and Artificial Intelligence Laboratory (CSAIL). He is a director of The Web Science Research Initiative (WSRI), and a member of the advisory board of the MIT Center for Collective Intelligence. In April 2009, he was elected as a member of the United States National Academy of Sciences, based in Washington, D.C.

Chapter 10

The Slow Advancement Of Freedom

Over the past 5000 years heretics of every type, philosophers, writers, scientists, artist, inventors and others, have continually challenged tribal and religious belief with sacrilegious and blasphemous ideas. These ideas have pressed forward slowly expanding people's choices and freedoms. This has resulted in polytheism changing to monotheism and Paganism morphing into the modern religions. These changes allowed man to think more clearly and create a separation in his mind between religion and the natural world. Slowly this caused a division in man's mind between religious belief and tribal rule. This allowed the tribal leadership to separate from the religious enterprise leaving a secular ruling function the principle societal power creating the modern State.

From the evolutionary prospective 5 millenniums is but a blink of an eye. But when viewed from the prospective of a human lifetime the advance of freedom has been a very slow process. Each time the heretics made some headway there was a reaction. There is always the push-back by the tribal powers. Even so freedom slowly advanced. We can see this very clearly as we study the lives of the great heretics. But only half of the difficulty is the obstinacy of the tribal power structure. The other great obstacle to freedom is the masses, the tribesmen, the slaves themselves. You might think that any and every bit of freedom promised by a new idea, discovery or invention would come as welcome relief. But you would be wrong. The theist theology that created the tribal power structure also programmed the people. As much as the tribe needs the slave, the slave needs the tribe. Freedom is what tribal peoples had when they were hunter-gatherers. They do not want freedom, it scares them to death! Tribalism is a co-dependent relationship. One aspect of tribalism is not more evil than the other. So each new bit of freedom initially meets with a combined resistance from the tribal rulers and from the public at large. This is why the process is so slow. Generally each new advance in freedom is resisted by the older members of the tribe and requires the younger generation to validate it and accept it. Then it passes into history, becomes the new baseline, and subsequent generations think it has always been so.

Another factor in the advancement of freedom that I suspect you may find difficult to believe, is the human brain! The basic human being, the hunter-gatherer type human being, untainted by education in theism, is fundamentally reasonable and peace-loving! So wherever given the chance even tribal peoples will accept bits of freedom that promote reasonableness and non-violence. Tribal leaders must therefore constantly point out the heretical and blasphemous nature of innovation in order to keep the masses stirred up over freedoms encroachment. I believe it is this basic decency of human beings which acts like a ratchet mechanism preventing the complete loss of progress made by holding onto at least a little of each heretics message.

This slow advance of freedom is like many trends in history, a back and forth process. Freedom advances, people get scared, tribesmen scare themselves back into their place for another while, and then freedom advances a little more. As freedom advances it has a humanizing affect upon the people, religion, the tribe, and its leadership. This humanizing influence that freedom has on tribalism has been called; humanism. But it is nothing more and nothing less than more freedom and less oppression of the masses. It means that the tribesmen are allowed to think about a few more things. It means the tribesmen do not have to be afraid to allow themselves a little more mental freedom. It means a little more 'space' in the mind in which to think!

The modern mind is a mixture of these directly opposite ideologies! Make no mistake about this. These are diametrically opposite philosophies. Pagan polytheism is the epitome of evil. It is a system designed to facilitate the exploitation of people by an elite. It is pure force, violence, slavery, and murder. Humanism promotes the exact opposites, namely reason, fairness, benevolence, love, forgiveness, non-violence, mercy, and equality. Science also promotes values contrary to Paganism, such as skepticism, questioning, empiric observation of an objective reality, logical deduction, physical examination and experimentation. Theism promotes the view that people are for the use of the Gods and the elites. Humanism recognizes the sovereign rights that each individual has over their own life! Humanism is the denial of the claims of the elite to the lives of the masses. These two opposing world views are irreconcilably different! The philosophical conflict between tribalism, all types of collectivism, and humanistic science could not be greater and cannot be resolved conceptually.

The work of many of the heretics was to directly challenge religion. That of others was to advance science. Science has been one of the most humanizing influences. The combined effect of science and humanism on the Pagan religion, the tribal mind, and the tribe has gradually created the modern world. It has created the contemporary mind. It has created the modern religions and it has created the contemporary State.

The Modern Mind

The great heretics, through their courageous thinking and fearless teachings have over the past 5000 years, gradually created a counter belief system, an alternative belief system to tribal collectivism or tribalism! Where Pagan polytheism once held an absolute monopoly over peoples minds, where Paganism maintained almost total control over thought for 15,000 years, these great thinkers have created an alternative! The total lock on peoples' minds once enjoyed by religion and the tribe, has been broken! Each heretic, adding to what those before him accomplished has gradually created an alternative to the brutal theology of polytheism! This alternative to theism, to Pagan tribalism, is humanism or egoism or deism or the scientific belief system. Label this new philosophy what you will, it is this benevolent, rational, maternal view of human nature and the universe, that is the first challenge to theism! Ideas like liberty, justice and equality have taken hold, and continue to spread around the globe. These ideas are being given a prominent place in more and more peoples' thoughts.

Yet as you survey the world today you find that most people, regardless of their culture, nation of origin, or the brand of religion in which they were raised, are theist. Even most nonbelievers were raised by theist and therefore display most of the psychological attributes of the religious adaptation. So most people today have a core commitment to theism. How then do they deal with the scientific and humanistic world forming around them? Most people around the world see the merits of the new scientific viewpoint. Except for a relatively few dihard clerics and reactionaries, people tend to at least partially embrace the new modern scientific and humanitarian world view. They see the benefits to themselves of a rational, benevolent, scientific world. They see this especially in terms of more effective health care, safer food and water and better transportation. They appreciate the improving standard of living that this modern scientific, rational way of thinking produces. Even when they are unable to embrace these new ideas for themselves they want them for their children. But this partial acceptance of humanism does not mean that people are abandoning their commitment to theism! People still desperately want to believe that their Pagan morality is a good thing! They cannot 'see' the evil of human sacrifice. So they hold on to their religious beliefs! Imprint programming is very tenacious! The modern mind, the modern religions and the contemporary State are a mixture of evil bloodthirsty Pagan theism on the one hand and maternal, merciful, forgiving, non-violent humanism along with rational, logical science on the other. So modern theism retains most of its Pagan elements. The contemporary theist continues to believe in supernaturalism, in the innate evil of man, and in the morality of human sacrifice .

But as slaves and tribal peoples everywhere began considering these 'new' heretical ideas many centuries ago, things gradually began changing in the world around them. More so in some places than others, but to varying degrees people everywhere began adopting these new humanist concepts. People at first put these ideas into practice partially, tentatively and cautiously. But wherever science and humanistic ideas are applied the plight of the masses improve. A revolution in thought and action began taking place almost everywhere throughout the world. Unfortunately this thought revolution also spawned a number of armed revolutions around the globe. Everywhere in the world the slaves have become restless and everywhere they are questioning obedience to tribal and religious authority. Tribalism is under siege!

So how can a human mind at the very same time embrace two diametrically opposite belief systems? How can a person believe, in the very same mind, the concepts of both humanism and theism? What is going on here? How can the human mind loyally believe Paganism and science? The reference standards which serve as the linchpins for these two belief systems are totally different! How can contemporary peoples believe both these dueling philosophies in the very same mind?

Contemporary people are now doing just as most of the heretics! Modern peoples are utilizing **compartmentalization!** The masses are now using the same psychological methodology that the heretics used to avoid a direct confrontation with the powerful institutions of tribe and religion! Yes the slaves, the commoners, the masses, the tribesmen, rather than reject theism, have simply pushed it to the side or bottom of their mind to make room for this new egalitarian philosophy of humanism and science! The 'have my cake and eat it too' approach which seemed the safest approach for the heretics appears the safest way to go for everyone!

The contemporary theist during childhood using compartmentalization forms two 'minds' or identities. We can label the identity he develops out of his commitment to tribalism his **theist identity**. He also develops an identity out of his commitment to the humanism and science that he learns. We can label this second more superficial identity the **secular identity**. In most contemporary people the **theist identity** will form their *super*concept. The heretics have created this modern, this third psychological adaptation!

Contemporary theist parents, now to a great extent unconscious of the true meaning of theism, teach their child one set of beliefs in infancy and early childhood, and then criticizes him for embracing these very same concepts and behaviors later in childhood! Conversing out loud with God, talking to long dead religious figures and 'seeing' angels and dead relatives may appear cute when it's a 4 year old. But when a teenager does the same thing during a job interview most modern parents do not think it cute!

This condemnation of the child in later childhood for his commitment to fundamental theism is a modern development. The tribal child met with no such hypocrisy. It is this hypocrisy which moderns have toward theism which confuses children and creates the mental schism that is so characteristic of the contemporary mind.

A person is only born with one **self**! It is a permanent structure located in the brain. In the process of creating a self-identity, an awareness of the **self**, a conceptual portion of the **self** is added through societal programming. The self-aware or conceptual portion of the **self** may be created multiple times. It should not be multiple, but in the modern world it usually is. The creation of multiple identities can occur because the conceptual part of identity does not have to be integrated into the inherited hard-wired **self**! A conceptual identity does not have to be integrated; it can be floating and unattached to perception. *The perceptual and the conceptual parts of the **self** should be seamlessly integrated to form a strong, singular self, but in contemporary individuals this seldom occurs!* So the **theist identity** is the one usually connected to the **self** through the imprinting process and is therefore the person's *super*concept and his default standard of reference. A **theist identity** will therefore form the foundation psychology of the contemporary mind.

Thus the foundation of the contemporary mind will have **all** the attributes of the tribal theist. This will include collectivists 'magical thinking' cognition and a morality of self-sacrifice manifest as sadomasochism. This *super*concept is psychotic, murderously violent and emotionally unstable. This is the contemporary person's psychological baggage from 15 millenniums of indoctrination in tribalism.

You will recall that the theist indoctrination of tribal infants falls into two stages. Stage 1 of infant indoctrination with theism consists of teaching the newborn of his innate **evil.** This Original Sin concept is taught mostly by the infant's parents and to a great extent it is taught non-verbally.

Then, in the training of tribal youth Stage 2 of the theist indoctrination program consist of teaching revelation, supernaturalism, force, human sacrifice and self-policing. There is no humanism taught here! The tribal message is clear and consistent. It's a terrible message, but it is consistent! The message is that you do not belong to yourself, that you belong to the tribe, that you are a slave! Furthermore you are to believe what you are told to believe and you are to engage in the behaviors you are told to perform! From the slave's *super*concept through his infant experiences down to his religious schooling to the organizational structure of his tribe nothing is telling him anything different. All of his programming, his entire **identity**, is based upon the belief and commitment to being a slave! Human sacrifice is the ultimate virtue, the sacrifice of others as well as self-sacrifice. The child's basic or fundamental theist indoctrination meshes with his secondary theist instruction. There is no hypocrisy or conflict for the tribal child.

This is not the case with the contemporary child. Stage 1 of theist indoctrination may not differ much from the tribal experience. The doctrine of self-evil or Original Sin may be well taught at home. Parents may begin the process during infancy of programming self **identity** with the theist belief of Original Sin or self-evil. Theist parents are diligent in teaching this evil because they believe it essential to preventing their child from becoming evil! Their understanding of evil in this regard is 180 degrees off the mark! But once the child reaches school age and begins public or private school, even in many religious schools, he meets a mixture of theist and humanitarian and scientific views.

At this second stage of theist indoctrination the modern child typically meets with humanistic and scientific ideas which conflict with the earlier learned tribalism. This is where the discrepancy and the cleavage begin. So the modern child develops a core value system that is identical with that of the tribal child. We have labeled this initial programming of the *self* the **theist identity.** Because the modern family milieu continues to be anti-self and collectivist, this core identity will have the *cognitive* and *moral* attributes of a **theist identity!**

After the child exits his home environment and enters school he typically runs into the contradictory humanitarian view! Here is where the contemporary child begins to construct a second identity, one that is very different from his **theist identity**. This second viewpoint or identity begins creating the psychic schism I refer to. This second identity is composed of concepts of scientific humanism. We have labeled this second identity the **secular identity.** The secular identity will produce a secular psychology. This **secular identity** will display the characteristics of scientific humanism: rationality, fairness, and compassion. This identity seeks self-gain, profit, success, and happiness. These two identities, theist and secular, and the psychologies they produce could not be more different! The theist identity is aggressive, believes that might makes right and has a sadomasochistic, self-sacrificing moral code. The theist identity is superstitious and psychotic and cannot separate its *self* from the collective. This is the cause of the mental conflict so apparent in contemporary peoples.

Once created the **secular identity** competes with the **theist identity** each trying to be the individual's correct, recognized, authorized and designated representative of the wired-in Hindbrain *self.*

As the contemporary child grows and matures he examines his **theist identity** and finds much to criticize. Modern society with its humanism and science, and its secular State, are opposed to the sacrificial murder of human beings. This is now illegal in most places. The child also listens as humanistic ideas tell him that he might not have been born so bad and evil after all! The young person is thus faced with this problem. *Many tribal theist values are unacceptable and/or illegal in the modern secular world!* So many tribal theist values, must to some extent, be **repressed** and hidden from consciousness and the contemporary world.

Those theist values which are socially unacceptable must be shoved out of sight, they must be repressed into the unconscious mind. So much of the contemporary child's **theist identity** must be repressed. Especially the sadistic aspects of a **theist identity** must be denied and repressed. The masochistic values of a **theist identity** are more socially acceptable in modern society and may remain in the conscious areas of the mind. But being repressed is not the same as being rejected and spit out of the mind.

Even though unacceptable theist values are repressed by the modern child's mind they are not gone! The contemporary child must still contend with his **theist identity**. He must still contend with the contradictions and conflicts between this collective tribal **theist identity** and his developing **secular identity**. So the conflict between the contemporary child's **theist identity** and his **secular identity** persist, it has not gone away, it has just gone into hiding. Whichever philosophy or identity wins this competition by adulthood will determine much of the person's life.

The selected identity may be the most consistent initiator of behavior in adulthood, but during childhood and adolescence behaviors may emanate from the **theist identity** or the **secular identity**. And of course the identity which initiates most of the successful behavior is strengthened by reinforcement.

Since theist tribal values are generally taught by the child's parents, and taught first, most of the adults in a theist culture will have a core **theist identity**. In other words the contemporary individual's *super*concept is usually tribal and collectivist. Then the secular, humanistic and scientific values are layered over and around the core of theism. So, usually the **secular identity** is superficial to the tribal identity, layered around it and answerable to it.

You can see this competing of ideologies clearly in contemporary countries that have established secular public schools. Sunday religious schools teach tribal lessons, while secular schools teach humanism and science. Even modern State governments have a division and separation between Church and State. This dual or split mental condition of modern man is apparent everywhere you look in contemporary society.The problems this condition poses are multiple.

So very little has been subtracted from the theist child's curriculum. But a good bit has been added to what a child of today must learn and do. The modern child is expected to learn and believe all the theist dogma his religion teaches him plus add to his mind all the science and humanism which his secular school requires that he learn! Religion may be filling the child's mind with ancient tribal Paganism , while the secular school system is filling his mind with reading, writing, arithmetic and science. These various sources of information and knowledge are headed in opposite directions! It is difficult to believe both the theist ideas and the scientific, secular information! Either Darwin is correct about evolution or Creationism is true? Either girls should have easy access to contraceptives or there use to prevent unwanted pregnancy is wrong? Either he was born evil or he was not. There is a long list of irreconcilable and conflicting beliefs between theism and scientific humanism. This contradiction is all around us but almost no one sees the conflict or dissonance! Do fish in the ocean perceive the water?

But the contemporary situation is actually more complicated than this. It is more complicated because both modern religions and the modern State are as mixed and conflicted regarding these ideologies as are the people. So it is not as clearly demarcated as religious schools teaching theism and the secular schools teaching humanism. It would be easier to see and to respond to if this were the case. But most schools, religious schools of most denominations and secular public schools, teach a mixed message. The general culture everywhere demonstrates this conflicting divide. Examples abound. You witness a convention of scientist giving a convocation of Pagan prayer at the opening ceremonies of a scientific forum! You see a meeting of religious leaders and clerics discussing the poor performance of students in their schools in the subjects of science and math! The church leaders have embraced science and humanism. Even the ruling elite have adopted humanistic beliefs. There are few clear-cut good guys or bad guys any more. The modern person, and most contemporary institutions are a mixture of both tribalism and secular humanism!

This is a classic mental conflict and there is no easy way to resolve it. To actually resolve such conflicts requires that the reality of the conflict be faced and then somebody gets hurt. The easiest way to avoid reality and a confrontation with its hurt, pain and rejection, is for the child to accept both belief systems! But unlike the menu in some restaurants he can't pick and choose. Religion demands complete fidelity. Science cannot compromise with the truth. His mind demands non-contradictory integration. The only way open to the child to be loyal to his religion and also be loyal to the reason of his own mind is through the process of **compartmentalization**. He does what many of the great heretics did. The child splits his mind in two. One mind for his parent's theism and one mind for reason and himself!

Chapter 11
Modern Religions

Everything about the modern world reflects this bicameral construction. Our cultures core of primitive theist evil is contained within a covering of scientific humanism. This fact effects everything in our culture and in our lives! For this reason almost every aspect of the modern world encourages compartmentalization and hypocrisy! No one wants to face up to this situation! Just like everything else this is true of the modern religions.

Over the past 5000 years, as a result of the work of the great heretics religion has changed profoundly. The separation of religion from the governing elite is the characteristic which most defines the State and contrast it with the tribe. The religious elite are cut loose from the direct support of the governing elite. Religion must then compete in the marketplace. Nothing about the modern religions has changed more than how it is now promoted in modern society. In countries that now enjoy religious freedom, religion must be marketed in a totally different way than in the primitive tribe. The different theist religions present in many countries must now compete with each other for members. In many countries there has developed competition among the various religious sects for converts. Modern religions have been forced to add a sugar or candy coating around its theist teachings so that it goes down better with freer, contemporary people. This sugar 'coating' is made up of maternal, humanistic and scientific ideas taken from the great heretics.

Before the age of heretics tribal leadership simply ordered all the slaves to believe whatever version of theism they decreed. Only one version or brand of Paganism was allowed in the tribe and it was chosen by the tribal elite. Those that refused to obey were murdered as sacrifices to the Gods. It was not necessary that religion show concern for the individual. It was just the opposite, the Pagan religions existed for the benefit of the ruling elite. They didn't apologize or make excuses for exploiting the masses, as they saw it their right. But the heretics gave the masses the idea that they had some rights, and that their needs were worthy of some consideration! This has placed religions on the defensive. As Pagan religions plowed the teachings of the great heretics back into their dogma they tried to do it in a manner that had the least effect upon their theist teachings. But over time, with the teachings of each successive heretic they were forced to accept more and more humanism. Slowly the contemporary religions morphed out of their Pagan beginnings. Now in more and more places around the world many religions must actively market themselves as they compete to recruit membership. Religions compete in the marketplace like everyone else by offering more and more services and benefits at less and less cost.

For religion to survive in a free market situation it had to become 'user friendly.' Religion learned that it was often better if it did not directly confront humanism and science. In the battle of ideas between religion and the heretics, the heretics too often won! Even when religion put a heretic to death for his ideas, the ideas often survived and gained believers! So religion learned that it had to apply a coating or veneer of humanism around its theism to stay in the competition for minds. This was something new for the clerics. To keep itself alive, to avoid extinction, religion has been forced to adapt. It had to add a veneer of altruistic and maternal concern for the masses. Religion had to add a fraternal commodore and a benevolent mutual help and assistance program to its teaching and indoctrination format to keep people interested. Nothing would seem stranger to the Pagan clerics or tribesmen of the earlier epoch.

But under the veneer of this sugar coating of humanism religion remains fundamentally unchanged. The theist core of modern religions remain brutal, violent, barbaric, psychotic, mind-controlling, child- abusing and exploiting machines. Indeed contemporary 'civilization' at its core, continues to retain its evil, barbaric, and pediophillic Pagan past. So dealing with people who have a choice of religions, and even the choice of non-belief, is something religion has only recently been forced to deal with. The religious institutions have been forced, by the impact of the heretics' ideas, and people's commitment to them, to add a sugar coating of maternal concern and empathy around their bitter and brutal ideologies and institutions. This has been so effective that we now observe members of the ruling elite themselves expressing guilt over the way the masses have historically been treated!

Another big change for religion in the State is that there is now little government assistance at persecuting and punishing heretics. This has become illegal in many States. The monopoly one religion had in the tribe is gone. The theist predilection to create ever more religious sects, denominations and religions was freed or unleashed when the tribe gave-way to the State. Now without the apprehension and execution of non-conventional believers the number of different religions within each State has increased dramatically. Religions divide, new groups re-name themselves, and go their separate ways without bloodshed. Any prophet now finds it easy to proclaim a new vision, receive a new revelation and found a new faith. So long as a new religion obeys the laws of the State they were largely free of State harassment and allowed to grow and expand. This increased the competition among religions and sects and sped-up the humanization of religion. This process continues at the present time.

The situation with religion in contemporary times is obviously complex. By perpetuating the basic Pagan concepts, especially the belief in innate human evil or Original Sin, and the morality of human sacrifice, religion does mankind a tremendous disservice. Yet all the major religions provide help to their communities at every level. These institutions, just like the modern mind, are wrapped in a humanist covering. The benevolent, maternal, caring, humanistic covering gets most of the press most of the time by providing food and clothing to the poor, medical clinics for the ill, counseling for every age group, and much secular education. They do all these good works from which society greatly benefits. And yet religions are the societal 'genes' that bring forward to each new generation the evil of Pagan belief. For it is religion which creates the evil in our young parents, which they then inculcate in their offspring. Evil with which religion and everyone else, must then contend.

Chapter 12
The State

The State developed out of the tribe through the psychological mechanisms of compartmentalization and repression. The totalitarian tribe is the direct societal expression of polytheism. The State is the direct societal expression of monotheism. Both forms of theism are a denial or rejection of Nature's God or objective reality. All forms of theism are a worship of man, or more specifically the man rulers of men. Both forms of theism have as their epistemological standard of truth, their *super*concept, the authority of humans, a collective of humans or its human rulers. Theist would not agree of course. They would claim their standard of truth to be God as revealed to them through special men, the Prophets. But Prophets are men. The 'truth' is second or third-hand at best. Under theism God must speak through multiple layers of human interpretation. The distortion of the interpreter, to say nothing regarding the Prophet's basic premises, make theism the worship of human interpretations, in other words a worship of human conceptualization. These theist beliefs are conceptualized in human language and recorded by men in written human language in Holy Books. Bottom-line theism is the worship of man. And you will notice that modern religions spend a lot more time worshiping their Prophet, a man, than they spend talking about God.

The ideas of the heretics have gradually changed the tribal mind to produce the modern mind. Scientific humanism created the need to compartmentalize and repress the Pagan values to produce the contemporary State. Note that theism was never rejected; the socially unacceptable parts of it were just repressed and hidden. The principal difference between a tribe and a State is the separation of religion and the political enterprise. The State has now made it illegal for religion to murder human beings in religious rituals of sacrifice. This affords more religious tolerance and less harsh retaliation against non-conformist belief. The result is more religious competition. But just like the modern mind, the modern State retains the evil tribal elements as its core with scientific humanism layered around it.

The other big difference between the tribe and the State is democracy. Democracy may have been what spawned separation of religion and State. All the conflicts and shortcomings inherent in the split modern mind are replicated in the modern State. The division of governing bodies into two or more assemblies is a reflection of modern man's compartmented and conflicted mental condition.

The tribe has gone through the same superficial 'user friendly' modifications to produce the modern State that religion has undergone. While the chief or ruling elite of a tribe would find the placating and ingratiating behaviors of modern politicians unbelievable it is a reflection of how societal power is now obtained. Contemporary masses have obtained considerable political power principally through the democratic processes which are themselves the result of the heretics. So the political elite in order to retain ruling control over the masses must add benefits and services to get votes. This is what has caused the ancient tribe to morph into the modern State. So where the tribal chief simply commanded the slaves or masses to do his bidding, contemporary politicians must obtain power by obtaining votes. This has required the State to develop a 'sugar' veneer or coating around its core identity. And as with the contemporary psychological adaptation, the compartmented theist mind, conflict in the modern State is unremitting.

Although the tribe or State is completely parasitic and actually offers little of material value, statists worship the State as a deity; it is one of their Gods. So they are unable to recognize the evil and unnecessary nature of the State. What actually exist, what is real, is a geographic area of the earth containing a population of people who are dependent upon authority, psychologically manipulated, tricked and terrorized out of protection money by a governing elite and their police. Due to their collectivist identity and its masochism the citizens are programmed for this exploitation.

Ultimately the only 'value' of a State for most citizens is imaginary. The Statist child incorporates the State into his collectivist mind as part of his identity. We can call these deluded lovers of the State: governmentophiles! Governmentophiles are those who love and trust the government. State citizens worship and love the State. These are the 'true believers[18]'. These are the people who are constantly complaining of the poor return from the State for their love and devotion! Governmentophiles assume each and every government scandal, failure and fiasco to be an exception. They are tenacious in their belief in the myth of 'good government.' These are the people the ruling elites refer to as 'useful idiots'. The contemporary theist mind cannot conceive of life independent of a collective and their preference is for a State. With their need for a **collective/self** they incorporate the State to become their **self!** They worship the State as a God so it is a kind of self-worship. They cannot separate their identity from the State. They are the great mass of governmentophiles.

Statists cannot imagine providing for themselves what they conceive as 'vital government services'. The fact that private citizens or private enterprise, absent the coercion, force, thievery and violence of a State, could and would provide all the needed services far more affordably and effectively is a belief totally beyond their comprehension! Citizens believe they must have a

State and psychologically this is true. But theists believe they must have a State, a government, to survive! What they actually want is for the State to do their stealing and killing for them. It is a core belief of theists that stealing and killing is necessary for survival. By having their leaders order and organize this behavior it becomes anointed and acceptable, even noble. Most can relegate the guilt; all can share in the booty. So they respect and obey the governing elite as their religious programming has trained them to do. Citizens need the State for the same reasons Shama needed the Tribe; as an instrument for stealing (taxation and wealth redistribution) and killing.

Therefore the **first** delusion required for a State to come into existence, and remain in existence, is the belief by a majority of the people that a State and its government are needed and necessary. The State's existence reflects the fact that a majority of the citizens cannot imagine surviving very well without State stealing and killing. This is the typical feeling of people amputated of their objective orientation to nature and imprinted on a collective. Having an epistemological standard of truth based on a ruling elite practically guarantees theft and murder. This is because the usefulness of slaves or citizens to the governing elite is limited to commercial and military purposes.

A citizen has a psychological need for the State to intercede between him and reality, or Ytilaer, Nature's God. Modern man just like Tribal man has been taught to fear reality and needs an institution to intercede between him and reality. While most of the citizens of a State receive some of the booty acquired from stealing and killing some of the time, the only people that consistently profit materially from the existence of the State are the governing elite and their entourage. The governing elite need the State for the same reasons Shama needed the tribe: to save them from, to intercede between them and the forest. It is the forest i.e. the unregulated marketplace in contemporary times, which the governing elite and many citizens so desperately fear. They must have a parental surrogate (government) intercede between them and reality and provide them with a guaranteed or protected source of food. In other words they cannot be mentally secure without having a government enforced commercial monopoly.

So the State, like the Tribe, provides the rational for murder and theft. Adherence and devotion to theism allows each citizen to guiltlessly collect his portion of the booty confiscated by the State. As with the tribe certain delusions allow the citizen to remain blind to these evils.

The **second** delusion required for maintenance of the modern State is the unquestioning belief in the efficacy of force. The force and violence theists utilize on themselves inside their own minds causes them to project a similar need for force and violence in their dealings with all aspects of reality. This is no less true in dealing with their social reality, in dealing with people. Indeed in theist societies there truly is a need for force! It is required to secure order!

Theism sets up a system of rules, regulations and expectations in the citizen's mind. He must then depend upon his will-power to force his adherence to this legalistic structure. He will then project onto the State a need for societal rules and regulations which he demands his government create and enforce. Since State citizens have no true understanding of the need for social rules and are operating solely out of will-power, which is weak and unreliable, social order is always precarious and problematic. For State citizens force is a much more reliable means of maintaining order. It is indoctrination with theism that creates the cognitive and emotional acceptance of intra-psychic force which in turn creates the need for so much force and a preference for force. Thus force and violence are the standard of practice throughout the modern State. This reinforces another rationale for the existence of the State: to maintain social order.

I am not saying that without theism there would be no need for law and law enforcement. I am saying that the theist tribal mindset is such that no amount of law enforcement short of totalitarianism will create law and order. Just as theist 'need' a punitive conscience to force them to 'be good' they need a 'strong man' (government) (rulers) (police) to make them behave. Theists are so committed to the use of force that it permeates every aspect of their existence. They have a strong commitment to sadomasochism, they do not question it! Theists need rulers and leaders to think for them, do for them and force them to behave and obey their rules. They use force to organize the insides of their minds; they use force to organize their 'loved ones,' so naturally they can 'see' only force as the means of organizing their State! Like the tribe the modern State will inextricably morph toward totalitarian dictatorship. It will do this even if no one in the State consciously wishes it to do so!

The State, like the tribe is always at war. Governmentophiles just like the tribesmen know that all humans are born evil. But the citizens of the State have received absolution from the State collective and are much less evil than foreigners. So the citizens of every State are suspicious of other States. Govermentophiles know that other States may attack them at any time and that they must remain perpetually on guard. War is almost continuous. War and military conquest is as important from the psychological standpoint as it is as a source of spoils, territory and taxpayers. Constant warfare, or the constant threat of it, serves to reinforce, for all the citizens of the State, the necessity of sticking together for their mutual defense and survival. Neighboring States are always planning to attack, so every State has to stay vigilant and ready to defend itself. The psychological dependency inculcated into the citizen is thereby reinforced by the reality of modern life. Theists will always work things around in their country to reflect their inner fears. So what starts out as a fear in their religious imagination comes to be created in their lives. Societal organization recapitulates psychological organization. It is another self-fulfilling prophesy.

The citizens, imbued as they are of the State, actually insist on government regulation, ownership and control! They perceive no contradiction between this view and their warped idea of freedom. Theism has taught them the sinfulness of commerce and business; they know that the pursuit of life sustaining money is evil. They reason that those who pursue wealth the hardest must be the most evil of all. So they look to their parental or tribal surrogates, government, to protect them from the evil marketplace (God). They want the State to oversee businesses and make businessmen behave ethically. Talk about having the fox watch over the chickens! But even citizens that expect nothing from government but corruption can find no alternative to government regulation. They willingly approve and comply with such evil schemes as the government regulation of commerce! Such is the blindness of the theist mind! Such is the theist fear of freedom!

The **third** delusion required of citizens in order to sustain the State is the unquestioning belief in the righteousness of the State and the actions of its government. Governmentophiles believe that those officials who are elected to leadership positions through the democratic elective process are imbued with a spiritual or supernatural power to govern morally and intelligently. To win an election is to be anointed by God and automatically made more intelligent and more morally righteous that any mere citizen! Many elected officials believe this about themselves! The truth is quite the reverse. Nothing stimulates motivation from the evil sadistic **theist identity** more than winning an election to public office!

Citizens are thereby relieved by this delusion of any guilt they should have for sanctioning, and participating in theft, war, murder and pillage. Like all sadomasochistic relationships both the governing elite and the citizens acquire emotional support for their evil by belonging to the State. Stealing is universally condemned by States; they do not want the competition! Simply by re-labeling stealing and calling it taxation the natives are mesmerized into accepting it as just! Taxation is clearly theft. Yet theists, because of their self/collective identity are as blind to this evil as most others. Taxation laws, particularly income taxation, give the State a 'blank check' on every person's efforts and information on every citizen's wealth and income. The governing elite learned long ago that 50% of the wealth of citizen's who think they are free amounts to a lot more wealth than 100% of the wealth produced by people who know they are slaves! Income taxation has the moral equivalency of slavery. Does it shock you that slaves accept it?

Laws of government regulation give the rulers inside information on everyone's business policies and methods. Government regulation is what keeps the citizens in harness and results in every person working for the State. As surely as residential real estate taxes effectively eliminate home ownership, income and other taxes eliminate the citizens' freedom more effectively than iron shackles! The citizen actually owns nothing, the State, the elites, own everything, including, of course, the citizens themselves! The State simply by changing the name of its stealing to taxation for example, allows it to steal with impunity. Governmentophiles have no moral argument against it.

Citizens of the modern State must accept massive deception and delusion in order to believe that murder and theft is morally justifiable. They must believe that anything and everything sanctioned by the State is acceptable and necessary. Theism is the pure essence of evil, and the State exploits this theology. But the citizens do not know this, they cannot recognize this evil. All of this stealing and killing is viewed as grand, glorious, Holy and noble.

The State is a very formidable machine. The word 'machine' is accurate here because all the citizens are expected to 'play their role'. Every member of a State is intimated by the expectations of every other member in the State, especially those of higher 'rank'. Everyone in the State, from top to bottom is required to be a 'cog in the wheel' that the State needs and demands. Force and violence permeate the State. This is a hold-over from the tribe. In a very real sense everyone in a nation is a slave because everyone must conform to 'political correctness'. Anyone who fails to live up to the State's expectations could find himself under investigation by the State's taxing authority or assassinated by its secret police.

Each State's belief system, indoctrinated into the citizens when they are children, is a special version of a generic theism. Nation specific taboos and ruler worship dogma is easily piggybacked onto this generic theism to suit the needs and specifics of the governing elite. Once theism is inculcated into a child's brain it is relatively easy to modify or change the details and particulars of the State's special dogma. The State may change its name, geographic boundaries or ruling families, but the fundamental delusions regarding self- sacrifice, murder, theft and the nobility of the State remain intact and persist. A **theist identity** is extremely difficult to change.

The ruling class has known from the invention of the first tribe that the key to its rule over the masses is the control of food. Following the invention of the wheel and the improvement in transportation, both agriculture and society became more complex. Soon another invention, metal money, made wealth easier to hide and keep. Gold and silver money enhanced trade and commerce adding to the wealth (food) produced. Unlike farm land money could be stored, hidden and kept secret from State authorities. Money, gold and silver, like food before it, became a source of independence and therefore a threat to State control of the masses.

The **forth** delusion required of State citizens is the belief that only the State can be trusted to mint coin and print money. Theism has always been anti-commerce. The purpose of theism is to produce slaves. The last thing a slave needs to know is how to earn money. Commerce is pro-life whereas the basic tenets of theism are all anti-life, so there is a basic antipathy. But the invention of metal money and the seditious opportunities it offered caused the Pagan clerics to reinforce their anti-commerce, anti-business and anti-profit taboos. The anti-money bias was therefore accentuated in tribal theology. Earning money, saving money, earning money on your money, engaging in trade, business enterprise and pursuing commercial interest were added to the list of sins the slaves must avoid. It was crucial to the survival of the State and its governing elite, that the vast majority of people remain financially inept and dependent upon them. Economic prosperity by another State or individual is routinely interpreted as evidence of sin. So whether thinking about one of the earliest and most primitive tribes that existed before the invention of writing, the wheel or money, or the modern State, we see the same effort on the part of the ruling elite to control as much of the means of production as possible. As long as land was the source of all wealth the ruling elite held monopoly control over all the land. The invention of metal money allowed another avenue to the attainment of personal wealth by people. This could not be allowed.

The only thing holding the State together is a belief, a flimsy, abstract little concept that the State is a necessity! Therefore everything must be done to insure that the people cannot manage their business or their money, or their life, without State oversight. For this reason one of the very first things that the State did upon the invention of money, was to confiscate the concept and establish monopoly control and regulation over money.

If control of money could magically be placed off limits to the State the result would be the collapse of the State. Upon the invention of metal money the governing elite of every State immediately realized this and took steps to see that this did not happen. They saw that only by the control of money could the State survive.

The invention of paper money was the State's salvation. Paper money allowed the governing elite to survive, and the ability to manipulate paper money provided them a new and improved means of exploiting the citizens. The last thing the State wants is for their citizens to gain any independence. The ultimate basis of dependency for a living organism is its dependence on food. Whoever controls your food controls you. So the State ruling class, for their own security, seeks control of the citizen's necessities. The easiest way for them to do this once paper money became the excepted medium of exchange, was to control the supply of paper money.

Once paper money was universally accepted, and the State's total control of paper money established, the temptation of the governing elite to print more of it could not be controlled. The opportunity to exploit the citizens in this way was enjoyed by the governing elites in every country. "If they think and act like sheep, why not fleece them?" has always been the mantra of the ruling elite. Sticking it to the masses in this way was made even sweeter by the fact that the rulers knew they were using money to further indenture the masses, something that the citizens might have used themselves to promote freedom. An extension of this forth delusion that the tribal rulers should control all the food is that the State should control all the money.

From the beginning of recorded history up to our present time, we read of one State after the other inflating its currency, bankrupting their State, going to war or armed revolution, decimating the people, and then moving on to the next governmental mess simply to repeat the process. I cannot find a single exception! The masses produce wealth, build up wealth reserves, and then the governing elites destroy it. And in spite of it all, citizens remain obedient to their masters just as their theistic programming demands.

Even with the ruling class firmly in control of the money supply, constantly printing more and manipulating it for their own benefit, they still dare not allow the masses to gain a modicum of independence. Utilizing their thugs or henchmen in their organization called 'government' they worm their way into every nook and cranny of the market place with laws and rules to control and regulate every business, commercial and professional enterprise the citizens manage to create. Whenever someone advocates government regulation of anything, we can be sure the speaker is a member of the ruling elite or one of its many sycophants. The royal road to totalitarianism begins with government regulation. The anti-business taboo is there to give the rulers of the State the religious cover it needs to take over and control land, labor, money, business and professional services. Aside from the greed the governing class has for the wealth citizens produce, their motive is to prevent the accumulation of wealth by others and to prevent citizens from developing an independent sense of mind. Should a citizen, by some near miracle, accumulate massive wealth, he will be invited to join the governing elite;

if he declines they will squash him. He might use his wealth in some liberating effort. Better he run for public office.

This anti-commerce theme was present in theist theology from the beginning, but before the invention of money it was expressed in terms of food. The slave was not to hoard his game but to share it with his tribe. The hunter-gatherer should be seen as a prototypical entrepreneur. Where he took up his bow and arrow and flint knife and went out into the forest to bring home food, the modern businessman takes up his brief case and cell phone and goes out into the market place to bring home money. So anytime we see anti-money, anti-business, anti-commerce and anti-profit bias in a religion or political philosophy it is a Pagan religious element. The anti- business bias of modern tribal theology continues the anti-hunter-gatherer tradition.

The State is a system created **solely** for the exploitation of the masses. The governing elite cannot allow the citizens to know this. The governing elite work constantly to keep the masses from learning how wicked and evil the government really is! They do not want to know this themselves! So the governing class works hard to convince themselves and the masses that the State and its government are essential to everyone's survival. Therefore anything that reduces or negates dependency of the citizens on the government threatens to expose this lie and could potentially dissolve the State. The governing elite must therefore eliminate anything which encourages independence because it is a threat to the State's survival!

Numerous ideologies have been formulated to rationalize and justify this stealing and killing by the State. The socio-economic-political systems called communism, fascism, Marxism, progressivism, socialism, welfare statism, and state capitalism, are just tribalism re-tooled, re-labeled and re-marketed as something new and desirable. They are all just slightly different versions of generic collectivism and they appeal to those with the tribal **theist identity**. Intellectuals formulate these enslavement plans to ingratiate themselves with the governing elite. They are just Shama in more modern clothes without his feathered headdress and painted face.

You might think that the State is very vulnerable to the truth. But the Emperors' nakedness is almost always interpreted as fine clothes because of the citizens' profound dependency. Governmentophiles' need to believe is so great that they see what they are told to see! The State's expectations and the citizen's 'conscience' control the behavior of all the citizens at all levels within the State. Theism produces mental 'blind spots' which creates the mind-set to accept propaganda. If a person can be so blinded by theism as to willingly send his own child to certain death for the good of the State, than believing the reality of empty political propaganda is no feat at all.

Reality thereby comes to vindicate the theists' worst fears and horrors. This is why the development of democracy which had the potential of liberation has not helped theist populations that much. They predictably and consistently vote themselves into greater bondage. I know of no example of a democracy consistently voting itself into greater freedom. No form of government will work for theist, except the one they are programmed for, the one they constantly unconsciously strive to establish, or re-establish, collectivism, totalitarianism!

The Governing Elite

Who are the governing elite? What are the motives of the governing elite? Every Nation has a group of people who control the apparatus of State and profit there from. That profit is among other things, material goods. The governing elite profit materially in a greater degree than their time and effort would justify. The governing elite for example may take 60 to 80 % of all production for themselves and their pet projects. While benefiting in this great amount the elites might contribute less than 1% to the total time and physical effort of the agricultural, industrial and military enterprises. Material gain is a very real and substantial lure for elites.

But another even more powerful motive for many governing elites is, as it was for Shama, the **control** of the masses. Controlling the government of State allows the elite to avoid the forest. This fear of the forest can be translated as fear of reality, or in hunter-gatherer terms, fear of God. This fear comes from a feeling of underlying incompetence at dealing with reality. This existential fear is a very profound psychological fear that results from the adoption of a collective human *super*concept and a rejection of nature or reality as their supreme standard of truth. Thus the governing elite's motivation is an existential fear of the forest. In modern terminology it is a profound fear and distrust of the free unregulated market place, the fear of competition. Everyone would like to have a guaranteed income, a guaranteed job, or a business that is a protected monopoly. The governing elite more than want this, they believe they have a right to it and do not hesitate to send hundreds of thousands of people to their deaths in order to obtain and maintain this for themselves and their families!

The governing elite are thus existentially afraid to compete fairly among the masses in a totally free, unregulated and open market. They seek government protection of their businesses through a State enforced monopoly. This explains the psychological pressure for a democracy to become fascist on its way to totalitarianism. The lassie-faire marketplace is equivalent to the untamed forest, or God, it gives no one a special privilege, and everyone must face reality every day. **The ruling elite demand a special dispensation from competition** and the necessity of fundamental competence. Only the **control** over those who are perceived as competent, i.e. those who are successful in commerce such as farmers, tradesmen, manufacturers, industrialist, and businessmen, etc. partially relieves this existential fear.

The masses also receive a psychological gain or benefit. They are relieved of responsibility for State actions. Citizens love to complain about 'the government' but they are not about to change anything! The government readily accepts responsibility, leadership and control, they want it desperately. This is what the masses shun. The governing elites do not get their hands dirty and the citizens are not responsible. It is thus an interdependency.

Theism is the belief system. Theism has an agenda. The agenda behind theism is the survival and maintenance of the governing elite by the domination, manipulation and control of the masses. Nothing about a modern democratic State changes this theist agenda.

Theists always have problems with their societal structure. They can never get any system to work. They create many problems with their theist mind-set that they cannot solve. Theist concepts act like blinders on a mule making many solutions simply out of their conceptual purview. The delusion of the State collective paints theist into a corner intellectually. Nearly all the problems faced by a State are self-induced. They invariably end up in a place from which they cannot extricate themselves. In such cases they almost always choose the course that reinforces their delusions rather than the course which provides a real solution. Their chosen course of action invariably leads to violence, often war. The cost of fanatic devotion to their theism is extremely high in both lives and resources. No amount of high-sounding rhetoric and convoluted theories can change theism and the State from what they are: rationalizations for theft. To maintain the delusion that they are anything more comes at a high price. But for millenniums humans have preferred to pay this cost rather than examine and reject their barbaric belief system!

To understand the evils of unbridled abstraction you need only study the history of the twentieth century. If profound failure would cure the theists of their cognitive reification disorder the first half of the twentieth century would have surely prevented the second half of the carnage. But the definition of a delusion is that it is an erroneous belief which is impervious to reason. Only by harnessing the cognitive processes to reality, or if you perfer, to the natural God, does man reap the rewards, and avoid the pitfalls, of abstract conceptual thought. It is floating contradictory abstractions held by force in the human mind, and then acted upon, that are are the root of all evil. And a very large portion of that root is theism.

Due to strategic military competitiveness, in order to survive, many tribes have adopted the State societal structure. So the tribal model changed into the State model and spread around the world. Many tribes morphed into a State. Every State maintains an active propaganda program to constantly generate and perpetuate its lies, myths, delusions and deceptions. The State rulers use news makers, media experts, reporters and journalist to continually create and promote the State mythology. Every State is actively engaged in this. They must also create the illusion of being special. Every State seeks to differentiate itself. Each State's unique and specific mythology necessarily makes each State appear very different from the others. Differences in religion, language, type of government and methods of governing create ample excuses for conflict and warfare. Warfare is something nations seek anytime they believe they have a military advantage. Wartime is a States finest hour.

Most governmentophiles unconsciously work and vote for laws and policies that gradually lead to dictatorship. The democratic State is therefore constantly evolving toward an absolute totalitarian social structure. This is its nature. Totalitarianism is a type of dictatorship which suppresses individualism through the use of terror. Totalitarianism is the ultimate anti-self societal structure! A State may vary from brutal absolute dictatorship to a more liberal commune that enforces compliance through social pressure. How brutal and violent or lenient and benevolent a State may be is at the discretion of the governing elite and dependent upon their beliefs and ambitions. But internal pressures are constantly pushing every State toward its true identity as a totalitarian dictatorship! I know of no democratic State that does not regularly and dependably vote itself into greater bondage! Huxley and Orwell described a mild version of this reality.

The core moral construction of contemporary civilization is unmitigated evil. But this evil is contained within a veneer or layer of humanism. As evil as much of the behavior in the contemporary world is, it is astounding that it is not more so. Incredibly most theists behave decently most of the time. Behaving decently is going against their upbringing but behaving decently is their nature! If you have trouble accepting these statements give a look around the world, read tomorrows newspapers and come to your own conclusions.

Chapter 13
The Contemporary Mind

I want to return to the question of how the human mind can at the very same time embrace two diametrically opposite belief systems. How does a modern contemporary person believe, in the very same mind, the concepts of both humanism and theism? How can the human mind loyally believe Paganism and science? The reference standards for these two belief systems are totally different! How do contemporary peoples believe both these dueling philosophies in the very same mind?

As I mentioned before modern peoples are utilizing the mental mechanism of compartmentalization! The masses are now using the same psychological methodology that the heretics used to avoid a direct confrontation with the powerful institutions of tribe and religion! Yes the slaves, the commoners, the masses, the tribesmen, the citizens rather than reject theism, have simply pushed it to the side or bottom of their mind to make room for this new egalitarian philosophy of humanism and science! Moderns have split their minds into two so that each of these competing belief systems will have a 'place.' The 'have my cake and eat it too' approach which seemed the safest approach for the heretics appears the safest way to go for everyone!

Earlier we labeled the identity which develops out of a commitment to tribalism the **theist identity**. The contemporary child also develops an identity out of the humanism and science that he is taught . We labeled this his **secular identity**.

The influence of scientific humanism can seldom extinguish the tribal theist motivation no matter how sincerely the theist wishes it out of his mind and out of his life. This is because theism makes up part of his imprinted *super*concept. Theism is his default value system. The best that a contemporary theist can usually do is to **repress** his **theist identity**! The mental mechanism of repression hides the theism from awareness but it does not eliminate it from the mind. The contemporary person usually has a **theist identity** repressed into his unconscious mind. This means that all the savagery we associate with tribalism resides in the contemporary theist's unconscious mind!

Repression of one's **theist identity** is necessary because contemporary people do not wish to discard their theism! They know that their tribal values are out of place in the modern world, but they want to hang on to them anyway. They do not wish to renounce supernaturalism and do not want to shun their faith. Few individuals or groups categorically reject the use of force or question the workings of their religious conscience! And almost nowhere do contemporaries recognize the evil of self-sacrifice and its masquerade as virtue! Almost no alternatives to theistic morality is offered from any source. For most people theism remains their standard of truth and virtue. Contemporary peoples simply cannot see that the sacrifice of their self as just as evil as the sacrifice of others!

Talk to a contemporary theist and ask him about two of his concepts. First ask him what he means by the term 'God'. This inquiry will likely bring forth a description of an alternative universe housing mythical supernatural personages who will include a human-like God, plus assorted angels and saints. This realm may be called heaven. Another area of this person's supernatural realm may be an area called 'Hell'. This area may contain a creature called the 'Devil' plus assorted 'sinners'. As an individual explains his God concept he will probably include admonitions to behave morally by doing 'good' for others and avoiding 'sinful' thoughts and selfish behaviors. By 'moral' the contemporary theist means that a person should be actively engaged in self-sacrifice for the benefit of other tribal members, for the State collective, or for the 'community'.

Next ask a modern theist what he means by the concept 'reality'. This question will usually bring forth a description of nature, with perhaps examples from astronomy, math and science. He may point to concrete objects in the environment around you.

You have now demonstrated the existence of both of the individual's 'minds.' You have demonstrated two very different identities or 'minds' with very different reference standards. Only one of these reference standards can be an individual's *super*concept. If the individual was raised a theist their *super*concept or default reference standard will most likely be their tribal theist beliefs. **It is the attempt to fully believe both of these directly opposite ideologies at the very same time that characterizes the modern mind.** This is the distinctive attribute of this third psychological adaptation.

But the complexity of the modern mind does not end there! You will recall from Section II that the **theist identity** has two components! A **theist identity** exists as an inverse duplication to produce both sadism and masochism. This is the source of the sadomasochism which permeates tribal life. Its two halves are opposites but they stem from the same commitment to human sacrifice. The slave 'virtues' exist inside the theist mind as **masochistic** traits. Then because of the dialectic effect these masochistic traits are inversely replicated as **sadistic** attributes! Both of these motivations are present in the **theist identity**. Both traits are evil! Masochism is in no way better than sadism!

The modern theist's mind is therefore initially divided into the **theist identity** and the **secular identity**. Then because of the dialectic effect the **tribal identity** is further cleaved into a **masochistic** and a **sadistic** portion! You will now observe that we have three functioning areas of the modern mind. This is just as

Sigmund Freud observed over a century ago. He labeled these three areas the Ego, Superego and the Id. This arrangement can be listed as:

1. Theist Identity

    a. masochistic identity-Superego

    b. sadistic identity-Id

2. Secular Identity-Ego

Freud had no idea what caused his patients to exhibit these three areas in their minds. His guess was that the human mind was just made that way. Psychoanalytic theory takes on new relevance and greater practicality when we know how these mental constructs are derived!

Mental Conflict

A mind divided into compartments is characteristic of people of our contemporary era. **Compartmentalization** and **repression** are the psychological mechanisms which allows the creation of this modern mental condition. So the modern mind has at least three areas. The first area we have labeled the **theist identity**. As you know a **theist identity** is a mixture of two groups of attributes. There are the (1a) **masochistic** attributes devoted to *self-sacrifice* (Superego). And there are the (1b) **sadistic** attributes devoted to the *sacrifice of others* (Id). The result is usually the sadomasochistic tribal psychology.

The third mental compartment is the (2)**secular identity** which contains a scientific or rational humanistic identity with a positive self-value and a commitment to reality and reason (Ego)! (As explained earlier the Dialectic Effect does not cause replication of the **secular identity** because it is not a floating abstraction.) This is the Freudian tripartite mind.

Rarely will a contemporary make a total commitment to one identity or the other. Most modern people are conflicted over this all their lives. I don't mean that most people do not come to some sort of accommodation. Most compartmentalize, then refuse to discuss or even think about their beliefs most of the time. They simply blank-out these issues and play-like they are settled and not to be revisited. But this effort to deny that a conflict exist leads to a repression of the conflict. The sadistic portion of one's **theist identity** is usually pushed down into the unconscious mind the deepest because it is the most socially unacceptable, but it has not gone away. In most individuals the masochistic traits of their **theist identity** are repressed as well. Although much of one's **theist identity** is repressed it does not mean that these mental compartments get along!

The **secular identity**(Ego) constantly reminds the **theist identity** that its beliefs do not stand up to reason, that they are not logical. The **secular identity** is critical of both masochistic and sadistic thoughts and behaviors.

The masochistic **theist identity** (Superego) in turn constantly points out to the **secular identity** how 'selfish' its beliefs and actions are. Tribal peoples kept a wary eye on themselves, and their fellow tribesmen, searching for evidence of selfish, un-tribal behavior, so they could punish the 'sinner' or report him to tribal authorities for punishment. Now the contemporary theist's **theist identity** does likewise, constantly analyzing and criticizing the **secular identity** accusing it of selfishness and causing much guilt, anxiety, shame and depression. On those occasions when the sadistic **theist identity**(Id) is stimulated and aggressive or violent behaviors considered or initiated the individual usually responds with regret, remorse and guilt.

The better the mind a theist has the less the possibility of some kind of detente. Contemporaries spend a lot of time debating with themselves over how to interpret events and trying to determine what course of action to take. This talking to oneself, sometimes even debating out loud between one's 'minds'was not invented by Shakespeare, it is a characteristic of the modern mind.

A result of this conflict is often the condition contemporaries refer to as 'mental disease[16].' Technically, as contemporaries will tell you, only those who are 'diagnosed' by a ' professional' have the condition 'mental illness.' For our purposes here we can assume that all contemporary theist have this mental condition to one extent or another. Here we can admit that most contemporary theist handle their mental conflict in socially acceptable ways most of the time while some do not. Sigmund Freud , contemporary psychoanalyist, psychologists and psychiatrists have accurately outlined the socially unacceptable ways theists react to this mental conflict. They have labeled these socially unacceptable behaviors the 'psychiatric syndromes'. Some of these syndromes are:

1. Syndrome of Major Depression
2. Syndrome of Bipolar Disorder
3. Alcohol Abuse Syndrome
4. Chemical Abuse Syndrome
5. Obsessive-Compulsive Disorder
6. Panic Disorder Syndrome
7. Post-Traumatic Stress Disorder
8. Schizophreniform Disorders
9. Somatization Disorders
10. Sexual & Identity Disorders
11. Syndrome of Multiple Personality Disorder

These psychiatric syndromes are simply descriptions of behavior and nothing else! They are the results of the creative ways contemporary people have tried to force the three identities to get along and the symptoms generated by their efforts. There are dozens of psychiatric syndromes in addition to these. Professionals refer to these types of diagnoses as a 'descriptive diagnosis.' Such 'diagnoses' have as little real validity as it is possible to imagine, but they are extremely popular! After receiving a 'diagnosis' the mental 'patient' is then entitled to prescription medication to 'treat' their 'condition'. How are such 'diagnoses' created or invented in the first place?

The first step in creating a psychiatric diagnosis is for a 'professional' to describe a socially disapproved pattern of behavior in one of the official 'professional' journals. Next he or another 'professional' gives the pattern of behavior a name. All psychiatrists and other 'professionals' read the journal and begin to look for similar 'patients' in their practices. They find and 'diagnose' similar behavioral patterns seen in their patients. They report this in subsequent issues of 'professional' journals. The circular reasoning doesn't stop there.

Next various medications and other 'treatments' are tried and the results published, again in the 'professional' journals. The pharmaceutical manufacturers make a great deal of money inventing chemicals to 'treat' these mental 'conditions.' Everyone profits, the 'patient' gets a good excuse, the 'professional' makes a good income, the psychiatric 'hospital,' the pharmacy and the pharmaceutical industry all make out really well! Who is there to be concerned with the truth when hypocrisy is so profitable?

This is not to say that people are not suffering. They are suffering terribly. The contemporary mental conflict is extremely painful and is causing an epidemic of suicides, especially among adolescents. Contemporary 'treatments', pharmaceutical and otherwise, do nothing but paste over and cover up the destructive effects of theism on a person's mind. Psychiatry and psychology are but another of the many contemporary means of evading this conflict. Hide, lie and make up stories seems to be the modern way to handle conflict. To really help people suffering with this mental conflict requires that the delusions of theism be faced. The fact that **both** masochism and sadism are equally evil has to be recognized. Harming your **self** is in no way less evil than harming someone else! Both parties are equally human beings! There is really no difference between masochism and sadism. The label is just changed depending upon the victim chosen for the sacrifice. Dying in combat, and suicide bombing are acts of masochism and sadism combined. Killing others, killing oneself, or killing both, how do you consider one behavior morally superior to the other?

Why can't contemporary theists face these facts? Because facing facts is not something contemporaries do.

A Troubled Adolescence

It is understandable that the most complex psychology mankind has so far created would be difficult to manage. The conflicted behavior we observe in contemporary people was not present among hunter-gatherers. It did not occur in hunter-gatherer or tribal youth. It is a modern problem. This conflict is rough on children and adolescents as well as adults. While most contemporary adults may have accommodated to some extent to the conflict between theism and scientific humanism, and the hypocrisy which characterizes modern life, the young person has not. The contemporary adolescent goes through a painful period of mental anguish as he tries to reconcile these cultural and societal contradictions. Contemporary adolescence is characterized by this psychological and moral battle of coming to grips with this conflict. This theist adolescent phenomenon is something new and unique to our modern contemporary period. It is a conflict between a malleable elastic subjective tribal *super*concept and an objective, fixed and in alterable scientific reference standard. It is a conflict between tribalism and humanism, between blind obedience to State authority and loyalty to one's own reason. It is a battle between collectivism and individualism, between the passive compliance to the rule of the elites, and the sovereignty of the **self**. How does the modern child cope with such massive and overwhelming cultural hypocrisy?

The contemporary theist adolescent was taught as an infant that he is innately evil. He was taught by his religion that the only way to redeem himself to some degree is by some ritual act of his religion or his State. He has been taught that the greatest virtue is to sacrifice himself for others. He can only ransom his **self** and own a life by some sacrificial service to the collective (community).

On the other hand our teenager is being taught to obtain a good education, earn a college degree, get a good job or profession and earn a lot of money. He is advised to find a mate, settle down, buy a house, raise a family and plan for a comfortable retirement. All of this advice is so that he may enjoy the good life. How does the contemporary adolescent meet these disparate goals? Can he or she resolve these conflicting values?

When the youngster considers acting on his **secular identity** and behaving rationally he is criticized by his **theist identity** as selfish and evil. So he hesitates to take action which may bring him success and pleasure. On the other hand behaviors based on either aspect of his **theist identity** will make his life miserable. His **secular identity** is telling him that sacrificing his **self** is not a smart behavior. Reason in his **secular identity** is telling him that acting on his tribal motivations is dumb, that it will cause him great problems. If he acts on his **masochistic theist identity** he knows it will hurt and he does not want to be stupid. If he acts on his **sadistic theist identity** and goes into a homicidal rage, something he often feels like doing, then he knows it may well end up badly for him, with him dead or in prison. **Either** theist motivation leads to self-destruction.

There appears to be no way out of this trap. In order to obey his theist values and be a good, moral person by the contemporary religious standard he must stupidly sacrifice himself! If he had only his **secular identity** to satisfy, life would be hard enough with his family clamoring for him to get an education and get on with his career. The lack of endorphin rewards is making his life almost unbearable. Aside from alcohol or drugs, about the only thing that can temporarily stop the adolescent's struggle and find some refuge from the tyranny of his **theist identity** is to make a deal. In order to escape one's **theist identity** and obtain some relief from the continuous pain it causes, the adolescent tries to find some kind of collective **redemption**!

Only by being 'reborn' or 'born-again' can the contemporary adolescent postpone total destruction of the **self** and put-off self-destruction to some future time. Perhaps in the future in response to great success and happiness he will have no choice but self-destruct, but he wishes to put it off. The opportunity to be re-born means that maybe, perhaps, this time the individual can enter the world without Original Sin and escape a theistic identity. Without the possibility of redemption, the theist will engage in the self-destruction that is now programmed into his mind. Without the extremely difficult task of repudiating theism, redemption is the only recourse short of drugs, alcohol, mental illness or suicide. Suicide, too often combined with homicide, is a frequent choice of contemporary youngsters.

Redemption is a much better option. At least it keeps the adolescent alive! The redemptive process must be 'extra self', beyond or above the **self.** Redemption must come from without, from a 'higher power'. Theist believe that nothing good can come from within, the **self** is bad, and cannot redeem anything. There are many religious, philosophical, political and social/occupational systems that offer such redemption.

How is redemption accomplished? One's **theist identity** must be tricked into believing that the youngster is going to sacrifice his **self** for the good of the collective! Most adolescents are able to do this by 'making a deal' with their **theist identity**. Most contemporary societies and religions have rituals to assist young people in making this contract. In exchange for the promise to pursue a 'Holy Tribal Cause' their **theist identity** allows them to think well of themselves. Their morality of self-sacrifice will not allow them to pursue a life selfishly for themselves, but their theist conscience will approve a life lived for others in some sanctioned tribal pursuit. Since they have been taught 'live not for yourself', but for 'others' they con their **theist identity** into a contract whereby it appears that they are sacrificing their life for others. Some youngsters become missionaries, embracing the very poison that has caused them to be sick in the first place. Some girls have a baby and devote themselves to their child's happiness. Many dedicate themselves to helping others as teachers, doctors or nurses. Some youths join the military or go to work for the government. Others look for unique ways to justify their existence and bargain with their theist conscience for a successful life. Their life is not theirs, it belongs to the tribe or State, so it must be ransomed in some way. In exchange for self-sacrifice in this life, they bargain for success in this world and they hope for the promised reward in a utopia somewhere in the future. Any deal that allows them to appear to dedicate their life to some cause other than their own success and happiness may be acceptable. The hypocrisy of redemption is preferable to the constant pain of psychic conflict.

Theist adolescents latch onto altruistic and self-sacrificing socially approved occupations and careers as a drowning man grabs a life raft. Career paths that 'care for others' or claim lofty or Holy goals are indeed true life rafts, as guilt, depression, drug abuse or suicide are sometimes the only alternatives. If their career choice or trade is not blessed by the tribe or State as a noble cause, the individual will have to find some extracurricular volunteer or charity work to assuage his guilt. Perhaps they can make a lot of money by promising to 'give back' philanthropically to the collective later in life. Theist feel guilty for being alive when their theist morality demands that they sacrifice their **self** for the tribal collective in some 'Holy Cause,' usually a war. Aztec children walked up the Temple steps to sacrifice their life for their tribe, contemporary youngsters register for the draft or report to the military recruiter. Each tribe conducts the sacrifice of its children in its own way.

Once the theist ideology is taught creating the **theist identity**, the least difficult compromise for the youngster is to find redemption through dedication to a 'Holy Tribal Cause'. The masochistic programming will still be there, and will result in self-destruction or suicide if a deal or contract with one's **theist identity** cannot be reached and maintained. One's own self-analysis could expose the hypocrisy and reveal the contract as a sham. The smarter the youngster the greater likelihood he will see through the contract. Even with a contract firmly in place, success generates a huge guilt which is a powerful stimulus to self-destruction. The adolescent, because of his **theist identity** cannot, will not, fight for his **self**, but he can fight, kill and die for his 'Holy Cause'. Even with a 'contract' in place self-destructive tendencies may only be somewhat muted. One of the results is the display of masochistic and self-mutilation tendencies we see so routinely during adolescence. Making a deal or contract with one's **theist identity** is a deadly serious enterprise. Parents and teachers who take this process lightly, or worse, make fun of it, are playing with their youngster's life, and maybe their own!

Masochism which we observe so commonly among adolescents such as cutting themselves, body piercings and other forms of self-flagellation is the taking of delight by the **theist identity** in the pain and humiliation inflicted on the **secular identity.** The **theist identity** takes masochistic pleasure in the pain caused the **secular identity**. Masochism revels in the debasement that can be brought on the **secular identity**. The **theist identity** is punishing the **secular identity** for being selfish. It is a personal sacrificial ritual. Teaching a child that self-sacrifice is a virtue pits the Forebrain against the Hindbrain. The child jumps on the bandwagon of condemning his **self**, so that he will fit in and gain a sense of belonging with his theist family, friends, and culture. The physical pain is offset by the endorphins attained in the behavioral reinforcement or authentication of his **theist identity**.

With sexual maturity contemporary peoples expect the adolescent to embrace reality. Contemporary theist culture sends a mixed message about this as it does about everything. Be virtuous, follow the theist morality of self-sacrifice, but also be financially successful, embrace reality and pursue your commercial interest. This split in society's demands simply reflects the dichotomy in societies' belief systems but it places even more pressure on adolescents. With maturity, the young adult is expected to employ his mind, learn to support himself, and begin to build his own career and family. The arrival of maturity makes it more difficult to hold off and ignore these pressures. The push to 'get out on one's own' places tremendous stress on the compartmented adolescent mind. Faced with the demands of adulthood, the youth is pressured to make a firm commitment exclusively to rationality and the modern world. Yet theism continues to beckon for an altruistic life of Holy Sacrifice. Young people want to be 'good' and lead a respectable life, but the directions on how to do this are conflicting.

Thankfully, most theist adolescents do embrace the 'real world'. That is to say they redeem themselves by making a deal with their **theist identity** that allows them to choose a 'worldly' path. They pick a practical educational path leading to a career that earns money. They reject criminal behavior and enter the world of commerce. They choose a spouse and start a family. As with all theists they will continue their secret inner life with its hidden tribal agendas; but, at least outwardly, they choose rational practical pursuits. Only a small percentage of adolescents reject a major commitment to rational (Ego) pursuits and identify more fully with their tribal beliefs.

Those adolescents who choose to strongly reinforce their **masochistic theist identity** (Superego) are the 'overly religious' kids. These children find it difficult to compromise. They are very idealistic and see making a deal with their conscience as the hypocrisy it is. They are bright and courageous and determined to make their theist beliefs work in the real world. Taken to the extreme, these are the people who become the fanatics of various 'causes'. These are the kind of people that fuel Inquisitions, Holy Wars, crusades of all kinds, and terrorism. And the more radical they become in their religious thoughts and actions, the more they stimulate their unconscious sadistic theism. Therefore the tension between their religious masochism and their unconscious sadism *grows greater and the chances of violence increase*. But these youngsters also form the ranks of discontented youth from which new and innovative *superconcepts* are created. And rarely, extremely rarely the social environment selects from this rich source a spiritual leader to follow and emulate.

Those theist adolescents who identify with their **sadistic theist identity** (Id) are rarely candid about it. Most contemporary adolescents hide their tribal theism, but those who make a strong sadistic identification go further and develop a secret undercover life. They become disillusioned with their life and see its unrealistic aspects. They tire of an endorphin deficient existence. They come to view the dominant culture as dumb and unattainable to them. They become astute at living a 'double life'. This is the group of kids who easily become psychopaths and criminals.

Some individuals can switch back and forth between their **masochistic theist identity** and their **sadistic theist identity**. All the while they are using their **secular self** to rationalize their behavior and try to fool everyone around them. Some individuals go even farther and develop many mental compartments. They have multiple selves or personalities. Such compartmentalization compromises cognitive integrity and destroys the possibility of endorphin mediated pleasure and self-esteem. This makes the urge for exogenous endorphins, like medication, alcohol or illicit drugs, even greater. All of these psychic phenomena are consequences of the childhood indoctrination with theism.

Compartmented individuals, in general, are potentially dangerous. This means that most theist people are prone to violent outburst. They tend to be angry and fearful. They must expend a great deal of energy maintaining their compartments by exercising will-power and determination. It doesn't take much to set them off.

The more a person engages in behavior that ignores a need, the more of an endorphin deficiency he will suffer. An individual that tries to deny his **secular identity** may therefore be enduring a bland, inhibited, pleasure-less life. He may possess an endorphin-starved brain. If, at this point, he should inadvertently have an exciting and pleasurable experience, it may well change the entire course of his life. Such an exciting and pleasurable experience, because of his endorphin starved state, may feel much more powerful than it would be otherwise; indeed, in relative terms, it is an extreme change. What might be pleasurable to anyone becomes, in our subject, a life-changing epiphany! He takes this great feeling as enlightenment. He sees it as a beacon, beckoning him toward his true identity! Exercising his 'magical thinking', he may take the experience, and the feeling, as indisputable proof of his real 'self'.

What this experience embraces is of crucial importance, since the individual may adopt it and its meaning with great fervor. The experience may be of a sick child taken in pain to a doctor's office. It may be a homosexual encounter, or a first visit to a saloon and his first drink of alcohol. It may be the first experience with a drug, looking through a peep-hole watching adults having sex, a beating or castigation by an authority figure, or his killing of a pet. What constitutes a life-changing experience under these circumstances is as varied as life and human nature. When it is combined with a collectivist identity and 'magical thinking', such abominations as serial killers and suicide bombers are sometimes the result.

The conflict in the mind of theists is between the true Hindbrain self and his tribal self. It is between the Forebrain which the tribe may program and the Hindbrain that is locked by genetics in the mode of self-preservation. It is the conflict between obtaining endorphin reward from satisfying their biological needs and obtaining endorphin reward from authentication of his **theist identity**. It is a torment that rages in the unconscious of theist continuously. Many theists are able to rigidly control these mental compartments and appear 'normal' most of the time, perhaps for life. But creativity requires more mental agility, more freedom and openness of mind. Therefore the more creative a theist is, the more likely he is to stray from a 'normal' behavioral pattern. His 'deviant' behavior may occur rarely, perhaps only under the influence of drugs or alcohol, or may express itself as a complete lifestyle. Heretics and geniuses in their quest for greater mental freedom through weaker walled mental compartments exhibit more deviance. If you will study the personal lives of the great heretics more carefully you will see what I mean. They are having less success at keeping everything neatly segregated and tightly walled-off. You see this mental torment and self-destruction also among the greatly talented for this same reason!

A massive amount of mental energy is required to control the compartmented mind. All this pent-up energy and anger can make the compartmented individual very dangerous. This situation has been described as being like two huge locomotives sitting on the same track, facing each other nose-to-nose. Both locomotives are running at maximum power. The noise, smoke and heat generated are enormous as the iron wheels spin at great speed against the iron rails. But neither locomotive is going anywhere! Nothing positive is being accomplished and there is a great danger of fire and explosion.

Whenever a person's *super*concept or core **theist identity** is challenged, this equilibrium or balance may be thrown off and this can cause a mental explosion. Tolerance of other sects and challenges to their own beliefs is very difficult for religious people. This is because their *super*concept is based upon collective acceptance and tribal approval. Theirs is a people centered universe. Alternative beliefs cause anxiety and anger to escalate very rapidly. This can cause a violent and often deadly outburst of behavior. While these traits are occasionally valuable to chiefs, kings, prime ministers and presidents in whipping up fervor for war, they are disturbing when they describe your fellow citizens and neighbors.

A strong religious identifier, for example, may experience the intrusion of a foreign culture, or the proximity of a different tribe or religion, or secular non-believers, as an attack on their **identity**. A theist identifier may experience the observation of a 6-year-old girl playing joyfully in her yard, as a stimulant to his sexual fantasies. He may become enraged at her for stimulating his sexual thoughts and feelings, feelings he is trying hard to extinguish or control.

The first example may lead to terrorism. The second case may result in molestation, rape and murder. A person's **identity** is so enormously important that people will kill and die when they perceive it to be threatened.

The creation of mental compartments housing absurd mystical concepts and sadomasochistic motives, usually kept secret but at other times proclaimed loudly to the world is the key to understanding the religious mind. Once a mind is unbridled from objective reality, from a natural God, free from the constraints of cause and effect, magical thinking runs amuck. Then superstition rules over logic and fear replaces reason. Whether it is the well-kept secret ideology of one person or the proselytized militant ideology of billions, the unifying concept is a rejection of the restraints of reason and the embrace of uninhibited, structure-less fantasy. Whether the 'floating mental construct' must be protected and reinforced by fantasy, perversion, serial molestation, or murder, or more simply by recruiting more converts, the failure of reasonable people to call absurdity what it is, only encourages increasingly absurd behavior. To quote Voltaire, 'those who believe absurdities may commit atrocities'.

Through theist indoctrination, a child suffers a damaged core sense of self-worth and is left with damaged cognition. These two consequences of theist teaching-- perversion of morality and perversion of cognition—can be observed to some degree in most theists. Programming one part of the brain to oppose another part is diabolical. The person is left with a damaged ability to apprehend reality and gullibility for life. All his life he will be prone to falling for one religious cult or political fad after the other. He may change tribes, he may change religions, he may change his political beliefs, he may even change his patriotism, he may change from one fanaticism to another, but he won't question the virtue of self-sacrifice. And a person willing to sacrifice himself is usually more than willing to sacrifice other people. Thus all fanatics get their start in a religion.

The reason why this is the case is that the theist, because of a *super*concept based upon human authority, develops a psycho-epistemology of force. He can believe anything his will power and self-discipline can force into his mind! Force thereby becomes the religious person's preferred approach to solving problems not only psycho-dynamically, but socially and politically as well! This is the source of the tribal admiration of force and denigration of reason and persuasion.

So each of these theist concepts: evil, revelation, force, supernaturalism, human sacrifice and self-policing attacks an area of the human brain. The evil morality of self-sacrifice and self-policing perverts Hindbrain function drying up much endorphin secretion. Force and supernaturalism pervert cognition forcing the acceptance of contradictions creating compartmentalization, self-deception and mental illness. This highly complex psychology results from our inability to make a firm and uncompromising commitment to reality and reason. This complex psychodynamics is the result of hypocrisy and results because we refuse to be honest with ourselves! Combined these mental manipulations retard maturation and produce individuals who are mentally conflicted and prone to violence. Drug abuse, mental illness, violence, murder, war and suicide are the consequences.

Our world reflects this conflict between colliding and incompatible ideologies. To the extent a theist is able to keep his tribal **theist identity**, with its savage psychotic sadomasochistic motivation tightly walled-off and contained, he is able to behave in a rational and benevolent manner. But his **theist identity** is always pressing for authentication, for action, for masochistic and sadistic violence! The contemporary world of mental conflict, war and violence is the direct result of our failure to deal decisively with these conflicting beliefs. Our modern contemporary world has come to reflect this mental schism in every aspect of our lives and institutions.

Self-Destruction

The contemporary theist child represses his **theist identity.** This means that he pushes his commitment to sadomasochistic tribal values with its violent aggression towards his **self** and others into his unconscious mind. He turns to his **secular identity** and tries to reinforce it by behaviors which a reasonable person would adopt. He pushes his evil **theist identity** out of consciousness as he reinforces his **secular identity** in his everyday behavior.

The modern person's conscious **secular identity** is pro-life and programmed to pursue success and happiness. His **theist identity** is not motivated toward the individual's success and life. It is programmed for sacrifice and service to the tribal elites. The modern citizen hasn't consciously dealt with these ancient and destructive beliefs, he has just pushed them out of consciousness.

The **theist identity** now resides in his unconscious mind. His **theist identity** is there lurking in his unconscious mind always ready to judge, condemn and punish. There in his unconscious mind his sadomasochistic **theist identity** is constantly pushing for authentication through tribal behaviors. This ancient insanity is now contained in a repressed mental compartment where it sits like a festering abscess trying to burst forth into his mind and behavior!

Even though much of the **theist identity** is repressed in the individual's unconscious mind it is still his *super*concept. And unconsciously the individual is still committed to these ancient evil beliefs. Unconsciously he believes that these tribal Pagan morals are 'more' moral than any of his secular motivations. The **secular identity** may be the identity the person consciously wishes to promote but it feels much less authentic than his default value system; his **theist identity**. Any time he is uncertain as to what he should do, especially where important life decisions are concerned, his tendency will be to fall back on his **theist identity.**

This repressed theist programming behaves like a destructive computer virus! Once this unconscious program of theist self-evil and self-destruction is installed in the child's mind, even pushed down and walled-off by repression, only will-power and self-discipline, or the development of mental illness, can prevent it from running its course and destroying the individual! This **theist identity** will press inextricably toward behaviors that will lead to his own destruction. There, cycling is his unconscious mind, theism becomes an unconscious software program of self-destruction!

Psychological maturity in the contemporary context simply means that the individual has learned to apply sufficient mental discipline to hold his tribal tendencies in check. Most of the time he chooses secular, more socially acceptable behavior. If enough will-power and self-discipline can be marshaled against his **theist identity**, then it might lie low and wait for success or happiness to give it more power and trigger it off. Repressed theism is directed specifically against the individual's **secular identity**. The 'broken-record' message playing in the background of the theist mind is: live not for yourself, but for others. Career, romantic and financial success is individual success and inimical to tribal theist morality. A tribal theist always shares his booty with everyone in the tribe! Tribal success is collective success. This is why many talented theist self-destruct in the mist of their successful careers.

Theist Morals in Contemporary Life[28]

Alice had planned a career as a commercial artist. But she fell in love with Roger. They eloped and her college plans were put on hold. Soon Eric, Rose and Teresa arrived and she reluctantly took up duties as a full-time housewife and mother. She had hoped that once the children were all in school she could return to her drawing and to school. But something, usually money problems always seemed to prevent this dream. So week-days for Alice all went the same.

She had to awaken early in the morning to get her husband and children off to work and school. She always awoke to the same thought: I am tired. I don't want to get up and do my duty, but my husband must get off to work. And my three children must soon leave for school. I must not be selfish and sleep in. I must get-up and make breakfast for my family.

As she dressed she always became anxious to get everyone off so that she could be alone. While alone all day she does a lot of shopping, cleaning, washing and straightening –up, but she also chain smokes and drinks wine constantly. Alice hates her life. She feels trapped and taken-advantage-of by everyone in her life. She feels like her potential has been lost and somehow life did this to her. She has sacrificed her life and happiness for her family.

Betty lives in the same small town as Alice and they are acquaintances. But Betty feels very differently about her life. Betty came from a very poor farming family and could never even think about a college education. Betty and Ben started dating in High School and neither ever dated anyone else. Shortly after graduation Betty and Ben were married. And like Alice Betty soon had three babies to care for. She wakes up about the same time each morning as Alice, but with very different thoughts in her mind.

When Betty awakens early in the morning she thinks: I am tired. I could use more sleep but I love my husband and I appreciate how hard he works to provide for our family. I am so proud of my kids and love them so much. They will be leaving for school soon and I won't see them all day. I want to see them off with a good breakfast.

Notice that the **behavior** of the two women is identical. If you were only observing what they **do** you would miss the point. It is their **attitude** that is 180 degrees different. In the example of Alice we have a person who has a low estimate of herself (she is a bad person; she is 'selfish.') This estimate is reinforced by the way she thinks! Only by the exercise of her **will** and her willingness to *sacrifice* her dreams and do her duty makes her behavior 'good.' Down deep she feels like her behavior is phony, that she is phony. She must fight against her fundamental, negative view of herself, force herself to act in opposition to her nature, and do what is right. Her thought processes **reinforce** a view of herself as a person who must fight her natural inclinations, in this instance to neglect her family. She must make herself do the right thing. She is a long-suffering masochist sacrificing her happiness to do her duty. Endorphin deficiency is assured by her attitude. The attitude she has is one of resentment. Her outlook will be pessimistic. How will this attitude affect her self-esteem? How will this attitude affect her sense of **responsibility**?

In the second example of Betty, she thinks only of what she wants. Her husband and her kids are hers to enjoy. She **owns** this situation! Therefore she thinks like an owner. She selfishly wants to do good for her family. She is **reinforcing** a view of herself as a person who naturally wants to care for those she loves. Her **attitude** may be one of pride and self-satisfaction. Her outlook will be optimistic. What will this do to her self-esteem? How will this attitude affect her sense of **responsibility**?

The individual who must 'force' himself to do what is 'right', to act in opposition to his 'selfish desires', out of moral duty, will assign responsibility to forces outside himself. 'They' caused this. 'They' have put him through this. It is 'their' fault. One advantage of a collective identity is that you always have plenty of people to blame. You *are* victimized by the collective!

An individual who is motivated by normal selfish desires, and does what he wants to do, will accept authorship and ownership of his actions. Such individuals tend to accept responsibility for their behavior.

The code of self-sacrifice is thus very destructive to family relationships. Theists almost always have trouble with their relationships. Masochistic self-sacrifice perverts motivation, attitude and outlook. It also encourages rejection of personal responsibility. Self-sacrifice means denial of one's needs. It forces a person to choose between behaving 'morally', and behaving in a way that is good for oneself. The morality of self-sacrifice makes a person choose between an endorphin deficient 'good' upstanding life, and a morally indefensible, but endorphin abundant one. The more altruistic and self-sacrificing an individual behaves, the more morally righteous he feels, and the more endorphin deficiency he endures.

Now let us examine Jim's life. Jim lives deep in the mountains, two hours from the nearest settlement and four hours by automobile from city services. He lives there with his wife who has to commute to her job as a secretary. Jim does not like nearby neighbors. When he lived in town, people constantly took advantage of him. A neighbor once borrowed Jim's chainsaw because he needed to cut down a dead tree in his yard. This neighbor kept Jim's saw for many weeks, forcing him to repeatedly ask for its return. When the neighbor finally returned the saw to Jim, it was broken into many pieces. The neighbor told Jim he was sorry but the tree had fallen on his saw. Jim humbly accepted the apology, went to town and purchased a new chainsaw.

A few days later the neighbor heard Jim cutting some firewood and came over. The neighbor admired Jim's new chainsaw, and said that he needed to finish cutting-up his downed, dead tree. Jim generously offered his neighbor the saw, along with extra gasoline. That is when Jim vowed to move away from people, off into the wilderness where he would be alone.

Jim does not want to loan his chainsaw. Instead of telling his neighbor that he does not loan his chainsaw he completely rearranges his life to avoid having neighbors! Why? Jim has a collectivist morality. The collective has a claim on him that he cannot deny. 'From each according to their ability-to each according to their need' is Jim's moral code. So how does he handle it? He doesn't question his moral code, he moves off into the woods where he will not see or hear of other people's needs!

Now let us travel to Palestine to meet Mr. Aldouri. Mohammad Aldouri was born in Palestine and has lived there all his life. He received all of his education at a school at his local Mosque. He is a devout Muslim. His hatred of the Jews is only overshadowed by his love of Allah. He has never actually met a Jew, but he has heard many awful stories about them. He has dedicated his life to the Palestinian cause. This morning he is dressing in the clothes of a Jewish cleric, under his gown strapped to his body is enough explosive to destroy a building. He kneels on his prayer rug; he prays earnestly for a long time. Then he rises and poses for a videotaping of his last words to his family. The video will be given to his mother after his death. He says good-bye to each of his loved ones and he leaves by car for the trip that will bring him nearer to his target.

Mohammad has a collectivist identity. He cannot separate his **self** from the Palestinian cause. 'Their' enemy is his 'enemy'. He needs to reinforce his collectivist identity through action. The closer he gets to his target the more powerful his emotional high.

Carl sits in his living room. It is late at night, his wife sleeps, and all is very quiet. Everything is very still, nothing seems to be moving. Yet inside Carl's mind there is a raging mixture of thought and feeling. As quiet and motionless as things appear on the outside, they are chaotic and violent inside of Carl's mind.

Like a broken record, Carl goes over and over every shortcoming, every failing, every mistake, every misstep that he has ever made, as far back in his life as he can remember. He calls himself every derogatory name and description he can think of. He is stupid, he is lazy, he is fat, he is ugly, he is worthless, he is selfish, he is rotten to the core, and he is a slob, without any saving virtue or redeeming value. This 'verbal' abuse goes on inside the privacy of his mind for hours on end. Upon arising, his wife comes into the living room, she finds Carl in a very black and foul mood. She suggests that Carl get back on his Prozac.

Carl's punitive conscience is as sadistic in its punishment of him as he is masochistic in his motivation. His collectivist conscience knows no mercy!

Lisa has been married three times. She is 43 years old and has a son, age 24, and a daughter, age 21. Both are now out on their own. Lisa is going through her third divorce and is living alone. She says that she has never been happy. She has spent her whole life trying to please others and make everyone happy. She has concluded that men cannot be trusted. Lisa finds many faults with herself. She is too selfish and thinks too much about her unhappiness. Yet she admits that pleasing others and avoiding disapproval is the central motivation in her life. She says she suffers from low self-esteem. She avoids confrontations and feels guilty if she sticks up for herself. She attributes her low self-esteem to being fat, ugly, and lazy, and she constantly reminisces about her personal failures. She has decided that she must be the type of woman that men always dump.

Lisa has lived for others all her life. She can't understand how others could find anyone they could appreciate more. She has always put her *self* last. It has been so long since she did something for her *self* she is not sure she even has one now.

Bill is in his garage. He tried this once before and messed up. He is determined to get it right this time. He has shut the garage door and is now sealing off all the edges of the garage door with duct tape. Bill is very angry. He is angry with his wife, but mostly he is angry with himself. He has always done everything others wanted him to do so that they would accept and love him. His wife has insisted that he pay off her friend's huge credit card debt. Although he despises debt, and dislikes this friend of hers, he has complied with her wishes. He is miserable when he refuses her and feels like he has betrayed himself when he complies. He can't win any way he goes. He views himself as defective, as 'mentally ill' and beyond hope or 'cure'.

This time he has made sure his car is full of gasoline and will run for a long time. After making sure the window and back door were also well sealed, he starts the automobile and sits down on a chair to await the inevitable. His last thoughts are of his father and the terrible way his father treated him.

When you have spent a lifetime slowly killing off your *self* to the point that you no longer have one, then suicide may seem like the next logical step.

Clyde sits in Cellblock 4 at Folsom Prison on death row. He has confessed to killing five people and has hinted there may be a lot more. Clyde hates people, he hates himself, and he hates every living thing. He will not admit to this hate. He may not be aware of it, since he has no point of reference. His hate is pervasive and complete throughout his being. He does not experience it as a separate or distinct emotion for he is the living embodiment of hate. When he was a child, Clyde tortured and killed animals and took delight in the agony of dying creatures. He believes that to destroy that which is evil that which one hates, is good. Clyde is not unhappy. Authentication of one's sadistic tribal identity is more important to Clyde than anyone's life, including his own.

 John tried to hang himself after his wife announced that she was leaving him. In his mind, *to take his life would show his wife and children how much he loved them.* John has never had any sense of 'self' and has always felt hollow and empty. When he met his wife to-be, he fell madly in love with her. She became his reason to exist; she became his 'holy cause', someone to devote his life to. When she quit him, he lost not only a wife, but also all reason to live. His *self* was 'emptied out' and he had nothing, nothing else to lose. With her leaving, he believes he has lost his soul, his being, his 'self' and his reason to exist. He was already dead even before he tried to kill himself. All that was left to do was the formality of putting a toe tag on his body and sending it to the morgue.

When you believe in the virtue of self-sacrifice you assume others place a high value on it as well.

Ben lives near a shopping center. He goes to this shopping center often, for groceries, to do his laundry, and for other purchases. When he leaves home, he must be careful not to step on a crack in the sidewalk. He feels sure that stepping on a crack in the pavement will precipitate some terrible accident that may kill him. He halfway knows that this is silly, but he's not sure; he's still afraid.

Placing oneself into elaborate causation fantasies is just one of the consequences of reified 'magical thinking'.

Brendan is a very fine physician. He has a huge practice and works incredibly long hours. He earns a lot of money. However, Brendan spends little money on his family and even less time with them. Brendan is heavily involved with a number of charities. He is a well known philanthropist and contributes a lot of money and time to several important not-for-profit organizations. But he refuses to spend money on private schools for his children.

Brendan and his wife argue frequently over money issues. She recently took a job as a librarian and uses the money to pay tuition for her children to attend private schools. This has angered Brendan as he feels his wife is trying to shame him as someone who can't afford to provide for his family.

Brendan actually loves his family greatly. Otherwise his denial of their needs would be no sacrifice. A theist's guilt is often commensurate with his financial success.

All of these fictional characters demonstrate how real people suffer from their unconscious **theist identity**. Alice, Jim, Carl, Lisa, Brendan, and all the other characterizations, demonstrate the collectivist or tribal method of 'thinking'. Theism is simply incompatible with a contented, happy life. These examples do demonstrate that compartmentalization can contain some of the damage caused by the childhood indoctrination with theism. But even when walled-off the damage done by this evil is not greatly reduced. The destruction caused to people's lives in every country around the world by theism is incalculable.

How Indoctrination and Commitment To Theism Leads to Murder

Now we come to the important question of how the indoctrination of children in theism creates the destruction, violence and killing apparent all around us! It has been my thesis throughout this book that evil is taught. Therefore the tremendous amount of murder we have observed down through history and continue to observe in the world today is the direct result of teaching theism to infants and children. I believe I have here presented a credible explanation of how this works. Now I wish to outline the psychological steps that lead to the many acts of 'senseless' murder. I am not talking about the murder of others in competition for money, fame or love interest: I am here addressing serial and mass killing, serial murder, genocide and warfare.

How does a person come to the decision to kill strangers? How does a person decide to kill people that they do not know and have never met and who have not done anything to harm them personally. The desire to kill these victims arises simply because they live in another country or belong to another group, race or religion? Imagine for example murdering athletes at the Olympics solely because they represent a nation you do not like! Or imagine killing thousands of civilians, children, women, old people, simply because they live in a certain country! What mental processes allow a person to make the decision to do such things?

STEP 1. Program a child with theism using the two stage process. The first stage is to teach an infant that his *self* is innately **evil**. This causes the infant to incorporate into his mind an identity of self-evil. Later as an older child the second stage of theist indoctrination teaches the Pagan morality of human sacrifice. These two childhood mental incorporations result in a defective way of thinking called reification and a morality of human sacrifice. This creates a collectivist sadomasochistic psychology.

STEP 2. Reified collectivists' abstractions such as 'State', 'ethnicity', and 'religion' are then processed mentally *as if* they are real! To the theist or anyone thinking incorrectly in this way, such abstractions as these seem very *real*. These reified collectivist concepts are incorporated into the individual's collective identity. The person then cannot differentiate between his collective, religion or State and his biological *self.*

STEP 3. The theist will then identify the collective in his identity as 'his' country, 'his' religion, or 'his' group. Then upon designation by 'his' tribal authority, or through his own analysis, he concludes that some other collective is an enemy of 'his country' or 'his religion', he responds just as if he had been personally attacked. He conceptualizes his **self** as a member of a collective and he conceptualizes as his enemy a collective 'State' or 'religion'. He then sees the war as his collective/self versus their collective. The war is between abstractions, one abstraction against another. Abstractions do not exist in reality! There is a group of people who make up the population of a tribe or State. This population of individuals has beliefs and opinions that vary across a wide spectrum. Some may condone the war others may not. Yet all are forced erroneously but conceptually into a single, undifferentiated collective label. Now the theist collectivist will fight and kill members of the enemy collective and if necessary die fighting for 'his' collective! This is how the ruling or governing elite convince the slaves or the public to fight their wars for them.

STEP 4. A **secular identity** is more perceptual, more 'biological' and more focused on objective reality and 'concrete' in its orientation. A **secular identity** is more of a commitment to the **self**. This is the principal reason the **secular identity** is rejected by theist, because it is committed to the **self**. The most important things in life, those things which provide the everyday endorphin rewards are perceptual and have to do with individuals. Individuals are real. A person's love partner, family and friends are individuals and they are real. These aspects of life are tangible.

Perception is required to inform one of the needs of one's lover, family and friends. Without the perceptual, without the secular mind functioning well, an individual will have difficulties with personal relationships. Without a strong **secular identity** endorphin rewards may be scarce.

A **theist identity** is a **self/collective** identity. Tribal identities are more abstract, floating and conceptual, as is any identity built through reification. A sadomasochistic **theist identity** is an abstraction or conceptual commitment. It is an ideology or theology, in other words it is a belief system of some sort! The theist may join with other religious individuals or organizations, or with members of his gang or fellow fanatics or criminals in the commitment to a 'Holy Cause'. The individual may become intensely committed to such a group or 'collective' of 'True Believers.' The individual may then begin living more and more in his *collectivist fantasy world*. The more time he spends in this fantasy world and the more behaviors he initiates trying to authenticate his theism, the more isolated he becomes from his family and love ones. Whether the individual is destined to become a renowned religious prophet, the leader of some glorious government reform movement, a serial murderer, or the kingpin of the criminal underground, he is first of all a tribal theist, a conceptualist and collectivist. His motivation will emanate from his sadomasochistic **theist identity**! The **theist identity** may be the masochistic variety and the individual will in time become a fanatic religious visionary. Or it may be the case that the individual is committed to reinforcing his evil tribal sadism, so that in time he may launch into a criminal rampage. In either case he shuns his **secular identity** for one that *feels* more authentic.

As an individual focuses more on the conceptual abstractions of his **theist *collective* identity**, such as commitment to a particular religious theology, a specific political-economic theory, or a specific criminal plan, as he becomes increasingly committed to his sect, cell, fantasy, gang or party something begins happening to his **secular identity**! A person's secular identity and perception begin to atrophy as it is increasingly ignored while his **theist**

identity, with its collective belief system, his Holy Cause or Utopian political theory, is reinforced and hypertrophies. The individual may become increasingly distant so far as love and family relationships are concerned. His increasing commitment to a conceptual construct at the expense of perceptual satisfactions pulls him from his love relationships as he is drawn more and more to his commitment to his conceptual collective madness. He is developing into a fanatic. As his personal relationships dry-up because he provides less and less feedback and decreasing emotional support to the people in his life, his commitment to his conceptual collective belief system becomes increasingly important. More and more he reinforces his tribalism, less and less does he reinforce his secular life. Reinforcement of his tribalism like a snowball going down a snow covered mountain grows larger and larger. It could be a secular snow ball but in this case it is the conceptual, abstract snow ball of tribalism. Eventually he is living totally for his Holy Cause. He becomes a 'true believer[18]' and grows into a totally committed fanatic.

If a fanatic has hatred for certain group or collective and especially if he should join a club or group that similarly hates a certain group, then such an individual may find support for his hate. Such a support group becomes for our fanatic his own tribe, his own tribal collective. With the support of a small but vocal 'collective' our subject's fantasies may escalate into action. He wishes to authenticate his belief system, gain the approval of his peers, and this may lead to action. This is the cognitive defect which causes a lot of the evil theists' behaviors

The terrible consequences of this reification error in cognition do not stop with war or with one large group against another. This same substitution of a collective abstraction for individuals is applied with devastating results in other areas of life. To understand this better let us consider two individuals.

First let us examine the beliefs of a young woman regarding abortion. We can call our first subject woman 'A'. Let us say that she is totally opposed to abortion and tells herself that she would never do or even consider doing such a thing. She knows that many women undergo abortions; some utilize the procedure frequently as a means of birth control. She does not believe she could be a best friend to a woman that did such a thing. Our subject however is not interested in supporting laws that would outlaw abortion and she is not about to picket abortion clinics. She doesn't consider it her responsibility to affect other people's behavior. She is not concerned with the moral stature of 'her country'. She does not relate to a collective of pregnant women. She doesn't believe in abortion but she believes others have the right to live their lives as they see fit.

Now let us consider another young woman. Let us label this individual woman 'B'. She too abhors abortion. She considers it a **sin** and believes strongly that there should be **laws** against abortion. She feels responsible for all the aborted fetuses and infants, as well as all the misinformed and exploited mothers. She is concerned that 'her country' may condone abortion by not having strong laws against the practice. She believes that 'her country' has a responsibility that goes beyond protecting her rights to include *maintaining the morality* of the national **collective**. She has a growing hatred for physicians who perform the procedure. She joins a militant anti-abortion group in writing public officials and law makers urging them to support strong laws and vigorous law enforcement to abolish this evil. She joins in the picketing of a notorious abortion clinic and meets many people who share her views. She hears that a notorious abortionist will be working at the clinic the following day. She becomes despondent over the realization that 'her government' will do nothing to stop the killing of the unborn. She steals her father's hunting rifle from his home. She hides in the bushes near the back entrance to the clinic. She guns the doctor down the next morning as he arrives for work.

How is this second woman's mind different from the first one? Woman 'B' has a protective attitude toward an abstraction. Her *self* is meshed with the concepts of 'State' and 'government'. She feels responsible for the State. The moral condition and commitment of 'her country' or 'her government' as portrayed in its laws are of extreme importance to her. She feels a very strong and intense responsibility for the condition of 'her State's' morality. She identifies so strongly with the State that the moral condition of the State is experienced as her own personal moral condition! There is very little or no separation between the moral condition of her **self** and that of the State **collective**! Her **self** and the State **collective** are almost one and the same! A moral threat to the State and its reputation is experienced by the second woman as a threat to her **self, a threat to her identity**!

So the second woman experiences no separation between her **self** and the **collective**. To her they are one and the same. She has a **self/collective** identity. Protecting the honor of the State or its government is the same as protecting her own personal honor and reputation.

Collectivism is the cause of *the syndrome of the inappropriate acceptance of responsibility.*

Once a crime is committed the members of the support club or collective will usually disavow all connection to the perpetrator. They may be among the first to define the criminal killer as 'crazy'. In some cases this club, support group, tribe or collective exists only in the perpetrator's mind! For many killers their reference authority, the tribal elites exist only in their fantasies. The tribal elites may only exist in their Holy Books! Theists are religion creators! They are true believers. And their tribal religions are seldom up to anything good.

The perpetrator of these horrible crimes sees the world as 'his' collective against 'their' collective. He is determined to get his 'high', his endorphin reward, regardless of its cost to him or others. Even when caught he is often smug with the self-satisfaction that what he did was morally righteous! He is certain that his behavior was morally righteous because it makes him feel so good! He is doing 'God's work'! When the perverted motivation of tribal theism is combined with collectivist thinking, such abominations as serial killers, school shooters and suicide bombers are sometimes the result.

It is the seeking of the endorphin reward for authentication of one's tribal collectivist identity that motivates most murderous behavior. The killer may choose to 'go out' in a burst of 'feel good ecstasy' or he can remain alienated, depressed and miserable. Many theists live on the edge of suicide a great deal of the time. Often it is when drug use ceases to ameliorate his pain that he turns to something more potent.

As a group of collectivist pilots fly an airliner loaded with enemy passengers toward a targeted skyscraper, or as a true believer approaches his target wearing a vest of dynamite, they experience an outpouring of endorphins into their brains which is so profound and so powerful that they are feeling a level of pleasure and ecstasy never before known to them! The experience provides a 'high' which is so intense and over-whelming that they cannot question the wisdom of their action. Even if they were able to take some time out and consider their actions they would not alter their course! Theist collectivism with its sado-masochism seldom provides this kind of Hindbrain reward. But when an action of masochistic self-sacrifice can be combined with a tribally approved 'Holy Cause' like the sadistic destruction of tribal enemies, the feeling produced is near *nirvana!*

Delusions Of The Contemporary Mind

Delusions caused by the collectivist identity:

Modern theists believe that together as a collective they can do things that they could not do as individuals. This belief in the power of the collective is much more than just the addition of individual talents and strengths. Theists believe a collective takes on a mystical, magical, near supernatural power which far exceeds the combined mundane powers of the group of individual humans. They believe that this is true of all collectives. But perhaps the best way to observe this contemporary superstition is by observing the modern theist's favorite collective; the State.

The belief that an election, the vote, bestows a miraculous ability upon the elected to govern intelligently and wise is no more apparent than in the area of finance. Where a particular theist may admit that 2 dollars plus 3 dollars equals 5 dollars when considering his own personal finances, he firmly believes that such laws of identity do not apply to large collectives like government entities. We can observe that a group of theist once designated as a governmental collective will take on projects that they would not undertake as an individual or as a group of individuals. This is because reality for theist is not objective, firm and fixed in and by nature. Reality for theists is the collective censuses. If enough of the right people believe it to be true, then by definition it is true! The theist firmly believes that the collective can actually change reality as it sees fit! Observe state legislatures, National Congresses and other governmental groups. Changing stealing into taxation and murder into warfare are believed by theist to be real changes! These are often cited as examples of how the State can do anything it wishes! One state legislature even tried to change the mathematical constant *pi* from 3.14159 to an even 3.0[27]! Homosexual behavior has been voted to be a biological condition! Next I expect hurricanes to be voted illegal and outlawed!

This 'magical thinking' is apparent wherever theists express themselves. They confuse behavior with objects. They pass laws outlawing firearms to stop killings, laws outlawing fatty foods to stop obesity, and raise taxes on cigarettes to stop lung cancer. These and many other examples are not just silly laws, they represent the illogical, delusional way theist think! They confuse nouns and verbs! Then they believe that as a collective they can change the behavior of the collective! Such 'social engineering' is not just ill-advised; it is delusional almost to the point of psychoses! It demonstrates a mechanical, pre-industrial, view of human nature!

This mechanical view of human nature is evident in all the behavioral 'sciences'. Again consider economics. While economics is obviously one of the behavioral sciences, it has to do with how humans choose to spend their money; you wouldn't know it by the mechanistic models of economics taught in colleges and universities[20]. All the major universities teach a completely untenable and fallacious model for how the financial markets work. Then when the markets develop a crisis we hear remedies proposed by politicians and their 'experts' and then laws enacted as absurd as 'spending ourselves out of debt!' Again, these are not simply mistakes; they reflect a pattern of fallacious thinking based upon theist delusions! These delusions make solutions nearly impossible to find, as correcting for unexpected conceptual failures invariably leads to even more delusional remedies!

Academia is ripe with such delusional thinking. Memorization and regurgitation are equated with intelligence by theist. Intelligence for theist is identical with susceptibility to indoctrination. These memorization abilities garner 'good grades' and mark such individuals as 'smart'. 'Smart' people advance up the corporate ladder and advance in government service whereas non-smart people do not. Such 'smart' people are better than their peers at memorizing the phony theist models for everything from economics to business management to political theory. They are therefore more dangerous in every area of endeavor since they tend to be 'true believers' in their 'field'. So we observe that the higher one goes up the chain of command in corporate management or government service, the more bizarre are the ideas and goals of the senior executives! So the people with the best sense of the real situation are often those lower down in the organization with the least 'education' and the least authority. The 'common man' often has a much better solution that the academics or the politically powerful. The 'educated fool' has become so ubiquitous in business and government as to represent the majority class!

Delusions caused by the morality of self-sacrifice:

The morality of human sacrifice turns the understanding of moral behavior on its head. The theist views anything which emanates from, or is of service to, the *self* as selfish, suspect and probably evil. Whereas activities which are anti-self, which or absent any benefit to the *self* tend to be considered worthwhile and good. Theist society is therefore very suspicious of private, self-serving, free enterprise! People engaged in productive activities trying to make a profit are considered extremely suspect! What could be more selfish than engaging in activities which promote life?

On the other hand government activity or actions, lacking a profit motive, are seen as benevolent and good. Non-profit corporations are accorded favorable review whereas corporations that turn in an exceptionally profitable annual report are accused of profiteering and calls go out to punish them with an excess-profits tax! Government workers are not viewed as tax feeders, rather they are accorded extra admiration because instead of pursuing a career aimed at profit they are perceived as sacrificing by choosing a career in 'public service'.

This denigration of private, free enterprise and the profit motive cast businesses, businessmen, private workers, private and public corporations in the role of exploiters, spoilers, and destroyers of society and the environment. While the truth is just the opposite the issue is complicated by the frequent merger or collusion between corporations and government. Business enterprises have a very difficult time in theist society growing larger than a small local presence without government influence and political power. In order to become a 'big business' it is usually necessary for a company to suck-up to government. Often a cozy relationship between the business and government grows to the point that such a corporation actually becomes a kind of hermaphroditic entity, half government agency and half private company. This usually gives such a bastardized 'corporation' a monopoly or near monopoly in its industry. This merger of government with business is called 'corporatism' which is the fancy word for fascist. The contemporary mind is an unhappy marriage between rationality and tribal savagery, and this is the make-up of these huge pseudo-corporations. These suck-up fascist 'corporations' are an abomination; having all the evil attributes of government entities yet masquerading as private enterprise.

The theist attitude that good people must be protected from bad corporations and bad people engaged in selfish business, leads to the whole mantra that government should provide 'consumer protection' for its citizens! This has been the rational for the massive increase in government regulation of commerce. Virtually no free market, no competitive free enterprise, no profitable business can exist for long with such prejudice against it. And this prejudice very obviously emanates from the theist ethic of self-sacrifice. This delusional bias against the *self*, against profit, against business and against business corporations is so powerful that all enterprise is often brought to a standstill while government regulators pour over their next move!

The rightful owners of most private businesses and business corporations are the business people who create them. But the prejudice against business is so great among theist that the property rights of business owners are routinely set-aside while collectives of workers, trade unions, have their political and legal powers enhanced. This bias against the *self*, the profit motive and free competitive enterprise is so great that theist will allow labor unions to completely destroy the market and devastate the economy, rather than uphold property rights and permit freedom.

The truth is that government produces nothing. Everything a people have is produced by productive people. The only reason anything is produced is because someone expects to profit from producing it. A system of morality which punishes productivity and rewards destruction is a suicidal morality. Governments are the biggest polluters on the planet and governments pay the largest number of people to be destructive and non-productive!

SUMMARY

I have in this section discussed how a few theist delusions affect government policy. Elsewhere I outline some of the ways theist delusions compromise interpersonal relationships. This section shows how theist delusions enacted into law are destroying Western civilization. While it is the clerics who keep tribal collectivism and the Pagan morality of human sacrifice alive, and the politicians who give this ancient evil its legal power, it is teachers, professors and intellectuals who are leading the suicidal charge toward the precipice. In the most prestigious academic institutions around the world a professor cannot voice support for competitive free enterprise. It would be received as blasphemy and evidence for disqualification. The Orient is temporarily spared this destruction because of the reduced influence and power of its academic and intellectual classes. This will change with affluence and will eventually be their economic downfall as well.

If you were to assert that these delusions have nothing to do with organized religion you would be wrong. Religion is the manufacturer or producer of theism, its promoter and distribution center. But if you were to assert that theism can exist and flourish in a Godless secular society you would be correct. In 1917 when the Soviet Union was established the communist undertook a serious and brutal effort to destroy the Russian Church. They were largely successful at eliminating organized religion in Russia. But theism was not destroyed. Far from it, the Russian people simply changed the **label** of their *super*concept from 'religion' to 'communism', leaving their theism untouched. But after a generation or two their communism began to fade without the support of religion.

We see this competition between the worship of the heavenly collective versus the State collective in the various political parties of most modern democratic countries. The party which claims to be conservative promotes itself as desiring to conserve traditional religious values. This means that it is more committed to belief in and worship of the heavenly collective. The party which claims to be liberal is the one most committed to the State collective having disavowed most of its 'religious' values. Both are actually totally committed to theism and the State collective, it is just a matter of labels. Most of the campaign debate is rhetorical rather than substantive.

'Treatment'

Imagine a doctor trying to treat patients without knowledge of, or belief in, the germ theory. Imagine trying to treat heart disease without understanding that the heart pumps blood through a system of tubes all over the body creating blood pressure! Imagine trying to fix fractured bones without an accurate picture of the human skeleton in your mind. Or imagine believing that draining part of the blood from a patient's body would improve their condition as they fought a contagious disease!

Medical treatment under these conditions was seldom of benefit and it often reduced the patient's chances of recovery. So it is with psychiatric and psychological treatments today. The mental conflict which is the cause of much guilt, anxiety, panic and depression is caused by religion. The etiology of 'mental illness' is theism. Brushing patient's and clients religious concerns aside by labeling them 'religiosity' and refusing to take them seriously is a disservice. The reason the behavioral sciences have progressed so slowly as compared to the hard sciences is because we declared religious beliefs out of bounds for scientific analysis. Instead of making psychotherapy about reality, facts, reason and logic, instead of making the therapeutic encounter about identifying and exposing delusions and 'magical thinking', the mental health professions have joined forces with those elements of society dedicated to hypocrisy and denial. Where is the courage in that?

"Treatment" of theism and its concomitant conflicts and symptoms requires that the individual's *super*concept be changed from a human collective standard of truth to an objective reference. So long as most people have human collectives as their epistemological authority conflict, 'mental illness', spiritual competition and violence will remain commonplace. Until an individual's *super*concept is changed to an objective and fixed reality they will suffer from the reification cognitive fallacy and the morality of sadomasochism.

To assist an individual in making this change does not have to be difficult. Many theists are afraid to question excepted dogma. Often the biggest obstacle to clear thinking is fear. The 'cure' is **education** not medication! Socrates should be our model of the correct approach. It is through the asking of questions that you lead yourself and others to greater enlightenment.

Section III
Summary

1. Cognition requires some degree of freedom within the mind. Invention and discovery are the results of this freedom.

2. Compartmentalization was used by tribal individuals to get some free 'space' in which to think. Some of these individuals became heretics.

3. Each heretic builds on what those before him have accomplished expanding the free mental 'space' available in which to think. Thus science, technology, and humanism slowly advance.

4. But even the heretics have difficulty extricating themselves from the totality of mind damaging theism. For theist, genius and insanity closely overlap.

5. The evil of theist concepts is accepted by most people as 'good'. They are unable to identity theist beliefs for the evil they are. They try to believe two opposite value systems. This is an incompatibility. A 'mind divided against itself" The modern age, the age of psychic conflict begins.

6. Compartmentalization was recognized by Freud. It has a number of consequences or side-effects. One side effect is the proliferation of mental compartments. Mental illness is another consequence of theist compartmentalization.

7. These mental compartments housing and hiding delusional fantasies account for a great deal of absurd, violent, perverse and psychotic behavior observed in theists.

8. The theist belief system, and the repression of much of it, and the creation of violent and perverse mental compart-

ments because of it, is the cause of most of the violence and destruction observed in contemporary human behavior.

9. This conflict is played-out in the mind of every theist adolescent and is the cause of his anguish.

10. Heretics, the modern mind, and civilization itself is characterized by this split between conflicting philosophies. A core of Pagan evil wrapped within a covering of humanism creates the irreconcilable conflict of the modern world.

11. So each 'mind' has its own reference standard of truth. Moderns give way to reality in matters of science but default to their theist collectivist *super*concept in matters of morals and in times of strong emotion.

12. The contemporary State just like the tribe has no actual or real reason to exist. It exists for psychological reasons. It exists as a projection of the contemporary mind. The State engages in but two behaviors, it can accomplish but two things, stealing and killing. From these activities it acquires loot and booty which it distributes to the elites, with some for the masses.

13. The contemporary world in all its elements is at its core; evil. The contemporary mind, the contemporary religions, the contemporary State and contemporary governments are at their core; evil. This is because they all emanate from, and developed out of, Pagan theism.

14. Contemporary theists behave decently through will-power and self-discipline to reinforce their secular self. In important matters they often fall back on their core theistic values.

15. The modern world reflects the values of its theistic inhabitants: a core commitment to stealing and killing.

16. The *super*concept of this third psychological adaptation is a combination of the self/collective and reality. Their ethic is a mixture of self-sacrifice and rational self-interest. Their cognition is a combination of reified abstraction and reason.

17. Surely, and hopefully, this third psychological adaptation of mankind is transitory.

SECTION IV
Deism

Mankind's Fourth
Psychological Adaptation
563BC-2010AD

Chapter 14

The Present

It is apparent from a number of vantage points that the current psychological adaptation is unsustainable. A species that adopts an anti-life philosophy can't expect to prosper well. And a species that adopts a suicidal philosophy can't expect to survive very long!

The hunter-gatherers were peaceful and prospered for over 180,000 years. Tribesmen pursued stealing and killing somewhat successfully for over 15,000 years. But with their low level of technology the tribes were never quite able to get their killing into a highly efficient, assembly-line mode. These superstitious barbarians did their best but without clear cognition their military technology remained backward and their killing remained retail and parochial.

The contemporary psychological adaptation combines aspects of two 'minds', two psychological adaptations, creating the worst possible synthesis. The tribal **theist identity** continues the motivation toward stealing and killing while the **secular identity** provides the clear-headed cognition to discover and invent. And this is what the past 5000 years has been. It has been an ever improving technology largely devoted to military enhancements. Now with the discovery of nuclear energy man has the weapons to totally destroy all life on earth taking not just himself, but every species to extinction! This current psychological adaptation is less than 5000 years old and already in deep trouble. It may prove to be a lethal synthesis!

In evolutionary terms a lethal synthesis is created when a combination of mutations occurs in a particular species which cannot be survived in any possible environment. Such an event inevitably leads to extinction!

The great heretics have taught us what we needed to know, but we refuse to listen. They have taught us that there is no supernatural realm or alternative universe where God lives. God is real, he is the natural world. Believing this God and studying him leds not to enslavement but to freedom. Theism is not the work of this natural God, it is the work of man and it is the essence of evil. Theism spins off evil in all directions, from the creation of slavery to the creation of the State.

Yet instead of eliminating this evil we accommodate to it! We have learned to live with evil everywhere in our lives. Many people are motivated more by their repressed evil than by anything else. Evil cannot perceive evil. It is now obvious that evil is not a dark and deep alley, it is a box canyon. We can only leave evil by going back the way we came. We must renounce theism explicitly and the many evil edifices we have built.

Are there any indications that humanity may be in the process of doing this? Is the era of the ruler coming to an end? Is the epoch of the State and governments sliding off into the dust bin of history? Are we transitioning to something different, hopefully better?

Such an event would be good for Homo sapiens as well as the planet unless theism and the State were replaced with something worse. Although it is difficult to conceive of anything worse, we can't underestimate the creativity of the human mind. Is there evidence that beneficial changes may be underway and is there any indication of the nature of such a future world? The answer is yes!

The Future

Over 200,000 years ago, upon evolving into existence, Homo sapiens adopted a simple and straight-forward epistemology. Our ancestors were in awe of the world around them, a world of nature. They adopted Nature as their measure of truth and they worshiped their natural world. This resulted in a simple and straight forward psychology.

After an unknown number of failures witch-doctors finally succeeded about 20,000 years ago in creating a psychology that provided slaves and permanent settlements. The invention of this slave psychology required a change in man's epistemology. The standard of truth had to be changed from nature to the collective of man. This worship of man manifests itself as worship of, and obedience to, a ruling authority. To pull man from his commitment to Nature as the ultimate truth and insert Man as his source of truth required the indoctrination of children with certain concepts. The Pagan religions and its religious schools were established to indoctrinate children with these concepts. The Pagan religions taught the theology of polytheism which created the collectivist mind, produced slaves and established the tribe as the premier societal structure. I have labeled this system of Polytheistic Pagan religion, the tribe, its slaves and their shared delusional belief system: tribalism. Tribalism is the prototypical collectivist totalitarian dictatorship.

About 5000 years ago the first heretics began to appear and question the validity of various aspects of tribalism. Abraham introduced the idea of one God and monotheism was born. Mankind's epistemology improved with the change from many 'truths' to but one tribal truth. But man's *super*concept or standard of truth remained the collective ruling authority. Many heretics tried to remain loyal to their revealed theist religious beliefs and at the same time embrace new ideas which questioned tribalism. This created two 'truths', scientific or natural truth and theistic 'truth'. The masses continued their dependence upon the ruling elite for 'truth', but a small minority developed reliance upon scientific truth, at least under certain circumstances. Belief in a scientific or natural standard of truth gained a growing number of adherents. Most people tried to have it both ways and hold both reference standards in their mind. The inherent contradictions between theism and the scientific, humanistic ideas created psychic conflict. Mental conflict and mental 'illness' have become the hallmarks of the modern contemporary mind.

Over these last few millennia mental freedom has continued to expand as each heretic added to the growing sum of scientific and humanistic knowledge. A new scientific humanism gradually began to grow until a new 'religion' began to emerge. It's not really a religion because it has few formal institutions. Yet it is replacing religion in the minds of many people.

This new 'religion' began as far back as 2500 years ago. At least by Buddha's time a few people had begun to embrace reason exclusively and completely reject revealed religion. This new 'religion' is developing from the merger of three traditions. Each of these three pathways adds important ingredients to this new 'religion'. What might this new 'religion' of scientific humanism hold for the future? Before trying to answer that question let's examine the history of each contributing branch of this new belief system.

The first line of thought which I wish to examine is science. We need to understand the history of science to appreciate how it contributes to this new 'religion'.

1. History of Science

Science is a body of empirical, theoretical, and practical knowledge about the natural world, produced by a global community of researchers making use of scientific methods, which emphasize the observation, experimentation and explanation of real world phenomena. Tracing the exact origins of modern science is possible through the many important texts which have survived from the classical world. However, the word scientist is relatively recent—first coined by William Whewell in the 19th century. Previously, people investigating nature called themselves natural philosophers.

In prehistoric times, advice and knowledge was passed from generation to generation by word of mouth. The development of writing enabled knowledge to be stored and communicated across generations with much greater fidelity. In Iraq from around 3500 BC Mesopotamian peoples began recording observations of the world with extremely thorough quantitative and numerical data. But their observations and measurements were taken for practical purposes rather than for scientific laws. A concrete instance of Pythagoras' law was recorded possibly millennia before he formulated an abstract theorem.

Significant advances in Ancient Egypt include astronomy, mathematics and medicine. Their geometry was a necessary outgrowth of surveying to preserve the layout and ownership of farmland, which was flooded annually by the Nile River. The 3, 4, 5 right triangle and other rules of thumb served to represent rectilinear plats.

The earliest Greek philosophers, known as the pre-Socratics, provided competing answers to the questions found in the myths of their neighbors: "How did the ordered cosmos in which we live come to be?" The pre-Socratic philosopher Thales, dubbed the "father of science", was the first to postulate non-supernatural explanations for natural phenomena such as lightning and earthquakes. Pythagoras of Samos founded the Pythagorean School, which investigated mathematics for its own sake. He was the first to postulate that the Earth is spherical in shape. Subsequently, Plato and Aristotle produced the first systematic discussions of natural philosophy, which did much to shape later investigations of nature. Their development of deductive reasoning was of particular importance and useful to later scientific inquiry.

The important legacy of this period included substantial advances in factual knowledge, especially in anatomy, zoology, botany, mineralogy, geography, mathematics and astronomy; an awareness of the importance of certain scientific problems, especially those related to the problem of change and its causes; and a recognition of the methodological importance of applying mathematics to natural phenomena and of undertaking empirical research. In the Hellenistic age scholars frequently employed the principles developed in earlier Greek thought in their scientific investigations. They, for example, applied mathematics and deliberate empirical research in their work. Thus, clear unbroken lines of influence lead from ancient Greek and Hellenistic philosophers, to medieval Muslim philosophers and scientists, to the European Renaissance and Enlightenment, to the secular sciences of the modern day. Neither reason nor inquiry began with the Ancient Greeks, but the Socratic Method did, and along with the idea of Forms, promoted great advances in geometry, logic, and the natural sciences.

The level of achievement in Hellenistic astronomy and engineering is impressively shown by the Antikythera mechanism (150-100 BC). The astronomer Aristarchus of Samos was the first known person to propose a heliocentric model of the solar system, while the geographer Eratosthenes accurately calculated the circumference of the Earth. Hipparchus (190 – 120 BC) produced the first systematic star catalog. In medicine, Herophilos (335 - 280 BC) was the first to base his conclusions on dissection of the human body and to describe the nervous system. Hippocrates (460 BC – 370 BC) and his followers were first to describe many diseases and medical conditions. Galen (129 – 200 AD) performed many audacious operations—including brain and eye surgeries— that were not tried again for almost two millennia. The mathematician Euclid laid down the foundations of mathematical rigor and introduced the concepts of definition, axiom, theorem and proof still in use today. His **_Elements_** is considered the most influential textbook ever written. Archimedes, considered one of the greatest mathematicians of all time, is credited with using the method of exhaustion to calculate the area under the arc of a parabola with the summation of an infinite series, and gave a remarkably accurate approximation of Pi. He is also known in physics for laying the foundations of hydrostatics and the explanation of the principle of the lever.

Theophrastus wrote some of the earliest descriptions of plants and animals, establishing the first taxonomy and looking at minerals in terms of their properties such as hardness. Pliny the Elder produced what is one of the largest encyclopedias of the natural world in 77 AD, and must be regarded as the rightful successor to Theophrastus. For example, he accurately describes the octahedral shape of the diamond, and proceeds to mention that diamond dust is used by engravers to cut and polish other gems owing to its great hardness. His recognition of the importance of crystal shape is a precursor to modern crystallography, while mention of numerous other minerals presages mineralogy. He also recognizes that other minerals have characteristic crystal shapes. He was also the first to recognize that amber was fossilized resin from pine trees because he had seen samples with trapped insects within them.

Indian astronomer and mathematician Aryabhata (476-550 AD), in his book <u>Aryabhatiya</u> (499AD) worked out an accurate heliocentric model of gravitation, including elliptical orbits, the circumference of the earth, and the longitudes of planets around the Sun. He also introduced a number of trigonometric functions including sine, versine, cosine and inverse sine, trigonometric tables, and techniques and algorithms of algebra. In the 7th century, Brahmagupta recognized gravity as a force of attraction. He also lucidly explained the use of zero as both a placeholder and a decimal digit, along with the Hindu-Arabic numeral system now used universally throughout the world. Arabic translations of these texts were soon available in the Islamic world, introducing what would become Arabic numerals to the Islamic World by the 9th century.

The first 12 chapters of the Siddhanta Shiromani, written by Bhāskara in the 12th century, cover topics such as: mean longitudes of the planets; true longitudes of the planets; the three problems of diurnal rotation; syzygies; lunar eclipses; solar eclipses; latitudes of the planets; risings and settings; the moon's crescent; conjunctions of the planets with each other; conjunctions of the planets with the fixed stars; and the patas of the sun and moon. The 13 chapters of the second part cover the nature of the sphere, as well as significant astronomical and trigometric calculations based on it

During the 14th-16th centuries, the Kerala School of astronomy and mathematics made significant advances in astronomy and especially mathematics, including fields such as trigonometry and calculus. In particular, Madhava of Sangamagrama is considered the "founder of mathematical analysis".

China has a long and rich history of technological contribution. The four great inventions of ancient China are the compass, gunpowder, papermaking, and printing. These four discoveries had an enormous impact on the development of Chinese civilization and a far-ranging global impact.

There are many notable contributors to the field of Chinese science throughout the ages. One of the best examples would be Shen Kuo (1031–1095AD), a scientist and statesman who was the first to describe the magnetic-needle compass used for navigation, discovered the concept of true north, improved the design of the astronomical gnomon, armillary sphere, sight tube, and clepsydra, and described the use of drydocks to repair boats. After observing the natural process of the inundation of silt and the find of marine fossils in the Taihang Mountains hundreds of miles from the Pacific Ocean, Shen Kuo devised a theory of land formation, or geomorphology. He also adopted a theory of gradual climate change in regions over time, after observing petrified bamboo found underground at Yan'an, Shaanxi province. If not for Shen Kuo's writing, the architectural works of Yu Hao would be little known, along with the inventor of movable type printing, Bi Sheng (990-1051AD). Shen's contemporary Su Song (1020-1101AD) was also a mathematician, and astronomer who created a celestial atlas of star maps, wrote a pharmaceutical treatise with related subjects of botany, zoology, mineralogy, and metallurgy, and erected a large astronomical clock tower in Kaifeng city in 1088. To operate the crowning armillary sphere, his clock tower featured an escapement mechanism and the world's oldest known use of an endless power-transmitting chain drive.

The Jesuit China missions of the 16th and 17th centuries learned to appreciate the scientific achievements of this ancient culture and made them known in Europe. Through their correspondence European scientists first learned about the Chinese science and culture. Western academic thought on the history of Chinese technology and science was galvanized by the work of Joseph Needham. Among the technological accomplishments of China were, according to the British scholar Needham, early seismological detectors (Zhang Heng in the 2nd century), the water-powered celestial globe (Zhang Heng), matches, the independent invention of the decimal system, dry docks, sliding calipers, the double-action piston pump, cast iron, the blast furnace, the iron plough, the multi-tube seed drill, the wheelbarrow, the suspension bridge, the winnowing machine, the rotary fan, the parachute, natural gas as fuel, the raised-relief map, the propeller, the crossbow, and a solid fuel rocket, the multistage rocket, the horse collar, along with contributions in logic, astronomy, medicine, and other fields.

With the division of the Roman Empire, the Western Roman Empire lost contact with much of its past. The Library of Alexandria, which had suffered since it fell under Roman rule, had been destroyed by 642AD. While the Byzantine Empire still held learning centers such as Constantinople, Western Europe's knowledge was concentrated in monasteries until the development of medieval universities in the 12th and 13th centuries. The curriculum of monastic schools included the study of the few available ancient texts and of new works on practical subjects like medicine and timekeeping.

Meanwhile, in the Middle East, Greek philosophy was able to find some support under the newly created Arab Empire. With the spread of Islam in the 7th and 8th centuries, a period of Muslim scholarship, known as the Islamic Golden Age, lasted until the 14th century. This scholarship was aided by several factors. The use of a single language, Arabic, allowed communication without the need of a translator. Access to Greek and Latin texts from the Byzantine Empire along with Indian sources of learning provided Muslim scholars a knowledge base to build upon. In addition, there was the Hajj, which facilitated scholarly collaboration by bringing together people and new ideas from all over the Muslim world.

Muslim scientists placed far greater emphasis on experiment than had the Greeks. This led to an early scientific method being developed in the Muslim world. The most important development of the scientific method was the use of experiments. Ibn al-Haytham, also regarded as the father of optics, was a strong advocate of experimentation. He depended upon his experiments for empiric proof of the intromission theory of light. Some have also described Ibn al-Haytham as the "first scientist" for his development of the modern scientific method.

Rosanna Gorini writes:

> "According to the majority of the historians al-Haytham was the pioneer of the modern scientific method. With his book he changed the meaning of the term optics and established experiments as the norm of proof in the field. His investigations are based not on abstract theories, but on experimental evidences and his experiments were systematic and repeatable."

Due to the development of the modern scientific method, Robert Briffault wrote in ***The Making of Humanity***:

> "What we call science arose as a result of new methods of experiment, observation, and measurement, which were introduced into Europe by the Arabs. Science is the most momentous contribution of Arab civilization to the modern world, but its fruits were slow in ripening. The debt of our science to that of the Arabs does not consist in startling discoveries or revolutionary theories; science owes a great deal more to Arab culture, it owes its existence....The ancient world was, as we saw, pre-scientific. The Greeks systematized, generalized and theorized, but the patient ways of investigations, the accumulation of positive knowledge, the minute methods of science, detailed and prolonged observation and experimental inquiry were altogether alien to the Greek temperament."

In mathematics, the Persian mathematician Muhammad ibn Musa al-Khwarizmi gave his name to the concept of the algorithm, while the term algebra is derived from al-jabr, the beginning of the title of one of his publications. What is now known as Arabic numerals originally came from India, but Muslim mathematicians made several refinements to the number system, such as the introduction of decimal point notation. Sabian mathematician Al-Battani (850-929AD) contributed to astronomy and mathematics, while Persian scholar Al-Razi contributed to chemistry and medicine.

In astronomy, Al-Battani improved the measurements of Hipparchus, preserved in the translation of Ptolemy's ***Hè Megalè Syntaxis*** (The great treatise) translated as Almagest. Al-Battani also improved the precision of the measurement of the precession of the earth's axis. The corrections made to the geocentric model by al-Battani, Ibn al-Haytham, Averroes and the Maragha astronomers such as Nasir al-Din al-Tusi, Mo'ayyeduddin Urdi and Ibn al-Shatir were later incorporated into the Copernican heliocentric model. Heliocentric theories may have also been discussed by several other Muslim astronomers such as Ja'far ibn Muhammad Abu Ma'shar al-Balkhi, Abu-Rayhan Biruni, Abu Said al-Sijzi, Qutb al-Din al-Shirazi, and 'Umar al-Katibi al-Qazwini.

Muslim chemists and alchemists played an important role in the foundation of modern chemistry. Scholars such as Will Durant and Fielding H. Garrison considered Muslim chemists to be the founders of chemistry. In particular, Geber is "considered by many to be the father of chemistry". The works of Arabic scientists influenced Roger Bacon (who introduced the empirical method to Europe, strongly influenced by his reading of Arabic writers) and later Isaac Newton.

Some of the other famous scientists from the Islamic world include al-Farabi (polymath), Abu al-Qasim (pioneer of surgery), Abū Rayhān al-Bīrūnī (pioneer of indology, geodesy and anthropology), Avicenna (pioneer of momentum and medicine),Nasīr al-Dīn al-Tūsī (polymath), and Ibn Khaldun (forerunner of social sciences such as demography.

An intellectual revitalization of Europe started with the birth of medieval universities in the 12th century. The contact with the Islamic world in Spain and Sicily, and during the Reconquista and the Crusades, allowed Europeans access to scientific Greek and Arabic texts, including the works of Aristotle, Ptolemy, Geber, al-Khwarizmi, Alhazen, Avicenna, and Averroes. European scholars had access to the translation programs of Raymond of Toledo, who sponsored the 12th century Toledo School of Translators from Arabic to Latin. Later translators like Michael Scotus would learn Arabic in order to study these texts directly. The European universities aided materially in the translation and propagation of these texts and started a new infrastructure which was needed for scientific communities. Europeans began to venture further and further east (most notably, perhaps, Marco Polo) as a result of the Pax Mongolica. This led to the increased influence of Indian and even Chinese science on the European tradition. Technological advances were also made, such as the early flight of Eilmer of Malmesbury (who had studied Mathematics in 11th century England), and the metallurgical achievements of the Cistercian blast furnace at Laskill.

At the beginning of the 13th century there were reasonably accurate Latin translations of the main works of almost all the intellectually crucial ancient authors, allowing a sound transfer of scientific ideas via both the universities and the monasteries. By then, the natural philosophy contained in these texts began to be extended by notable scholars such as Robert Grosseteste, Roger Bacon, Albertus Magnus and Duns Scotus. Precursors of the modern scientific method, influenced by earlier contributions of the Islamic world, can be seen already in Grosseteste's emphasis on mathematics as a way to understand nature, and in the empirical approach admired by Bacon, particularly in his ***Opus Majus***. According to Pierre Duhem, the Condemnation of 1277 led to the birth of modern science, because it forced thinkers to break from relying so much on Aristotle, and to think about the world in new ways.

The first half of the 14th century saw much important scientific work being done, largely within the framework of scholastic commentaries on Aristotle's scientific writings. William Ocham introduced the principle of parsimony: natural philosophers should not postulate unnecessary entities, so that motion is not a distinct thing but is only the moving object and an intermediary "sensible species" is not needed to transmit an image of an object to the eye. Scholars such as Jean Buridan and Nicole Oresme started to reinterpret elements of Aristotle's mechanics. In particular, Buridan developed the theory that impetus was the cause of the motion of projectiles, which was a first step towards the modern concept of inertia. The Oxford Calculators began to mathematically analyze the kinematics of motion, making this analysis without considering the causes of motion.

In 1348, the Black Death and other disasters sealed a sudden end to the previous period of massive philosophic and scientific development. Yet, the rediscovery of ancient texts was improved after the Fall of Constantinople in 1453, when many Byzantine scholars had to seek refuge in the West. Meanwhile, the introduction of printing was to have great effect on European society. The facilitated dissemination of the printed word democratized learning and allowed a faster propagation of new ideas. New ideas also helped to influence the development of European science at this point: not least the introduction of Algebra. These developments paved the way for the Scientific Revolution, which may also be understood as a resumption of the process of scientific change, halted at the start of the Black Death.

The renewal of learning in Europe, which began with 12th century Scholasticism, came to an end about the time of the Black Death. The Renaissance followed and showed a decisive shift in focus from Aristoteleian natural philosophy to chemistry and the biological sciences (botany, anatomy, and medicine). Thus modern science in Europe was resumed in a period of great upheaval: the Protestant Reformation and Catholic Counter-Reformation; the discovery of the Americas by Christopher Columbus; the Fall of Constantinople; but also the re-discovery of Aristotle during the Scholastic period presaged large social and political changes. Thus, a suitable environment was created in which it became possible to question scientific doctrine, in much the same way that Martin Luther and John Calvin questioned religious doctrine. The works of Ptolemy (astronomy) and Galen (medicine) were found not always to match everyday observations. Work by Vesalius on human cadavers found problems with the Galenic view of anatomy.

The willingness to question previously held truths and search for new answers resulted in a period of major scientific advancements, now known as the Scientific Revolution. The Scientific Revolution is traditionally held by most historians to have begun in 1543, when the books ***De humani corporis fabrica*** (On the Workings of the Human Body) by Andreas Vesalius, and also ***De***

Revolutionibus, by the astronomer Nicolaus Copernicus, were first printed. The thesis of Copernicus' book was that the Earth moved around the Sun. The period culminated with the publication of the ***Philosophiæ Naturalis Principia Mathematica*** in 1687 by Isaac Newton, representative of the unprecedented growth of scientific publications throughout Europe.

Other significant scientific advances were made during this time by Galileo Galilei, Edmond Halley, Robert Hooke, Christiaan Huygens, Tycho Brahe, Johannes Kepler, Gottfried Leibniz, and Blaise Pascal. In philosophy, major contributions were made by Francis Bacon, Sir Thomas Browne, René Descartes, and Thomas Hobbes. The scientific method was also better developed as the modern way of thinking emphasized experimentation and reason over traditional considerations.

The 17th century "Age of Reason" opened the avenues to the decisive steps towards modern science, which took place during the 18th century "Age of Enlightenment". Directly based on the works of Newton, Descartes, Pascal and Leibniz, the way was now clear to the development of modern mathematics, physics and technology by the generation of Benjamin Franklin (1706–1790), Leonhard Euler (1707–1783), Georges-Louis Leclerc (1707–1788) and Jean le Rond d'Alembert (1717–1783), epitomized in the appearance of Denis Diderot's Encyclopédie between 1751 and 1772. The impact of this process was not limited to science and technology, but affected philosophy (Immanuel Kant, David Hume), religion (notably with the appearance of positive atheism, and the increasingly significant impact of science upon religion), and society and politics in general (Adam Smith, Voltaire), the French Revolution of 1789 setting a bloody climax indicating the beginning of political modernity.

The Scientific Revolution established science as the preeminent source for the growth of knowledge. The early modern period is seen as a flowering of the Renaissance, in what is often known as the Scientific Revolution, viewed as a foundation of modern science. During the 19th century, the practice of science became professionalized and institutionalized in ways which continued through the 20th century. As the role of scientific knowledge grew in society, it became incorporated with many aspects of the functioning of nation-states.

Science has to be seen as a great threat to religion and a stimulus toward atheism. Science is making it increasingly difficult to blame disease and natural disasters on Gods anger toward mankind. Revelation is less appealing all the time. Every tiny increment of knowledge breaks loose one more shackle. Science is a powerful force behind this new religion.

Now I wish to turn to the second line of thought which I believe may be merging into a new religion. We need to know something about the history of secular humanism.

#2 History of Humanism

Contemporary humanism can be traced back through the Renaissance to the Islamic Golden Age and then to its ancient Greek roots back to the time of Gautama Buddha 563–483 BC and Confucius 551–479 BC. The sixth century BC pantheists Thales of Miletus and Xenophanes of Colophon prepared the way for later Greek humanist thought. Thales is credited with creating the maxim "Know thyself", and Xenophanes refused to recognize the Gods of his time and reserved the divine for the principle of unity in the universe. Later Anaxagoras, often described as the "first freethinker", contributed to the development of science as a method of understanding the universe. These Ionian Greeks were the first thinkers to recognize that nature is available to be studied separately from any alleged supernatural realm. Pericles, a pupil of Anaxagoras, influenced the development of democracy, freedom of thought, and the criticism of superstitions. Although little of their work survives, Protagoras and Democritus both espoused agnosticism and a spiritual morality not based on the supernatural. The historian Thucydides is noted for his scientific and rational approach to history. Many medieval Muslim thinkers pursued humanistic, rational and scientific discourses in their search for knowledge, meaning and values. A wide range of Islamic writings on love, poetry, history and philosophical theology show that medieval Islamic thought was open to the humanistic ideas of individualism, occasional secularism, skepticism and liberalism. Humanism's divergence from orthodox Christianity can be identified with the condemnation of **Pelagius** by Jerome and Augustine. Pelagius perceived humans as possessing inherent capacity for developing the qualities that the church perceived as necessitating the gift of grace from God. Pelagius rejected the Doctrine of Original Sin. The Humanists likewise recognize humans as born not with a burden of inherited sin due to their ancestry but with potential for both good and evil which will develop in this life as their characters are formed. The Humanists therefore reject Calvinistic predestination, and understandably therefore aroused the hostility of Protestant fundamentalists. Renaissance humanists believed that the liberal arts of music, art, grammar, rhetoric, oratory, history, poetry, should be practiced by all levels of wealth. They also approved of the self, human worth and individual dignity. Noteworthy humanist scholars from this period include the Dutch theologian Erasmus, the English author Thomas More, the French writer François Rabelais, the Italian poet Francesco Petrarch and the Italian scholar Giovanni Pico della Mirandola. The term humanism was coined in 1808, based on the 15th century Italian term umanista, which was used to designate a teacher or student of classic literature.

One of the earliest forerunners of contemporary chartered humanist organizations was the Humanistic Religious Association formed in 1853 in London. This early group was democratically organized, with male and female members participating in the election of the leadership and promoted knowledge of the sciences, philosophy, and the arts. In February 1877, the word "Humanism" was publicly used, apparently for the first time in America, to apply to Felix Adler, pejoratively. Adler, however, did not embrace the term, and instead coined the name "Ethical Culture" for his new movement – a movement which still exists in the now Humanist-affiliated New York Society for Ethical Culture.

Active in the early 1920s, F.C.S. Schiller considered his work to be tied to the Humanist movement. Schiller himself was influenced by the pragmatism of William James. In 1929 Charles Francis Potter founded the First Humanist Society of New York whose advisory board included Julian Huxley, John Dewey, Albert Einstein and Thomas Mann. Potter was a minister from the Unitarian tradition and in 1930 he and his wife, Clara Cook Potter, published **_Humanism: A New Religion._** Throughout the 1930s Potter was a well-known advocate of women's rights, access to birth control, civil divorce laws, and an end to capital punishment.

Raymond B. Bragg, the associate editor of **The New Humanist,** sought to consolidate the input of L. M. Birkhead, Charles Francis Potter, and several members of the Western Unitarian Conference. Bragg asked Roy Wood Sellars to draft a document based on this information which resulted in the publication of the ***Humanist Manifesto*** in 1933. The Manifesto and Potter's book became the cornerstones of modern humanism. Both of these sources envision humanism as a religion

In 1941 the American Humanist Association was organized. Noted members of The AHA included Isaac Asimov, who was the president from 1985 until his death in 1992, and writer Kurt Vonnegut, who followed as honorary president until his death in 2007. Robert Buckman was the head of the association in Canada, and is now an honorary president.

After World War II, three prominent humanists became the first directors of major divisions of the United Nations: Julian Huxley of UNESCO, Brock Chisholm of the World Health Organization, and John Boyd-Orr of the Food and Agricultural Organization.

The term 'secular humanism' was first used in the 1930s, and in 1943, the then Archbishop of Canterbury, William Temple, was reported as warning that the "Christian tradition... was in danger of being undermined by a secular humanism which hoped to retain Christian values without Christian faith."

During 1960s and 1970s the term was embraced by some humanists who considered themselves anti-religious as well as those who, although critical of religion in its various guises, preferred a less strident approach.

The release in 1980 of ***A Secular Humanist Declaration*** by the newly formed Council for Democratic and Secular Humanism ("CODESH") gave secular humanism an organizational identity. Now named the Council for Secular Humanism (CSH), it lays out ten ideals:

1. Free inquiry as opposed to censorship and imposition of belief
2. Separation of church and state
3. The ideal of freedom from religious control and from jingoistic government control
4. Ethics based on critical intelligence rather than that deduced from religious belief
5. Moral education
6. Religious skepticism
7. Reason
8. A belief in science and technology as the best way of understanding the world
9. Evolution
10. Education as the essential method of building humane, free, and democratic societies.

Secular humanism further describes a world view with the following elements and principles.

1. Need to test beliefs – A conviction that dogmas, ideologies and traditions, whether religious, political or social, must be weighed and tested by each individual and not simply accepted on faith.
2. Reason, evidence, scientific method – A commitment to the use of critical reason, factual evidence and scientific methods of inquiry, rather than faith and mysticism, in seeking solutions to human problems and answers to important human questions.
3. Fulfillment, growth, creativity – A primary concern with fulfillment, growth and creativity for both the individual and humankind in general.

4. Search for truth – A constant search for objective truth, with the understanding that new knowledge and experience constantly alter our imperfect perception of it.

5. This life – A concern for this life and a commitment to making it meaningful through better understanding of ourselves, our history, our intellectual and artistic achievements, and the outlooks of those who differ from us.

6. Ethics – A search for viable individual, social and political principles of ethical conduct, judging them on their ability to enhance human well-being and individual responsibility.

7. Building a better world – A conviction that with reason, an open exchange of ideas, good will, and tolerance, progress can be made in building a better world for ourselves and our children.

Every tiny increment of knowledge breaks another shackle. You can see the clash between the secular humanist views and religion. The secular humanist movement is quite large and it is a significant threat to religion.

Before turning our focus now to the third line of thought which I believe is producing a new religion I need to explain how adopting a religion differs from merely accepting a theory, ideology or intellectual belief system.

Chapter 15

Conceptual Model VS *Super*concept

There may be a problem with the scientific and the humanist belief systems serving as a *super*concept for most people. These belief systems seem to lack the emotional power to supplant the authoritarian and anthropomorphic theist *super*concept. The scientific and humanist belief systems therefore tend to remain intellectual ideologies compartmentally isolated from the individual's core commitment to theism. This is not true of all scientist and humanist of course but these belief systems have thus far failed to 'take off' winning the allegiance of millions of people over a short span of time.

For a new *super*concept like that represented by science and/or humanism to strike an emotional chord with the masses, and push commitment to theism aside, it has to be accepted **emotionally**. It must trigger a connection with early childhood **percepts** and satisfy an emotional need that resides deep in the individual's mind. Those individuals with their *super*concept in flux, those who have not been able to make a firm closure on their current belief system, must make an emotional commitment to a new 'religion' before it can be adopted as the answer to their spiritual quest. They must 'see' the new belief system **emotionally** as providing a better 'truth', a superior answer to their serious and important questions.

You can tell if a belief system has this kind of emotional power in this way. If a new belief system is embraced emotionally, reverently and fervently it will then be taught to the children of its devotees. These children will then be imprinted on the new *super*concept, and they thereby will become powerfully and emotionally attached to it. If the new belief system is but their parents' intellectual preference, then that is what it tends to remain.

Some scientists and secular humanists are 'true believers'[18] and believe very powerfully and emotionally in their commitment to these belief systems. But these scientific and secular humanist belief systems are apparently too intellectual and esoteric to appeal to huge numbers. So while there are many people who are intellectually committed to the scientific and the secular humanist view, not so many have accepted these concepts as their *super*concept. This may not be the case with the third tradition of thought I now wish to discuss.

The third tradition of thought which I believe is contributing to the emergence of a new 'religion' is Deism. Deism has shown that it has the power to heal the conflicted contemporary mind! Deism apparently also has the capacity to serve as a very powerful *super*concept, capable of sweeping theism aside!

3 # History of Deism

The term "Deism" originally referred to a belief in one deity, as contrasted with the belief in many Gods (Polytheism). During the late 17th century, the meaning of "Deism" began to change. It originally referred to forms of radical Christianity - belief systems that rejected miracles, revelation, and the inerrancy of the Bible. Currently, Deism is no longer associated with Christianity or any other established religion. Deism is not a religious movement in the conventional sense of the word. There is no Deistic network of places of worship, nor any priesthood or hierarchy of authority.

Deism was influential among politicians, scientists and philosophers during the late 17th century and 18 century, in England, France, Germany and the United States. Early Deism was a logical outgrowth of the great advances in astronomy, physics, and chemistry that had been made by Bacon, Copernicus, Galileo, and others. It was a small leap from rational study of nature to the application of the same techniques in religion. Early Deists believed that the Bible contained important truths, but they rejected the concept that it was divinely inspired or inerrant. Deists were leaders in the study of the Bible as a historical (rather than an inspired, revealed) document. Lord Herbert of Cherbury (1648) was one of the earliest proponents of Deism in England. In his book ***De Veritate*** published in 1624, he described the "Five Articles" of English Deists:

1. Belief in the existence of a single supreme God
2. Humanity's duty to revere God
3. Linkage of worship with practical morality
4. God will forgive us if we repent and abandon our sins

5. Good works will be rewarded (and punishment for evil) both in life and after death

Other early English Deists were Anthony Collins (1676-1729AD), Matthew Tindal (1657-1733AD). In France its leaders were Rousseau (1712-1778AD) and Voltaire (1694-1778AD). Many of the leaders of the French and American revolutions followed this belief system. Among the U.S. founding fathers, John Quincy Adams, Ethan Allen, Benjamin Franklin, Thomas Jefferson, James Madison, Thomas Paine, and George Washington were all Deists. Deists played a major role in creating the principle of separation of church and state, and the religious freedom clauses of the Constitution of the United States.

The contemporary American definition of Deism is well expressed by The World Union of Deists founded in Charlottesville, Virginia, U.S.A., on April 10, 1993 by Robert Johnson. "Deism is the recognition of a universal creative force greater than that demonstrated by mankind, supported by personal observation of laws and designs in nature and the universe, perpetuated and validated by the innate ability of human reason coupled with the rejection of claims made by individuals and organized religions of having received special divine revelation."

www.deism.com

This new 'religion' can be called scientific humanism or it can be given this more formal name: Deism. "Deism is belief in God based on the application of reason to the designs/laws found throughout nature. Deism is therefore a natural religion and is not a "revealed" religion. The natural religion/philosophy of Deism frees those who embrace it from the inconsistencies of superstition and the negativity of fear that are so strongly represented in all of the "revealed" religions. Most contemporary religions are called revealed religions because they all make claim to having received a special revelation from God which they believe their various and conflicting holy books are based on. Deism is not supported formally by any institution so technically we can't call it called a religion".

Whatever you wish to call this new ideology, it has developed from the full, uncompromising embrace of science, reason, humanism and the rejection of theism. This new 'religion' rejects evil, force, supernaturalism, human sacrifice and the punitive theist conscience. Deism and scientific humanism reject a human-centric, intervening God and a flexible, malleable subjective truth. As a result of the acceptance of 'reality' or 'Nature's God' there is no need for compartmentalization. The adherents of this new religion avoid compartmentalization and the hypocrisy inherent in that psychology. The epistemology of Deism utilizes a *super*concept based on Nature, not on a human ruling authority. Deists believe that Nature or reality and not man is the power and ruler of the universe!

With the refusal to take the 'have my cake and eat it too' approach to the conflict between theism and reality, compartmentalization is avoided and with it all the cognitive and emotional consequences. Lying to oneself, the hypocrisy, the mental conflict, the anguish and the rage disappears with the theistic epistemology and psychology. The collectivistic *super*concept is extinguished. Reality and God are one! The mind-dichotomy is healed.

Do these individuals who embrace Deism foretell the future of mankind? With a cognitive standard based upon an objective, disinterested and non-interventionists truth, there is no worship of or dependence upon man. There is no dependence upon leaders or rulers and for many Deists no need for, or loyalty to the State. Many of these new thinkers see no need for a State. They see the fallacy of 'good' government. If the Paganism of many modern religions could be magically distilled away would what remains be Deism?

Deism
A new 'religion'
Mind of the Third Millennium

The Homo sapien brain is uniquely designed to function properly when used the way it is used by Deists. Their spirituality is not riddled with superstition, and compromised by the conflict and savagery associated with theistic belief. Moderns of this new era view Nature, the world or the universe, as reality or God. Their conceptual standard of truth is not man, it is a non-interfering, non-interventionist[1], consistent and dependable objective standard. Their God is, in many ways, interchangeable with the concept some label 'reality'. The Deist God is tangible, visible, knowable, and in constant direct view. There is no supernatural or other-worldly aspect to the Deist God. The God of the third millennium is perceptible and understandable. Thinking, the type of thinking Deist, or scientific humanist do, where concept building starts with percepts and is integrated without contradiction, is called percept-driven or 'empirical'. This type of thinking is often called 'common sense'. These new thinker's psychological method of dealing with contradictions is through rejection.

I believe this 'empirical' type of thinking to be characteristic of this new psychology that I see developing. Such thinking is 'bottom-up', from perception to conception. Such thinking begins at the ground level so-to-speak. An empirical thinker does not allow concepts to overrule perception. So imagination and conceptualization is kept in check by a diligent, non-contradictory

---

[1] Deism is so free of man-made dogma, Deists are free as individuals to believe God intervenes in human affairs or not. Deists who do believe in intervention realize the truth to what the Deist George Washington wrote, that divine providence, if it does exist, is "inscrutable". This is a permanent block to clergy or anyone else who wants to control others by pretending they can control Providence.

cognition. A cognition that is kept subservience to objective reality, or as the modern Deist might explain, to God. The scientific humanist or Deist is not 'up in his head' rather he is focused on God, nature, or objective reality. He is looking out at the world, out at the environment, out at an objective and Natural God. He is looking for 'signs' from God, for God to 'talk' directly to him. God 'talks' to the scientist, and the Deist, by 'what works'. The Deist is trying to understand God's laws, nature's laws. 'What works' tells him he is closer to understanding God's laws. He is curious about his world, his God, his reality.

These free thinkers are all around, though we scarcely notice them. From some of the freethinkers that I have been fortunate to meet and get to know, I can tell you that they are highly skeptical and percept-driven. They are hyper skeptical with little gullibility or suggestibility. Some view the State as parasitic and phony as its fiat money. Many do not vote, and they do not join the military. They agree with Thomas Paine who said "It is the duty of the patriot to protect their country from its government".

Seeking Freedom From The Theist World

I made an appointment to meet and interview Mr. and Mrs. Jake Appleton at their home. I got their names from an individual who stated that he was a Deist and who offered to introduce me to others of the persuasion. I had to promise to keep their real names and address secret.

This family lives out from Seattle, Washington. Yesterday I arrived at Sea-Tac Airport and have been driving for three hour this morning following directions. I am far from any town and now I see the road I have been instructed to take. I make a left turn onto this narrow, blacktop road. I now have 12 more miles to go.

The secrecy I am sworn to in order to meet Mr. Appleton is incredible. I would not even have known about him if I had objected to a very rigorous interrogation. Only after circulating the early proofs of this book was I able to make contact with some of these people.

The 12 miles seem to be longer than usual. Partly it's because there is no radio reception out here and partly because the road is monotonous and rough. Finally I arrive at my turn. The next leg of my journey is 3 miles long. But now the road gets even narrower, rougher, curvier and it's climbing up the mountain continuously. I noted my odometer and sure enough exactly 3 miles and I come upon a driveway entrance. As instructed I turn in and stop before a rusty old iron gate that looks like it hasn't been opened in decades. I honk my horn as told to do and the gate opens.

Now I drive less than a hundred yards to see a house built of dark brown logs standing at the apex of the mountain. Several large 8 foot satellite discs are partially visible behind the structure. A number of solar panels can also be seen. As I drive up to the front of the house a tall sandy haired man, appearing to be in his early thirties comes out the front door and greets me as I exit my vehicle.

"Hi! I am Jake Appleton." He says in a deep masculine voice.

"You are Doctor Jones, I presume?" he asked.

"Yes. Thank you so much for agreeing to see me" I said.

"Come in. I want you to meet the rest of my family" He said as he motioned me up the walkway toward the door to his home. He opened the door for me and I entered a brightly lighted great room of perhaps 24x 35 feet in size. A large L-shaped sectional sofa dominated the room.

A beautiful woman, blonde hair with very, very green eyes approached me with her hand extended.

"My name is Heather Appleton" she said, "Welcome to our home Doctor Jones, I enjoyed reading your book *The Devastating Delusions of the Contemporary Mind*."

"Thank you very much. I am happy to make your acquaintance" I responded.

"You and Jake have a seat here at the coffee table. May I get you something to drink?" Heather said.'

"Thank you. I could use a drink of water. That is quite a trip here from Seattle. Do you drive it often?" I asked.

"No, we do not make that drive much. If you keep going where you made that left turn 15 miles back there is a small town that has all the essentials we need. I am afraid we are pretty much home-bodies." Jake replied.

We sat in the great room. I sipped my glass of water. Jake and Heather and I talked for the next 2 and one half hours.

I learned that there exists a fair size community of Deists in the United States and around the world. These people communicate via the Internet. This community exists as a virtual society almost entirely on the World Wide Web. Deist websites are the rallying point for this community.

I knew that Jake and Heather Appleton is not their real names, but the names by which everyone in their present location know them. They have two daughters age 8 and 10 and they are home schooled by both their parents. Their family lives almost completely below the radar of all government authority. They are off the electric grid as well, producing all the power they use with a small hydro-electric plant which Jake build and installed himself. I saw several solar panels and they showed me a room full of 12 volt batteries. Jake and Heather chose their piece of property because of the year around stream that runs through it, and the opportunity that affords to generate their own electricity.

Jake and Heather file no Federal Income tax, Washington State has no income tax, and they are very secretive regarding their finances. I did learn that Jake is a computer programmer and freelance software engineer. He finds work over the Internet. Heather is a copy-editor and freelance writer. She also finds work over the Internet. Both are avid gardeners.

"So neither of you participate in the politics of the country?" I responded at one point in our conversation.

"That's right. We do not register or vote. The State, the Federal Government has nothing to offer us; and we want nothing from the State." Heather answered.

"You see no need for protection, national security, that sort of thing?" I asked.

"No. We believe that most of the international tension is one government playing one-upmanship against another. We make a very clear distinction in our minds between a nation's government and the people who live in a particular government's jurisdiction. Governments are constantly creating a sense of emergency among their citizens." Jake replied.

"So you see no benefit of national security?" I reiterated

"If it were strictly a defensive security force perhaps that would be acceptable. But States, especially the United States right now, goes all over the world meddling in every nation's affairs, starting or at least joining in every conflict it can find, all to promote the commercial interest of its pet businesses and its religious crusade for democracy. This creates enemies by the bucket –load" Heather answered. "We are pacifist and do not condone firing rockets from 30,000 feet into private homes inside a sovereign nation, because terrorists are thought to be inside. That to us is cowardly and it is murder, and we disavow any support of it." She continued.

"We do not want to support all the State's evil with tax money. We believe it is against God's law to pay taxes of any kind, in any amount, to any government." Jake added.

" How do you respond to the criticism of your position which holds that not every citizen may choose for himself what national policy shall be, for how could you maintain a country if everyone made up their own mind like you do?"

Jake responded, "Exactly!"

"I see your point." I indicated.

"What do you say to those who value the protection the State provides in the form of clean water, fire protection, medical care, roads, airports and so forth?" I ask.

"We have no problem with municipalities. Communities can run things as they wish. If local government goes too far a person can pull up and leave. But county, state and Federal governments do much more harm than good and we don't need them and don't want them."Heather answers.

"All the supposed 'protection' is really just the protection of government jobs. They do not do anything we can't do better and cheaper ourselves. And we do. We don't look for government to protect us from anything or provide us with anything. It works out a lot better without them" said Jake.

"The people who believe in 'government regulation' are the worse 'true believers' of all. It is governmentophiles like that which drag free countries back into fascism and communism in the name of 'safety'. Heather adds.

"You can tell I read your book very carefully and have adopted some of your terminology!" Heather continued.

"Well most of the ideas and terms I use in the book I borrowed from other people, so I am glad to pass them on." I respond.

"Doctor Jones. This is Abby, she is 8 years old and this is Lilly who is 10." Heather said introducing her two daughters.

"I am very pleased to meet you" I answer. Both girls are blonde with their mothers green eyes. "May you join us? With your parents' permission I would like to ask you both some questions."

"Sure", all four responded, and the girls sat down and joined us on the big sofa.

Heather got up from the sofa and went into the kitchen. It is just off the great room. She retrieved a large tray with coffee, cups, hot tea and cookies. "Everyone help yourself." She announced.

"Abby, how do you find living in so remote a location?" I ask.

"The way we live is the only life I know. I have lots of friends on the Internet. I have met some of them on our travels. I love the country, my dog Sunshine, and all our animals. I am very happy here." She gave a big smile.

"How about you Lilly, do you not feel lonely for other kids your age?" I ask.

"I can remember when we lived in the city. There were so many people, so much noise. Everyone was in such a hurry! I like the life I have now much better. But it wouldn't be nearly as good without the Internet. I have friends all over the world and I visit some of them almost every day over the World Wide Web." She stated.

"Well, that certainly answers that question." I offer.

"Jake, how do you and Heather see politics, especially the way things work here in the U.S.?" I ask.

Heather beats Jake to answer, "A State is just a huge mob. It's just a mob that condones and relishes murder and theft. Then they just divide up the loot that's been stolen. Every few years they fight over what faction of the mob gets to divide up and distribute the loot to the mob members. So they hold an election. We don't condone murder and stealing so we have opted out of the State."

"I see. What aspect of the State do you think I failed to cover sufficiently in my book?" I asked looking at Heather.

"I think you did a really good job in your coverage of the evils of the State. But it's difficult to say too much about the immorality of State citizens. The upper and middle class see what welfare and government handouts do to the lower class. But they are blind to what participating in the stealing and killing does to them!" Heather responded.

"You are so right. I am constantly astounded at people's blindness to the evil of theism and its societal expression." I comment.

"Doctor Jones it is getting late. The night comes early here in the mountains. We want to invite you to have dinner with us and to spend the night here. The road back to Seattle is very dark and winding as you know. They are much more difficult to drive at night, especially if you are unfamiliar with them. If your travel plans will allow please stay with us tonight?" said Jake.

"My flight does not leave from SeaTac until tomorrow afternoon. I accept your gracious invitation." I replied.

"Fine. Before the sun goes down we want to show you around our little homestead." Jake said motioning me to follow him and the girls.

I followed them out the back door and immediately see a large red barn. We entered the barn to the smell of cow manure and the sound of chickens clucking. I noted several cows, many chickens and several cats and dogs. All the animals were thankfully very friendly. I also noticed a lot of machinery and equipment, a tractor and several implements, a motor driven tiller, and assorted hand tools.

We then exited the rear of the barn to a large and beautiful vegetable garden. Peas, beans, greens, potatoes and several types of berries grew in abundance along the long rows. Apple trees, plum, cherry and pear trees lined the back of the garden in front of the trees of the forest.

We stayed outside talking about organic gardening until the darkness overcame us and we returned to the house. Inside the table was set. I was shown the bathroom. Then we were all seated at the table. I waited to see what kind of meal ritual might occur.

Everyone sat with their hands in their lap and Jake said, "Thank God for this Earth and for its bounty. You provide us with all we need. We are truly loved. We truly love Nature's God and seek to understand you better every day."

With that everyone began passing the food around the table. Fried chicken, mashed potatoes and gravy, peas and brown bread were outstanding. Tea, milk and grape juice were drink choices with apple pie and ice cream for dessert.

"I detect some Southern influence in your cooking." I enjoined.

"Yes, my Mother was from Hattiesburg, Mississippi. She moved to California to attend UCLA where she met my Dad." Heather offered. "They now live near Modesto. We see them at least once a year." She added.

"Grandpa and Grandma have an almond orchard." Lilly added.

These girls are definitely not the 'be seen and not heard' variety. Both Lilly and Abby are very well informed and very opinionated! They speak their mind and can hold their own in adult conversation. I was very impressed with these two sisters.

"I guess from your feelings regarding the State, that you are anarchist?" I ask the Appleton as we sit in the great room after dinner.

Heather answers first "were not what you would call 'pure anarchist' as we think law and order are necessary. But by contemporary standards I think most people would view our beliefs as anarchy. The 1960's television series 'Gunsmoke' stared James Arness as Marshal Matt Dillon of Dodge City, Kansas of the 1800's. That show depicted about how much law I think we need."

Jake laughed. "I think you have it right in your book, theist can't make any organizational structure work. A moral people could make many different societal structures work".

"Suppose the U.S. Congress passed a Constitutional Amendment establishing separation of Commerce and State?" I asked.

"Well, that would really be something. But of course it could never happen. That would negate everything the U.S. Government, every government stands for!" Heather answered.

"If such an amendment to the Constitution could be passed and ratified by a majority of states it would say volumes about the country and the enormous strides it would have made to get there. But it would take a sea change in the populations' attitudes." Jake responded.

"So what I hear from both of you is that the theist 'morality' is so strong it prevents any worthwhile improvement in society, in its government?" I ask.

"Yes, that is the way we see it. But I wouldn't call it theist morals; theism is the total absence of morals! Just because they claim their evil to be moral virtue doesn't make it so. Sure they try to maintain a monopoly over the term 'morals' but it's to keep inquisitive minds from noticing that theists have none!" Jake said.

"Theist Clerics work very hard to keep their members as separated from God as they possibly can. And since God is available 24/7 to be seen and heard, it takes an enormous effort to maintain blindness to him!"Heather adds.

"We seldom watch television anymore." Heather continues, "People are so obsessed with the worship of celebrities and politicians! We just can't relate to it. I think you have to have a people-dominated mind to find such people interesting!"

I respond, "How does that song by Streisand go? 'People who need people are the luckiest people in the world'?

"Yeah. That's it. It certainly appeals to theists." Jake adds

"Theists are obsessed with what 'other' people think of them." adds Jake.

"Yes, they constantly worry about 'what other people think. How other people feel about them. They really do worship 'other people' the ubiquitous 'they." I add.

"Let's turn on the television set and look at the programs offered tonight. We'll look for just for a few minutes. This will interest Abby and Lilly as well."Heather said.

With that Heather picked up the remote and turned on the television. The first program was CNN a news show. The commentator was reporting on the Israeli -Palestinian conflict. Next Heather changed the channel to a police detective show. The detectives were hunting down a murderer. Then several more police shows. Then she reached a channel that was over the next few days going to show all the James Bond Agent 007 movies. Heather clicked off the TV.

Heather spoke, "First we saw CNN reporting on two groups of religious fanatics in the Middle East. They are so busy hating and killing each other they really don't even hear each other. Without their two respective governments the people themselves could probably work things out, but neither group believes it can exist without a State and a government. Without a State how could either expect to efficiently steal from and murder the other!?"

"The Palestinian problem does appear insolvable" I add.

"Neither side has any morals or any moral leader. They desperately need a Gandhi to show them the moral path." Jake commented.

"Then we change the channel from real killing and stealing and go to the fictional stealing and killing depicted on one cop show after the other." Heather states.

"What does that mean to you Heather?" I ask.

"It's all about people, it's all about men. It's all about killing. It's how to take advantage of other people, how to exploit people, how to manipulate and kill people and take from them. The criminals are detestable but the police are no stalwarts of honesty. They lie, manipulate, break the law and kill without hesitation. These shows completely lack any educational or moral value, except they teach you that only those on a government payroll have any chance coming out alive!" Heather finishes.

"The next channel was having what they call a 'Bond-a-Thon' the entire James Bond movie series, one after the other for days!" said Jake.

"Jake what do you think of the James Bond character." I asked.

"James Bond sort of epitomizes the culture" Jake begins, "He is a government man, a G-man. He is licensed by the State to kill. That alone speaks volumes. The State, as the repository of human authority, is able to bestow the right of executioner upon its employees!"

"Well Jake, that's what they do with soldiers!" I respond.

"Yes, but supposedly, at least originally, that was to defend the country. Now it's come to mean anything the ruling class wishes to define as 'in the National interest.'"

"True" I answer.

"So the stealing and killing by the State is made very clear and very attractive in the James Bond stories, and people love the stories!" Jake concludes.

"Bond is a good example of the collective standard of truth to which governmentophiles subscribe. Their epistemology holds that the government is the source of truth. They truly do worship the State as their God!" I respond.

"And you'll notice in the movies the women agents freely use sex as a tool and bribe against their opponent. So it's clear that in all matters of morals the State is the supreme authority. It is as you say in your book, humans, especially those in authority, are the theists' source of truth." Heather says.

"You have to figure if you will take another humans word for reality, after having an alleged exclusive interview with God, then human authorities and 'experts' become your measure of all things." I add.

"Just look at their heroes. I do proofreading and copy-editing, so I read a lot of fiction. Everyone is writing a novel or at least their memoirs. They are all about whining and blaming everybody else. But their notion of a hero is someone with a special knack for getting people killed!" said Heather.

"Well when somebody as awful as Abraham Lincoln is admired and venerated what else can we expect?"Adds Jake.

"Yes, from Alexander Hamilton to Abraham Lincoln, from Woodrow Wilson to FDR, from Adolf Hitler to Joseph Stalin, and from LBJ to Bush, the tyranny of Utopian fanaticism always seems to follow when theists gain political power." I add.

"Now it's time for me and the girls to go to my office and work on our studies. Jake wants to show you his office. So we will say good night Doctor Jones. We will see you in the morning at breakfast."Says Heather.

I said good night to Heather, Abby and Lilly.

Jake escorted me to his office in the northeast corner of the home. I have never seen so many computers in one room before. He builds his own computers. He designs and manufactures specialty and custom software. He has more business than he can handle, but he does not want to expand, he does not want employees.

We sat down in two identical brown leather chairs.

"Jake, your office and your obvious skills are very impressive. How did you come to the Deist orientation?" I asked.

"It was not much of a problem for me. My parents were the easiest going, most rational people you can imagine. My siblings and I had a most idyllic childhood. My father avoided clerics. He would joke when asked to attend church, 'I received my smallpox vaccination the same week I was baptized. They both took. So I have no need to go back for a booster!"He would answer.

He always made a joke of religious ideas and helped us not to take religion seriously. When I discovered Deism through Robert Johnson's website I knew I had found a home. It made it easier for me to profess Deism knowing that such beliefs would not bring condemnation from Mom and Dad."He said.

"They sound delightful. How do they feel about your Deism?" I said.

"They find it intriguing. I think at heart they have been Deist all their lives. Now they just have a name for it." Jake said.

"I wanted to ask you about your security. How do you make sure the authorities don't learn about you? With the efforts worldwide to track down terrorists, aren't you concerned about being accidentally exposed?" I asked.

"Most terrorists get caught planning an attack or purchasing weapons or explosives. Since I am a pacifist I don't purchase guns or explosives and I don't plan terrorist attacks. But you are correct. The terrorists and I use many of the same techniques to stay out of the State's grasp. But I have only myself and members of my family to hide, I do not have to expose myself trying to recruit members or communicate over the Internet with cell members." Jake answered. "Also, the man that I obtain my identification documents from is an absolute genus. His passports, birth certificates and driver's licenses are better, I repeat better, than the real ones! I have traveled to many countries and proper documentation has never been a problem."

"That's good to know. I am gratified that you don't take any more chances than necessary." I respond. "But I am sure you do not want to use an alias with your bank accounts?" I ask.

"We keep our long term savings in gold. It's deposited outside the United States. And yes it's under our true identity complete with fingerprint and retinal identification. But we have several bank accounts under different identities here in the States, mainly for business purposes." Jake answers.

"I guess you have doubts about the U.S. dollar, that's why the gold?" I say.

"Yes, all the major nations now issue fiat money. It's just paper. The fiat money scheme is the latest way governments have come up with to steal from their own citizens as well as foreign investors. That said the U.S. dollar is the worst. The United States is broke, insolvent, they owe several times what they have available. The U.S. Government has borrowed trillions to the point now where it literally can never re-pay its debts. It has worked its Ponzi scheme for over forty years while wildly spending everything it could steal or borrow. So default is inevitable." Jake ends.

"That's very interesting. How did the U.S. Government work a Ponzi scheme?" I ask.

"Throughout each year the Federal Government issues interest bearing treasury notes and bonds which it sells to people, institutions and governments all around the world. When it's time to pay the interest on those notes, it sells more notes to more people to get the money to cover the interest payments. If it can't sell enough of the worthless notes to cover the promised interest, it prints money to pay the interest!"Jake explained. "That's just one of the Governments Ponzi schemes, the biggest, it has many others. The Social Security Administration is another."Jake said.

"The monopoly over the country's currency which most governments force on their citizens is just another way in which governments exploit the people." I add.

"Yes. Exactly. Every possible source of exploiting the people is utilized." Jake adds.

We talked for another 45 minutes then Jake showed me to my bedroom. I procured my bag from my rental car. My only problem falling to sleep was the quiet of their place. I mean quiet!

I was awakened at 6:10 literally by the crowing of a rooster. By the time I finished my morning routine, dressed and made it to the kitchen it was obvious that I was the last to arrive.

"Good morning Doctor Jones" said Lilly handing me a cup of coffee.

I sipped the coffee then said, "Good morning to you. The coffee is fixed just the way I like it. How did you know how I liked it?" I asked.

"I watched as you prepared your cup yesterday." She responded. "Mom and Dad will be back in just a minute. They went out to the garden for something."

Heather, Abby and Jake came in with a large basket of potatoes and other garden produce.

"Good morning Doctor Jones, hope you slept well?" Heather said.

"Too well maybe, must be the mountain air."I answer.

"Please call me Henry; Doctor Jones is too- stuffy." I enjoin.

"Abby and Lilly smile, Jake nods and Heather indicates that we should all sit down for the breakfast that is prepared and ready to eat.

Fresh eggs and I do mean fresh, hash brown potatoes, sausage, milk, apple juice, and coffee.

The same prayer is said before breakfast.

As everyone was completing their meal I said, "What a wonderful breakfast. I feel like I should go out into the woods and fell a tree or cut up some logs or something!"

Jake says, "That won't be necessary. Beside you have a long drive ahead of you this morning?"

"True" Maybe I can impose on your time a bit more this morning with a few more questions?" I ask.

"Of course. Let's take our coffee into the great room where we can sit and talk." Says Heather.

Abby and Lilly excuse themselves. They leave, I believe to tend the animals. We have a seat on the sofa.

"I have a few more questions about Deism. First, do Deists make up the majority of the atheist and freethinker community?" I ask.

"Oh, no. Deists are the minority. Most are secular humanist and libertarians."Heather answers. "But Deists are the fastest growing group!"

"Are most people in your Internet community atheist?" I ask.

"Yes I think that is true as all Deists are atheist and many secular humanists and libertarians are atheist. But some are Unitarians and some are Christians and Buddhist and other religions. Many are non-religious and a few are anti-religious." Jake responds.

"Do most Deists share our views of the State?" I pose.

"Probably not. Many do, but many are politically active believing that the State concept may be saved." Jake said.

"I see that attitude more among those who identify themselves simply as atheist. Many of these people are very active in party politics believing that atheist activism can have a positive impact on government and the State." Heather adds.

"But as I understand it you and many Deists do not hold that view." I retort.

"Correct. We believe the State, all States. The United States for example will collapse in bankruptcy. I personally believe the United States will collapse just like the Soviet Union within our lifetimes." Jake commented.

"So if theism remains unchallenged something as evil as the present U.S. Government or Soviet Government will be reconstructed. So our task is education. This is where we are so hopeful regarding your book."Heather exclaimed.

"My last question is this: as you must see yourselves as living among savages, do you not feel lonely and socially isolated. And as an appendage to that question do such thoughts not make you depressed?" I ask.

"Henry, the world, the universe, God is wonderful. Everything in the universe is going just swell. Our planet is surviving very well. The only problems are human problems. Most of the problems humans have are self-inflicted. We do not view our contemporaries as savages, quite the contrary. Considering how theists are raised and the ideas in their minds, they are remarkably peaceful. But with billions of them, some are always loosing it, falling back on their theism and committing atrocities. We just have to be careful who we trust; everyone must be careful who they trust."Jake states.

"Theism is on its way out. God is working on this 24/7/365. Where, as you reported, the hunter-gatherer lasted 180 millennia theism began to fall apart in 15 millennia. In just 15 millennia the great heretics began pointing out its flaws and its evils. The case against theism grows stronger all the time. So whether it takes a few hundred years or several more millennia, theism will fade from the scene and one day cease to infect humankind." Jake finished.

"So you feel little pressure to proselytize?" I ask.

"Other people's happiness, other people's enlightenment is not a requirement of our happiness. Sure we would prefer a world where most people were cooperative and murder and stealing not so prevalent. But we believe that to a very great extent we make our own happiness. It does require a degree of secrecy. It has for 20 millennia. But it is not that difficult. It is a very satisfying, a very happy life." Heather added.

"We do have to keep that television mostly turned off as it tends to bring all the problems theists create and dump them in one's living room." Heather added.

I said goodbye to Abby, Lilly, Heather and Jake. I didn't notice until I was a mile down the road the sack of fresh vegetables and fruit on the floor of the car. I had an uneventful trip back to the Seattle area and to the airport. The flight back home was smooth. I had time to reflect on my meeting with the remarkable Appleton's. I relish any opportunity to see them again.

Religion in the New Millennium

The Deist 'religious' institution of this new millennium is the college and university. The research institute is a Deist religious institution because it is a place devoted to the pursuit of truth. The Temple of the Deist Age is the library, for it is where knowledge is stored and studied.

Worship for Deist also has another name, it is called scholarship. Study, learning, observation, experimentation, research, knowledge, and skepticism make up the lexicon of the Deist Religion. Scholarship is not just for its practical benefits, it is a way of showing respect for God, a way to worship Nature and reality.

The State in the New Millennium

Moderns carry repressed in their unconscious minds the evil savagery of tribalism wrapped within a veneer of rational humanism. This is called theism. They are so conflicted that no system of government can work very well. A government contributes nothing positive to a country. Deist and scientific humanist of the future will enjoy just the opposite. Almost any simple and inexpensive organizational model will provide the very minimal structure required to establish and maintain society.

Book Summary

I have two reasons for writing this book. My first purpose was to elucidate the processes of psychological evolution. I have explained the way mankind changes its psychology to adapt to environmental change. I described 4 great epochal adaptations mankind has created to survive in a changing world.

Upon evolving into a natural environment Homo sapiens first adaptation was to adopt Nature as their standard of truth. This allowed them to efficiently solve the problems they faced in the natural world. But after 180 millenniums they still had not found a way to create and maintain permanent settlements. Permanent settlements require agriculture and there was simply no power source available at that time which could be applied to farming.

That is to say that no power source was available at that time other than humans themselves. Witch-Doctors used the human imagination and conceptual ability to invent the collectivist psychology which would embrace slavery. People were taught to disbelieve their own perception and rely upon Pagan Gods as their standard of truth. This created tribalism which solved man's problem of creating permanent settlements. But this second psychological adaptation created the worse problem of totalitarian mind control, lack of creative freedom, terror and violence.

After 15 millenniums of brutal tribalism and coinciding with the invention of writing, heretics began to appear. They utilized the mental mechanisms of compartmentalization and repression to create some 'space' in which to think and invent. They gradually developed a competing ideology to theism. Splitting their minds into functionally separate compartments allowed one part to remain loyal to Tribalism while another part of their mind could venture to create a new ideology from science and humanism. The theist part of the mind continued to accept the Pagan Gods as the standard of truth and virtue. But the questioning, inquisitive, freedom seeking portion of the heretics' minds chose a new standard of truth. They embraced objective, non-interventionists, stable truth commonly referred to as Reality.

The tripartite mind with its inherent conflict is mankind's third psychological adaptation. This compartmented, troubled mind was ushered in about 5 thousand years ago as the heretics began their work. This has created the terrible combination of evil theist motivation and clear-headed thinking which characterizes the minds of contemporary peoples. This has resulted in mass murder on an increasingly efficient scale. With this adaptation evil is now mass produced!

Then about 2500 years ago the forth psychological adaptation which humans have so far created began to appear. It is created by eliminating commitment to the Pagan Gods and theism as a source of truth and virtue. The mind is then unified under one objective, immutable and non-interventionist truth. This is Nature's God of the Deists or the scientific and humanistic God: Reality.

The second reason I had for writing this book was to point out that evil was invented by man to create the collectivist mind and thereby solve the problem of constant nomadic trekking. This allowed the creation of tribes and slaves which solved the problem of building permanent settlements. The collectivist psychological adaptation however came with its own set of disadvantages. The evil it produces, especially when combined with the better cognition of the modern mind now threatens the extinction of our species.

Has the hunter-gatherer God Ytilaer been found by the heretics' and brought back to man's perception? Does Deism have the spark and emotional appeal to ignite a much needed, revolutionary and final assault on theism? Is science, humanism or **Nature's God** here to save the human race from itself?

Postscript

Theism is *not* the worship of God. Theism is the worship of coercive authority. If you don't believe this, read or re-read, the Holy Books of the great religions and see for your self.

- If an ancient philosopher must grant that freedom is wise
- If great men must write a constitution to recognize your rights
- If you are told what you may use for money
- If you are willing to be taxed in any way, in any amount, by any entity
- If you, or your children are forced to attend any school
- If you must get a prescription from a doctor to purchase pain or other medication

Then you are not free! You are not following the laws of God Reality. Perhaps you are not allowed to obey God Reality. God Reality did not design a single one of us to live in fear, cowardice and enslavement. You are a child of Nature and a citizen of the universe. You need not apologize to anyone for your life, joy and happiness and no group except your children has any claim to your dreams, your energy or your property. You can only be enslaved by your own mind, you must agree to it! The cure is always freedom, the disease always too little of it. Evil cannot exist without your cooperation. To be free only requires that you demand it.

Section IV
Summary

1. The divided and conflicted mind of contemporary man provides a degree of clear-headed cognition promoting discovery and invention. This is combined with an evil theist value system at the core of contemporary man's mind which threatens to destroy all life on earth.

2. The heretics have generated at least three lines of thought and tradition that challenge theism. One of these, or some combination of them, may provide a *super*concept emotionally powerful enough to unseat theism and heal man's mental schism.

References

Reference #1 Hutton, James ,M.D. 1726 – 1797.

1785 *Abstract of a dissertation read in the Royal Society of Edinburgh, upon the seventh of March, and fourth of April, MDCCLXXXV,* **<u>Concerning the System of the Earth, Its Duration, and Stability</u>**. Edinburgh. 30pp.

1788<u>. ***Theory of the Earth****; or an investigation of the laws observable in the composition, dissolution, and restoration of land upon the Globe.</u>* Transactions of the Royal Society of Edinburgh, vol. 1, Part 2, pp. 209-304.

Reference #2 Lyell, Charles 1797 – 1875

Principles of Geology 1st edition, 1st vol. Jan. 1830 (<u>John Murray</u>, London).

Reference #3 Wegener, Alfred 1880–1930

Wegener, Alfred (July 1912). "Die Entstehung der Kontinente" (in German). *International Journal of Earth Sciences*. doi:10.1007/BF02202896.

Wegener, Alfred (1966). ***The Origin of Continents and Oceans.*** New York: Dover.- (Translated from the fourth revised German edition by John Biram)

Reference #4 Darwin, Charles (1859)

On the Origin of Species by Means of Natural Selection, *or the Preservation of Favoured Races in the Struggle for Life* (1st ed.), London: John Murray, http://darwin-online.org.uk/content/frameset?itemID=F373&viewtype=text&pageseq=1, retrieved 2008-10-24

Reference #5 Ghiselin, Michael

1992. With Leda Cosmides and John Tooby, co-editors. ***The Adapted Mind: Evolutionary Psychology and the Generation of Culture***. New York: Oxford University Press.

Reference #6 Barkow, Jerome H.

1992. "Beneath new culture is old psychology." ***In The Adapted Mind. Evolutionary Psychology and the Generation of Culture***. J. H. Barkow, L. Cosmides and J. Tooby. New York, Oxford University Press: 626-637.

Reference #7 Cosmides, Leda

>Cosmides, L. & Tooby, J. (in press). ***Universal Minds: Explaining the new science of evolutionary psychology***(Darwinism Today Series). London: Weidenfeld & Nicolson.

Reference #8 Tooby, John

>Tooby, J. & Cosmides, L. (in press). ***Evolutionary psychology: Foundational papers***. Cambridge, MA: MIT Press.

Reference# 9 Gibbons, Ann (2003)

>"Oldest Members of Homo sapiens Discovered in Africa". ***Science*** **300** (5626): 1641. doi:10.1126/science.300.5626.1641. PMID 12805512. http://www.sciencemag.org/cgi/content/summary/300/5626/1641. Retrieved 2006-04-11. (abstract)

Reference # 10 McNeill, Willam H. (1999)

>"In The Beginning". ***A World History*** (4th ed.). New York: Oxford University Press. p. 8. ISBN 0-19-511615-1.

Reference #11 Shalins, Marshall David

>***Evolution and Culture*** (ed., 1960)

>***Stone Age Economics*** (1974: ISBN 0422745308)

>***The Use and Abuse of Biology*** (1976: ISBN 0472087770)

Culture and Practical Reason (1976: ISBN 0226733599)

The Western Illusion of Human Nature (2008: ISBN 13-9780979405723)

Reference #12 Hoeber (1979).

Selected Writings of Sir Charles Sherrington: A Testimonial Presented by the Neurologists Forming the Guarantors of the Journal "Brain" Oxford University Press

Reference #13 Rand, Ayn

(1966) ***Introduction to Objectivist Epistemology*** New York, The Objectivist, Inc.

Reference # 14 Lorenz , Konrad

***Here Am I - Where Are You?* - The Behavior of the Greylag Goose** (In collaboration with Michael Martys and Angelika Tipler). (1988). Translated by Robert D. Martin from Hier bin ich - wo bist du? ISBN 0151400563

Reference # 15 Andersen, Hans Christian

(2008). ***The Annotated Hans Christian Andersen***. New York and London: W. W. Norton & Company, Inc.. ISBN 978-0-393-06081-2.

Reference #16 Szasz ,Thomas, M.D.

    1974 (1961). **_Myth of Mental The Illness: Foundations of a Theory of Personal Conduct._** Harper & Row.

Reference #17 Snyder, Solomon H.

    (1974) **_Madness and the Brain_** New York, McGraw

Reference # 18 Hoffer,Eric

    (1951) **_The True Believer: Thoughts On The Nature Of Mass Movements_** ISBN 0-06-050591-5 Harper & Row

Reference # 19 Honneth, Axel

    (2008) **_Reification: A New Look At An Old Idea_** New York, Harvard University Press

Reference # 20 Taleb, Nassim Nicholas

    (2007). **_The Black Swan: The Impact of the Highly Improbable._** New York: Random House. ISBN 978-1-4000-6351-2.

Reference # 21 Chamberlain, David

    The Mind of your Newborn Baby ISBN 1-55643-264-X North Atlantic Books

Reference #22 Foulkes, David

> (1985). ***Dreaming: A cognitive-psychological analysis***. Hillsdale, NJ: Lawrence Erlbaum Associates.

Reference # 23 Moffitt, Kramer, and Hoffman

> (1993). ***The Functions of Dreaming***. Albany, NY: SUNY Press.

Reference # 24 U.S. Surgeon General's Scientific Advisory Committee on Television and Social Behavior

> (1972)***Television and Growing Up: The Impact of Televised Violence: Report to the Surgeon General, U.S. Public Health Service.*** Rockville, MD: National Institute of Mental Health; Publ. No. HSM 72-9090

Reference # 25 Pearl, Michael & Pearl, Debi

> (1994)***To Train Up A Child*** ISBN 1-892112-00-0 Pleasantville, TN No Greater Joy Ministries

Reference # 26 Maybury, Richard J. 1946

> Mr. Maybury is the publisher of *U.S. & World Early Warning Report for Investors*. He has written several entry level, common sense, books on United States economics, law, and history from a libertarian perspective. He has written these things in epistolatory form, usually as an uncle writing to his nephew, answering questions. Maybury was a high school economics teacher. After failing to find a book which would give a clear explanation

on his view of economics he wrote one himself. Some of his books include *Uncle Eric Talks About Personal, Career & Financial Security*; a book that is basically the foundation for his other books about the model perspective and Higher Law, *Whatever Happened to Penny Candy?*; a book that explains the history of the [United States] economic model and how it was based on free-market Austrian economics, *Whatever Happened to Justice?*; a book about his juris naturalist philosophical viewpoints regarding the foundations of America's legal system, British Common Law, the law of theFranks, and earlyChristian Ireland.

Maybury's Two Laws:

- Do all you have agreed to do
- Do not encroach on other persons or their property.

The first law is related to contract law. A contract is an agreement between two or more parties, in which they promise to perform certain actions for and recognize certain rights of the other parties. The second law is related to some criminal law and tort law.

Reference #27

In 1897 House Bill #246 was introduced in the Indiana House of Representatives. Although the attempt to legislate pi was ultimately unsuccessful, it did come pretty close to passage.

Reference # 28

>All of the individuals in this book referred to as 'patients' or 'clients' are fictional. These characterizations do not refer to any real human being, alive or deceased. I read and studied thousands of clinical case histories over the course of my years in psychiatric practice. When I sat down to write this book I used my imagination to create the fictional 'patients' represented here. Real people are much more complicated that the short sketches provided which are intended only to suggest some of the ways theistic belief damages people's lives. Certain beliefs can make a person just as sick as infection with certain microbes.

Henry E. Jones, M.D.

Glossary

Abstraction- the act of considering something as a general quality or characteristic, apart from concrete realities, specific objects, or actual instances.

Authentication-the act of establishing or confirming the self as authentic and who they claim to be.

Cognition-the act or process of knowing. The exercise of awareness, imagination, perception, conception, reason and judgement.

Contradiction - a proposition that contradicts or denies another or itself and is logically incongruous.

Egocentric, over-inclusive categorization- this phrase became popular for a time in psychiatry to label the type of 'magical thinking' seen in 'schizophrenics'. Then it was observed in 'neurotics'. I have found it typical of collectivist theist 'thinking'.

Empirical- Derived from, or provable by, observation and experience or experiment.

Homo erectus-an extinct species of the human lineage, formerly known as *Pithecanthropus erectus,* having upright stature and a well-evolved postcranial skeleton, but with a smallish brain, low forehead, and protruding face.

Homo sapien-human beings, our species- are bipedal primates belonging to the species Homo sapiens (Latin: "wise man" or "knowing man") in Hominidae, the great ape family. They are the only surviving members of the genus Homo.

Identity-The conceptual programming in the Forebrain which defines the **self** for the **self**. This leads to self-awareness and is well underway by the time an infant knows his name.

Integration- combining into an integral non-contradictory whole.

Millennium- n. *pl.* mil·len·ni·a or mil·len·ni·ums -A span of one thousand years

Mutation-a sudden departure from the parent type in one or more heritable characteristics, caused by a change in a gene or a chromosome

Prejudice- an unfavorable opinion or feeling formed beforehand or without knowledge, thought, or reason. Any preconceived opinion or feeling, either favorable or unfavorable.

Rational self-interest-The ethic of rational self-interest has been advocated by many philosophers and heretics down through history. Contrast with the masochistic ethic of self-sacrifice. The basic premise is that every man should have the freedom to use his mind for his own benefit. Most powerfully advocated and popularized by the philosopher-novelist Ayn Rand.

Reinforcement-In psychology it means the strengthening of a response through repetition so that the likelihood that it will reoccur increases.

Sacrifice-willingly surrendering of a higher value in exchange for a lesser value.

Self- One's consciousness of one's own being or identity, the ego. That which knows, remembers, desires and suffers.

Spiritual- of, or pertaining to, the mind or the intellect.

Stereotype- a simplified and standardized conception or image invested with special meaning and held in common by members of a group.

***Super*concept**-Some of the first concepts conceived by the child's mind are those regarding epistemological **identity**. One of those concepts is the standard of truth which the child accepts for non-contradictory cognitive integration. This is the *super*concept. Thus the *super*concept is part of **Identity**; it is some of the earliest, imprinted or core part of self **identity**.

Thinking- to exercise the power of reason as by conceiving ideas, drawing inferences, and using judgement.

Unconscious- the part of the mind containing psychic material that is only rarely accessible to conscious awareness but which has a pronounced influence on behavior.

Ytilaer the Great Spirit- This is 'reality' spelled backward. There were so many different names early man gave to his notions of nature's God that I chose to create this generic name to cover them all.

About The Author

Dr. Henry E. Jones was born in New Orleans, Louisiana in 1939. He was raised in the small Louisiana town of Wisner where he finished high school. Dr. Jones attended Louisiana State University in Baton Rouge, graduating with a BS in zoology. He then attended L.S.U. School of Medicine in New Orleans, receiving his M.D. Degree in 1965. He did a mixed pediatric internship at Charity Hospital of New Orleans.

Dr. Jones finished the first two years of a psychiatric residency at Warren State Hospital in Pennsylvania. He completed his psychiatric residency at Herrick Memorial Hospital in Berkeley, California in 1969.

Dr. Jones opened a private psychiatric practice in Los Angeles in 1970. He also worked as an instructor in psychiatry at University of Southern California-Los Angeles County Hospital. He practiced psychiatry exclusively for a few years, and then entered a Family Practice Residency at the Santa Monica Medical Center. In 1974 he completed this training. For several years, he limited his practice to psychosomatic disorders.

In 1976, Dr. Jones returned to Louisiana to practice in Ouachita Parish. During the next 25 years he practiced both psychiatry and general medicine. He served as Chief of Staff of one hospital and as chief of the Department of Family Practice, as well as Director of the Inpatient Psychiatric Unit of Saint Francis Medical Center in Monroe.

For almost 40 years Dr. Jones interviewed and treated patients suffering 'mental illness'. Over this period, he has evaluated and analyzed well over 2000 psychiatric patients. Much of this book is the result of what he learned from these individuals.

Printed in Great Britain by
Amazon.co.uk, Ltd.,
Marston Gate.